NONSENSE MUTATIONS AND tRNA SUPPRESSORS

Proceedings of the EMBO Laboratory Course and Aarhus University 50th Anniversary Symposium held in Aarhus, July 9th–21st, 1978

NONSENSE MUTATIONS AND tRNA SUPPRESSORS

EDITED BY

J. E. Celis

Department of Chemistry
Aarhus University
Denmark

J. D. Smith

Medical Research Council
Laboratory of Molecular Biology
Cambridge, UK

1979

ACADEMIC PRESS
London New York San Francisco
A Subsidiary of Harcourt Brace Jovanovich, Publishers

ACADEMIC PRESS INC. (LONDON) LTD.
24/28 Oval Road,
London NW1

United States Edition published by
ACADEMIC PRESS INC.
111 Fifth Avenue
New York, New York 10003

Library of Congress Catalog Card Number: 78–75276
ISBN: 0–12–164550–9

Printed in Great Britain by
Whitstable Litho Ltd., Whitstable, Kent.

CONTRIBUTORS

Abelson, J. Departments of Chemistry and Biology, University of California, San Diego, California, USA.

Altman, S. Department of Biology, Yale University, New Haven, USA.

Beckmann, J.S. Departments of Chemistry and Biology, University of California, San Diego, California, USA.

Brenner, S. MRC Laboratory of Molecular Biology, University Medical School, Cambridge, UK.

Bruton, C.J. Department of Biochemistry, Imperial College, London, UK.

Campbell, J.M. The Howard Hughes Laboratory for the Study of Genetic Disorders, Departments of Medicine and Biochemistry, Baylor College of Medicine, Houston, Texas, USA.

Caskey, C.T. The Howard Hughes Laboratory for the Study of Genetic Disorders, Departments of Medicine and Biochemistry, Baylor College of Medicine, Houston, Texas, USA.

Celis, A. Division of Biostructural Chemistry, Institute of Chemistry, Aarhus University, Denmark.

Celis, J.E. Division of Biostructural Chemistry, Institute of Chemistry, Aarhus University, Denmark.

Clark, B.F.C. Division of Biostructural Chemistry, Chemistry Department, University of Aarhus, Denmark.

Cotton, R.G.C. MRC Laboratory of Molecular Biology, University Medical School, Cambridge, UK.

Coulondre, C. Department of Molecular Biology, University of Geneva, Switzerland.

Cowan, N.J. MRC Laboratory of Molecular Biology, University Medical School, Cambridge, UK.

Fenwick, R.G. The Howard Hughes Laboratory for the Study of Genetic Disorders, Departments of Medicine and Biochemistry, Baylor College of Medicine, Houston, Texas, USA.

Fuhrman, S.A. Departments of Chemistry and Biology, University of California, San Diego, California, USA.

Galas, D. Department of Molecular Biology, University of Geneva, Switzerland.

Gauss, D.H. Max-Planck-Institut für Experimentelle Medizin, Göttingen, West Germany.

Gesteland, R. Cold Spring Harbor Laboratory, Cold Spring Harbor, New York, USA.

Grüter, F. Max-Planck-Institut für Experimentelle Medizin, Göttingen, West Germany.

Hofer, M. Department of Molecular Biology, University of Geneva, Switzerland.

Johnson, P.E. Departments of Chemistry and Biology, University of California,

San Diego, California, USA.

Kaltoft, K. Division of Biostructural Chemistry, Chemistry Department, University of Aarhus, Denmark.

Karn, J. MRC Laboratory of Molecular Biology, University Medical School, Cambridge, UK.

Kjeldgaard, N.O. Department of Molecular Biology, University of Aarhus, Denmark.

Knapp, G. Departments of Chemistry and Biology, University of California, San Diego, California, USA.

Konecki, D. The Howard Hughes Laboratory for the Study of Genetic Disorders, Departments of Medicine and Biochemistry, Baylor College of Medicine, Houston, Texas, USA.

Kruh, G. The Howard Hughes Laboratory for the Study of Genetic Disorders, Departments of Medicine and Biochemistry, Baylor College of Medicine, Houston, Texas, USA.

Kurland, C.G. Department of Molecular Biology, Wallenberg Laboratory, University of Uppsala, Sweden.

Marcker, K.A. Department of Molecular Biology, University of Aarhus, Denmark.

McLeod, A.R. MRC Laboratory of Molecular Biology, University Medical School, Cambridge, UK.

Miller, J.H. Department of Molecular Biology, University of Geneva, Switzerland.

Milstein, C. MRC Laboratory of Molecular Biology, University Medical School, Cambridge, UK.

Ogden, R.C. Departments of Chemistry and Biology, University of California, San Diego, California, USA.

Ono, B. Department of Pharmaceutical Technology, Faculty of Pharmaceutical Services, Okayama University, Okayama, 700 Japan.

Philipson, L. Department of Microbiology, University of Uppsala, Sweden.

Piper, P.W. Imperial Cancer Research Fund Laboratories, London, UK.

Sakano, H. Departments of Chemistry and Biology, University of California, San Diego, California, USA.

Schmeissner, U. Department of Molecular Biology, University of Geneva, Switzerland.

Schmitz, A. Department of Molecular Biology, University of Geneva, Switzerland.

Secher, D.S. MRC Laboratory of Molecular Biology, University Medical School, Cambridge, UK.

Sherman, F. Department of Radiation Biology and Biophysics, University of Rochester School of Medicine and Dentistry, New York, USA.

Smith, J.D. MRC Laboratory of Molecular Biology, University Medical School, Cambridge, UK.

Sprinzl, M. Max-Planck-Institut für Experimentelle Medizin, Göttingen, West Germany.

Stewart, J.W. Department of Radiation Biology and Biophysics, University of Rochester School of Medicine and Dentistry, New York, USA.

Waterston, R.H. Department of Anatomy, Washington University of St. Louis, St. Louis, Mo. 63110.

Wills, N. Cold Spring Harbor Laboratory, Cold Spring Harbor, New York, USA.

PREFACE

This book is a record of the proceedings of the EMBO Laboratory Course and Aarhus University 50 Years Anniversary Symposium on "Nonsense mutations and tRNA suppressors" held in Aarhus in July 1978.

Nonsense mutations and their suppressors have played a key role in the genetics and study of gene expression in bacteria and yeast. Also our knowledge of transfer RNA genetics, biosynthesis and structure-function relationships has depended to a large extent on the tRNA suppressors. The possibility of developing similar suppression systems in higher eukaryotes has stimulated extensive research on the characterisation of nonsense mutants and attempts to isolate nonsense suppressors in animal cells.

The main aim of the book is to introduce the reader to the field of translational suppression, specifically to nonsense mutations and tRNA suppressors. The book covers classic work on nonsense suppressors in prokaryotes and yeast as well as the latest developments in the search for nonsense mutations and tRNA suppressors in higher eukaryotes. To help the reader the book contains a few general chapters dedicated to tRNA and its role in protein synthesis as well as a compilation of wild type and mutant tRNA sequences.

We are grateful to EMBO, the Aarhus University and the Danish Natural Science Research Council for providing generous financial support and to the staff of Academic Press for their aid and cooperation in the production of this work.

The Editors.

CONTENTS

STRUCTURE AND FUNCTION OF tRNA

B.F.C. CLARK

Division of Biostructural Chemistry, Chemistry Department, University of Aarhus, Denmark.

Introduction

In the late 1950s no one could envisage how nucleic acids could programme the synthesis of protein by direct structural interactions with amino acids. However, Crick (1957) proposed his 'Adaptor Hypothesis' whereby an adaptor molecule (rather vaguely defined in terms of structural properties) would mediate between the amino acid and the nucleic acid which carried genetic information for specifying the amino acid sequence during polymerisation into protein. Shortly afterwards the Adaptor Hypothesis was confirmed by the discovery in rabbit liver extracts of a class of small RNA molecules capable of specifically binding amino acids (Hoagland *et al.*, 1959). Each example of this class of RNA molecule consists of a polynucleotide chain about 80 units long starting with a 5′-phosphate at one end and ending with a common sequence CpCpA at the other (3′) end. Later this class of small RNA molecule became known as transfer RNA describing its role in protein biosynthesis.

Transfer RNA functions in protein biosynthesis by carrying esterfied, activated amino acids to the ribosomal site in the correct order for peptide bond formation. During this central biological function the tRNA plays its role as an adaptor by decoding the genetic information carried by mRNA. Indeed, in modern terminology one could consider the tRNA as a molecular interface between the protein and nucleic acid languages.

The total tRNA in a cell makes up about 1% of dry weight in the case of a bacterial cell so that there are about 4×10^5 tRNA molecules of perhaps 55 different types in such a cell. Thus tRNA content corresponds to a concentration of about 0.5 mM, a figure to be borne in mind when trying to relate *in vitro* results to reality. Estimates of how much of the cellular tRNA is charged with an amino acid are notoriously difficult to obtain because of the lability of the aminoacyl bond to the tRNA. Probably estimates such as about 80% of the tRNA being charged (Lewis and Ames, 1972) are on the low side since it is likely that the catalytic activity of the adequate amount of aminoacyl-tRNA synthetase (activating enzymes) molecules in the cell keeps the tRNA fully charged provided that sufficient free amino acids are available.

Now that a three-dimensional structure is known for one tRNA species and more than 100 primary structures all with predictable secondary structures in clover leaf forms (Holley *et al.*, 1965) there is considerable motivation for attempting to explain tRNA function in terms of structure. At present, naturally, most of this work concentrates on explaining tRNA function during protein biosynthesis because so much more has been discovered about the biochemical processes involved, thanks to the research stimulus during the elucidation of the genetic code in the 1960s. Although the availability of the crystal structure of a tRNA provides a new meaningful basis for discussion of function it must be remembered that this structure gives only one conformation or view of the tRNA in its most stable or resting conformation (Ladner, 1978). How this conformation changes during biological function especially during protein biosynthesis is an intriguing and tantalizing problem currently engaging many different types of research workers using X-ray crystallography, chemical modification studies, enzyme kinetic measurements and physical chemical techniques such as proton magnetic resonance, electron spin resonance, laser light scattering and fluorescence.

In this chapter I shall concentrate on the role of tRNA in protein biosynthesis, but in the last section it will be pointed out that tRNA is becoming more and more fascinating as it is being implicated in more and more biological activities other than its traditional adaptor role. Possibly due to evolutionary pressure, biological economy and tRNA probably being as old as life itself, tRNA has become imbued with its multifunctional role.

This chapter does not aim to be all embracing of the extensive tRNA literature so I apologize for being arbitrarily selective in reference citations. A much more extensive literature survey will be found in a recent review by Rich and RajBhandary (1976).

Role of tRNA in Protein Biosynthesis

A summary of the current knowledge about the biological function or, more precisely, biological activities of tRNA in protein biosynthesis is given in Table I. Although much is known about the role of tRNA in protein biosynthesis even in this area some of the activities listed have not been characterized adequately enough to elevate them to functions.

TABLE I

Biological Activities of tRNA in Protein Biosynthesis

Elongation	*Initiation*
1. Activation of amino acids	
2. Recognition by EF-Tu	6. Recognition of initiator tRNA by IF
3. Location in A-site	7. Location in I-site (part of P-site)
4. Decoding mRNA	
5. Signal for 'magic spot'	8. Recognition by transformylase
9. Regulation: a) Repressor; b) Feedback Inhibitor; c) Suppression	

In summary for normal protein biosynthesis each tRNA species is charged with an amino acid (activity 1 of Table I) by an aminoacyl-tRNA synthetase ('activating enzyme'), and then the charged species is carried to the ribosome in the form of a ternary complex made with the elongation factor Tu (EF - Tu) and GTP (activity 2). In similar fashion the unique tRNA species, the initiator tRNA, a special class of methionine tRNA, is thought to be carried to the ribosome by an initiation factor and probably GTP as well (activity 6). The aminoacyl-tRNA is located by an uncharacterized mechanism in the A-site of the ribosome (activity 3) where it decodes mRNA via its anticodon triplet (activity 4). In contrast the initiator tRNA, formylmethionyl-$tRNA_f^{Met}$ in prokaryotes, and methionyl-$tRNA_f^{Met}$ sometimes called Met-$tRNA_i$ in eukaryotes, is located in the initiation (I)-site on the small ribosomal subunit (activity 7) for decoding the initiator triplet codon. This site becomes part of the ribosomal P-site. Since the prokaryotic Met-$tRNA_f$ has to be formylated for its initiator activity, it is also recognized by a special enzyme for this, the transformylase (activity 8).

When prokaryotic cells are starved for amino acids an unusual role has been detected for uncharged tRNA (Pedersen *et al.*, 1973). The uncharged tRNA is bound to the ribosomal A-site as if in mRNA decoding, but sets off a signal for the formation, by the so-called stringent factor, of unusual guanosine nucleotide derivatives, ppGpp and pppGpp, called 'magic spots' (activity 5).

A group of assorted roles for tRNA in the regulation of protein biosynthesis has been collected as activity 9. These include bacterial roles as a repressor of the histidine operon (Lewis and Ames, 1972) a less defined regulator of amino acid biosynthesis (Allende, 1975) and the well characterized suppressor of nonsense mutations (Smith, 1972 and this volume) and a possible role as a feedback inhibitor in yeast (Bell *et al.*, 1976). In eukaryotes, there are additional non definitive roles relating to evidence on the binding to tryptophan pyrrolase in Drosophila (Rich and RajBhandary, 1976) and the inhibition of protein synthesis in virally infected animal cells by the degradation of one or more essential tRNA species, a process that seems to accompany interferon production (Revel, personal communication).

Another set of phenomena concerning amounts of various isoaccepting species existing in different types of cells at various stages in growth or transformation can be classified under a general regulatory role. Obviously restricting the amount of one specific decoding aminoacyl-tRNA will control protein biosynthesis at this point in translation (Smith, 1975; Sharma *et al.*, 1976). However, despite the large literature on the subject, especially in connection with control of protein synthesis in carcinogenic states the regulatory role of tRNA has not been clearly characterized (Rich and RajBhandary, 1976).

Because of our present structural knowledge of tRNA it is now appropriate to study the molecular mechanism of protein biosynthesis in structural terms. We wish to know what happens to the tRNA structure during charging of the tRNA with an amino acid, what happens to the tRNA on the ribosome and how the tRNA gets to the ribosome.

A tRNA is enzymically linked to its specific amino acid at the 2′ or 3′ hydroxyl group on its 3′ - terminal adenosine, by its specific (cognate) aminoacyl-tRNA synthetase (activity 1 of Table I). Another chapter in this volume (by Bruton) describes what is known about the structure of this class of enzymes. Figure 1 summarizes the 2 step reaction concerned in charging the tRNA.

Since there is only one activating enzyme for charging several different tRNAs (isoaccepting species) with the same amino acid it was hoped that a knowledge of the several tRNA primary structures would reveal how the activating enzyme recognized them. Unfortunately, however, these studies have not been very helpful. Similar

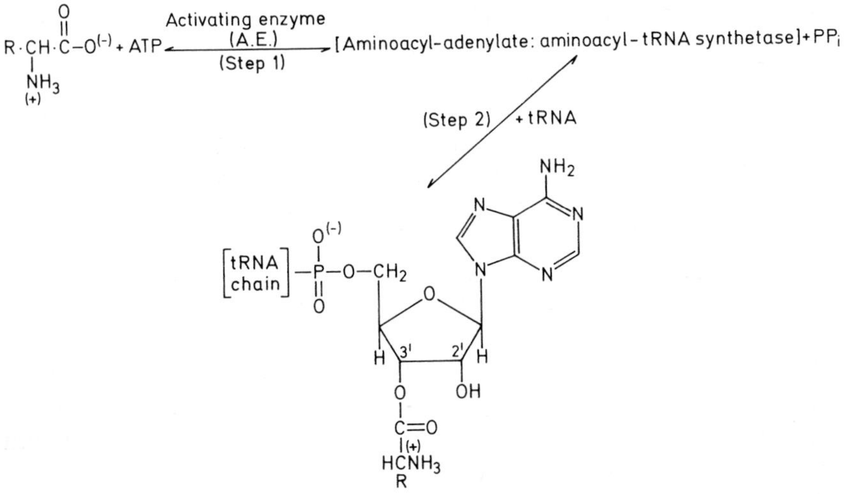

$$R\cdot CH(NH_3^{(+)})\cdot C(=O)-O^{(-)} + ATP \underset{\text{(Step 1)}}{\xrightleftharpoons[\text{(A.E.)}]{\text{Activating enzyme}}} [\text{Aminoacyl-adenylate: aminoacyl-tRNA synthetase}] + PP_i$$

(Step 2) + tRNA

Fig. 1 A two-step scheme for charging a tRNA with an amino acid by an activating enzyme (an aminoacyl-tRNA synthetase).

feelings have been expressed for the reciprocal comparative studies of the enzymes' amino acid sequences. Clearly the folded tertiary structure provides the important points of contact for recognition and it is still an intriguing problem how such a synthetase recognises a particular tRNA when we believe that many of the tRNA tertiary structures are similar. This enzyme specificty for tRNA and amino acid clearly is an important feature in ensuring fidelity of translation of the genetic code. What we really do need to solve this problem is an X-ray crystallographic analysis of a complex of an amino-acyl-tRNA and its cognate synthetase. So far no crystals containing such a complex have been reported so other physical and chemical methods are being used to gather useful information.

When the recognition site of a tRNA for its activating enzyme is discussed an uncertainty arises. There is no guarantee that every tRNA is recognized in a similar fashion by some property of a similarly located set of nucleotides. Perhaps the recognition of particular tRNAs by their specific activating enzymes has evolved differently, so that by now there are different classes of tRNA recognition features. If this is correct no generalization about the recognition process will be possible when details of recognition of one tRNA by its activating enzyme are known.

The possible different recognition features have even been observed in the different points of attachment of different amino acids to their specific tRNAs (for a review, see Sprinzl and Cramer, 1978). In recent years the problem of the point of attachment has been dramatically solved for a series of synthetases. At first it was thought that the amino acid was attached to the 2′ -OH group of the 3′ terminal A, as for example is the case for methionine and phenylalanine. The story has been complicated by other studies where the attachment point is the 3′ -OH group (as shown in Figure 1) as in the case of amino acids such as lysine and serine. In addition there are a few cases where the attachment point occurs at both 2′ and 3′ -OH groups as in the case of tyrosine. A reasonable explanation is that the stereochemical details of the substrate binding site of a cognate tRNA:synthetase pair have evolved independently of other cognate pairs. Indeed, the actual point of attachment may not be very important because in solution aminoacyl migration can occur between the two 2′ and 3′ -hydroxyl groups of the terminal A about 1000 times faster than peptide bond formation on the ribosome; the latter rate is about 10 per sec. There is good biochemical evidence that the amino acid must be attached to the 3′ -OH group for peptide bond formation on the ribosome so the amino acid ends up in this position even if it was originally attached to the 2′ -OH.

Our knowledge of what happens to the tRNA on the ribosome (activities 3 and 4 of Table I) is rather meagre in spite of a great body of results in this field, but it is likely that with our increasing knowledge about the ribosome there will soon be some clarification.

After polypeptide chain initiation a series of peptide chain elongation steps occurs before the process of chain termination releases the completed protein.

The generally accepted scheme for peptide chain elongation which involves relative movement of the mRNA and ribosome is shown as a cyclic scheme for peptide bond formation in Figure 2. This scheme is simplified in that a static view of the ribosome involving special tRNA binding sites is assumed. At present there is no strong evidence to persuade us to drop this view and think in terms of a more dynamic situation involving activated binding states.

The cyclic scheme shown is quite self-explanatory, starting with a situation in state A with a growing peptide attached to $tRNA_n$ in the peptidyl-tRNA binding site (P-site) of a 70 S bacterial type ribosome decoding codon n of the mRNA and, an aminoacyl-$tRNA_{n+1}$ decoding codon n + 1 in the aminoacyl-tRNA binding site (A-site). The mRNA is bound to the 30 S subunit and the tRNA stretches across both 30 S and 50 S subunits. The peptide bond is made by the enzyme peptidyl

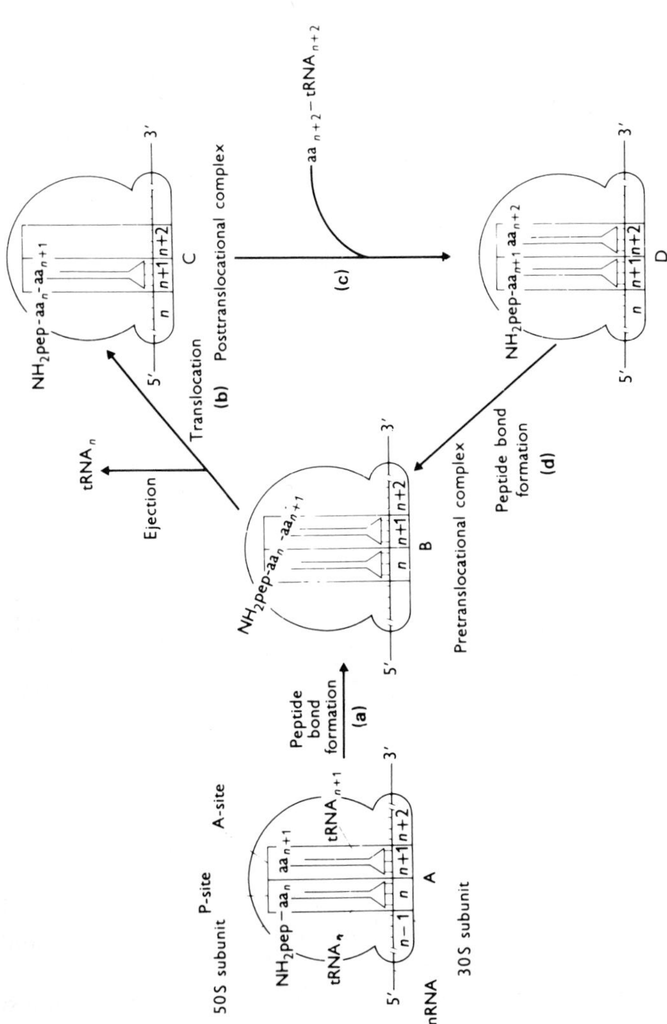

Fig. 2 Cyclic scheme for peptide bond formation on the ribosome in terms of two tRNA binding sites.

transferase on the 50 S subunit in step (a) leaving, in state B, an uncharged $tRNA_n$ in the P-site and a new peptide extended by one amino acid, aa_{n+1}, attached to $tRNA_{n+1}$ in the A-site. Movement of the $tRNA_{n+1}$ and mRNA occurs in step (b) to free the A-site state C for a new incoming aminoacyl-$tRNA_{n+2}$ in step (c). Step (b), involving movement of $tRNA_{n+1}$ with concomitant ejection of $tRNA_n$, is usually called translocation. What causes the ejection and relative motion is a stimulating area of research. When the new aminoacyl-$tRNA_{n+2}$ is bound in the A-site as in state D, the ribosome is back to a state equivalent to A ready for a new round of peptide bond formation and translocation giving the cyclic feature to the scheme.

Much work is in progress using physical chemical techniques, affinity labelling and chemical crosslinking to identify the protein and nucleic acid environments for the tRNAs bound in the different ribosomal sites.

Actually the situation is more complicated than in the simplified version of protein synthesis shown in Figure 2 because in this scheme the role of supernatant factors is ignored. Two protein elongation factors EF - T and EF - G are involved in the elongation step. It is believed that a subunit of EF -T called EF -Tu actually carries the aminoacyl-tRNA to the ribosomal A-site in the form of a ternary complex composed of aminoacyl-tRNA, GTP and EF - Tu. The cyclic scheme shown in Figure 3 is generally accepted (see a review by Miller and Weissbach, 1977) for the way in which EF - Tu plays a role in the elongation step. There is currently much research into the biochemistry and structure of EF - Tu.

EF - Tu provides a good contrast in the problem of specific recognition of aminoacyl-tRNA to an aminoacyl-tRNA synthetase because EF - Tu recognizes all aminoacyl-tRNAs whereas a synthetase recognizes at most a set of isoaccepting species. The details of how the ternary complex binds to the ribosome and how the EF - Tu:GDP is released after hydrolysis of the GTP are not known. However, the regeneration of EF - Tu:GTP is thought to be brought about catalytically by another protein factor, EF - Ts, which binds to EF - Tu to make the bimolecular EF - T. Interestingly there has been found a large amount of EF - Tu in bacterial cells - about 5% of total protein giving a concentration of about 0.3 mM which is of the same order as the aminoacyl-tRNA concentration. Perhaps the high amount of EF - Tu is needed to sequester the aminoacyl-tRNA during protein biosynthesis. Although no X-ray pictures of ternary complex crystals have yet been produced there is good progress towards solving the three-dimensional structure of EF - Tu:GDP.

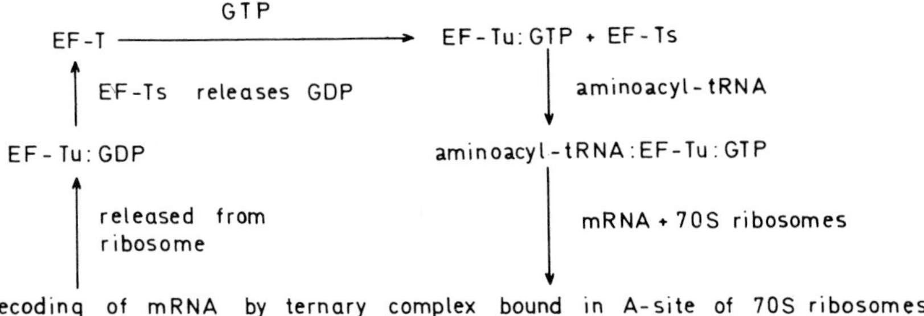

Fig. 3 Cyclic scheme for role of EF-Tu in protein biosynthesis.

A 6Å low resolution model has been described by Kabsch *et al.*, (1977) and a 2.7Å high resolution map showing the position of GDP and part of the polypeptide backbone has been reported by Morikawa and coworkers (1978).

Very little is known so far about how the tRNA could be complexed with EF - Tu (for review see Miller, 1978). However, an interesting speculation on the role of EF - Tu in providing a checking mechanism to increase fidelity of decoding has been proposed by Sprinzl and Cramer (1978). Based on some biochemical studies with analogues of normal aminoacyl-tRNAs to prevent aminoacyl migration, they have suggested that EF - Tu holds the aminoacyl-tRNA in the ternary complex with GTP so that the amino acid is linked to the 2′ -OH group of the 3′ -terminal A. Only after codon checking has taken place would aminoacyl migration occur prior to peptide bond formation.

Generalized Primary Structure of tRNA

The information from 101 different tRNA sequences (primary structures) known in July 1978 and listed in Table II (Barrell and Clark, 1974; Clark, 1977 and 23 new structures, Sprinzl *et al.*, 1978 and 10 structures, Sprinzl *et al.*, this volume) has been conveniently incorporated into standard 'cloverleaf' forms as shown in Figure 4. This remarkable feature of all the primary structures was first proposed by Holley and his colleagues (1965) and is based on Watson-Crick base pairing which forms double stranded helical stretches (the secondary structure). The simple classification shown in Figure 4 is based on size (see Table II for species).

Thus we have small and large tRNAs dependent upon the size of the extra arm (see also Figure 5 and Table II). The 10 new structures not listed by Sprinzl et al., (1978) are T4 Arg (Mazzara *et al.*, 1977)

TABLE II

Classes of tRNA according to Arm Sizes and Structure Correlations

Class 1 (86)	
A (50)	4 base pairs in b-stem (D stem) 5 bases in extra loop and containing m^7G
(and with A9)	Ec Ala_1, Sw Ala_1, Sw Ala_2, Ec Arg_1, Ec Asn, Ec Asp_1, Ec Gly_1, Ec Gly_3, Ec + Sal His_1, Ec Ile_1, Ec Lys, Bsu Lys_1, y Lys, Rbl $Lys_{2A(2B)}$, Rbl Lys_3, Svt Lys_4, An Met_f, Ec Met_m, y Met_3, Mye + Rbl Met_4, Ec Phe, Mp Phe, Bs Phe, Bsu Phe, y Phe, Wg + Ps Phe, Egc Phe, Bc Phe, Rbl + Hup Phe, T4 Pro, Ec Thr, Bsu Thr, T4 Thr, Ec + su^+ Trp, Ec Val_1, Ec Val_{2A}, Ec Val_{2B}, Bs Val_{2A}, Mye + Rbl + Hup Val_1.
(and with m^1G9 or G9)	R1 Asn, y Cys, Ec Met_{f1}, Ec Met_{f2}, Bsu Met_f, Tt Met_f, Nc Met_f, Mye + Rbl + St + X1 + Smg + Hup Met_f, yp Phe, y Trp, Chi + B1 + Rbl Trp_1.
B (7)	4 base pairs in b-stem (D stem) 5 bases in extra loop III without m^7G
	y Ala_1, Tu Ala_1, y Arg_3, Hay Lys, Mp Met_f, B1 + Rbl Trp_2.
C (10)	4 base pairs in b-stem (D stem) 4 bases in extra loop
	T4 Arg, y Asp_1, Ec Glu_1, Ec Glu_2, Ec + Sal Gly_1, Sta Gly, Wg Gly, y Gly, Sw Gly_1, Sw Gly_2.
D (19)	3 base pairs in b-stem (D stem) small extra arm with no. of bases (3-5)
	y Arg_2 (5), Ec Cys (4), Ec Gln_1 (5), Ec Gln_2 (5), T4 Gln (5), y Glu_3 (4) Ec Gly_2 (4), T4 Gly (4), Tu Ile (5), Ncm Met_f (4), Mul Pro (5) y Thr_{1A}(5), y Thr_{1B} (5), y + su^+ Tyr (5), Tu Tyr (5), y Val_1 (5), y Val_{2A} (5), y Val_{2B} (5), Tu Val (3)
Class 2 (15)	
	3 base pairs in b-stem (D stem) large extra arm with no. of bases (13-21)
	Ec + Sal Leu_1 (15), Ec Leu_2 (15), T4 Leu (14), y Leu_3 (13), y Leu_4 (13), Ec Ser_1 (16), Ec Ser_3 (21), T4 Ser (18), y Ser_1 (14), y Ser_2 (14), R1 Ser_2 (14), R1 Ser_3 (14), Ec + su_3^+ Tyr_1 (13), Ec Tyr_2 (13), Bs Tyr (13).

Footnote to Table II:

su	= Suppressor	An	= *Anacystis nidulans*
Bc	= Bean chloroplast	Bl	= Beef liver
Bs	= *Bacillus stearothermophilus*	Bsu	= *Bacillus subtilis*
Chi	= Chicken	Ec	= *Escherichia coli*
Eg	= *Euglena gracilis*	Egc	= *Euglena gracilis* chloroplasts
Hay	= Haploid yeast	Hup	= Human placenta
Mp	= Mycoplasma	Mul	= Murine leukemia virus
Mye	= Myeloma	Nc	= *Neurospora crassa*
Ncm	= *Neurospora crassa* mitochondria	Ps	= *Pisum sativum*
Rbl	= Rabbit liver	Rl	= Rat liver
Sal	= *Salmonella typhimurium*	Sf	= *Streptococcus faecalis*
Smg	= Sheep mammary gland	St	= Salmon testis
Sta	= Staphylococcus	Stf	= Starfish
Svt	= Svt 2 cells	Sw	= Silk worm
T4	= Bacteriophage T4	Tt	= *Thermus thermophilus*
Tu	= *Torulopsis utilis*	Wg	= Wheat germ
Xl	= *Xenopus laevis*	y	= Yeast, undefined and *Saccharomyces cerevisiae*
yp	= Yeast, *Saccharomyces pombe*		

Rl Asn (Chen and Roe, 1978), Ec Cys (Mazzara and McClain, 1977) Rbl $Lys_{2A(2B)}$, Rbl Lys_3, Svt-Lys_4 (Gross, personal communication), Sf Met_f (Delk and Rabinowitz, personal communication), ypPhe (McCutchan et al., 1978), Mul Pro (Dahlberg, personal communication) and T4 Thr (Guthrie *et al.*, 1978).

In Figure 5 the information from the small class 1 sequences has been incorporated into a standard generalized cloverleaf. To obtain this information 74 of the 86 class 1 species indicated in Table II have been used: clear exceptions based on functions, such as the initiator Met_f species and the cell wall Sta Gly, to the generalized form were omitted.

As shown in Figure 5 the Watson-Crick base pairs give rise to four double helical stem regions a, b, c, and e, three of which are closed by non-base paired loop regions I, II and IV. These four stems give rise to the clover leaf arrangement of the secondary structure. Whether there is a functional role for the secondary structure is still unclear. Another point of nomenclature illustrated in Figure 5 is that a stem plus a loop is also called an arm. Most of the tRNA cloverleaf forms have remarkably constant regions. There is a phosphate at the 5′ -end, and at the 3′ -end where the amino acid is attached there is a common sequence CpCpA. In addition stems a, c, and e contain 7, 5 and 5 base pairs respectively and loops II and IV each contain 7 non-base paired nucleotides.

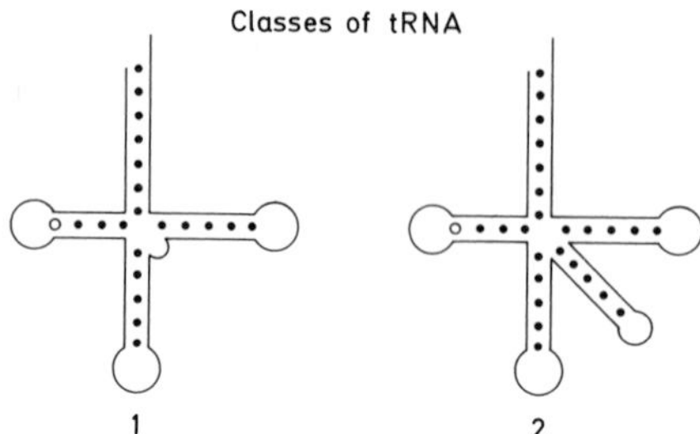

Fig. 4 Classes of tRNA. The open circle indicates that a Watson-Crick base pair is not always found in this position.

Stem b contains 3 or 4 Watson-Crick base pairs (Figure 5, open circles in stem b show the position of the possible non-Watson-Crick base pairs, and see Table II) but our knowledge of the three-dimensional structure of tRNA (Robertus et al., 1974a; Kim *et al.*, 1974) permits us to propose that the various non-Watson-Crick base pairs occurring in the fourth position of the stem b can be accommodated without distortion of the tertiary structure (Clark and Klug, 1975). There is then no point in differentiating between tRNAs on the basis of whether this position is a Watson-Crick base pair or not. The variable regions are confined to loops I and III and stem d. The arms can be also referred to for convenience by trivial historical names as shown in Figure 5, e.g. loop I + stem b as the D arm since it usually contains some D bases: loop II + stem c can be called the anticodon (ac) arm since the loop contains the anticodon; loop III + stem d, the variable finger or extra arm; and loop IV + stem e, the TΨC arm. Stem a is also called the amino acid (aa) stem since this is where the amino acid is attached.

There are also many invariant and semi-invariant nucleotide positions in the generalized structure. The invariant positions are shown by nucleoside letters whereas the semi-invariant positions are shown by R (signifying a purine nucleoside A or G) or Y (signifying a pyrimidine nucleoside C or U). The dotted stretches in the diagram are regions of variable length. For example, and including class 2 tRNAs, the extra arm can vary from 3 to 21 nucleotides in length. In contrast the D loop I is much less variable in length - from only 7 to 10 nucleotides long when the base pair b4, even though not of the Watson-Crick type, is considered as part of the D stem. For convenience the variable dotted regions can be named α, β and γ.

Until recently what the invariant and semi-invariant nucleosides represented was an interesting but open question. It was not possible

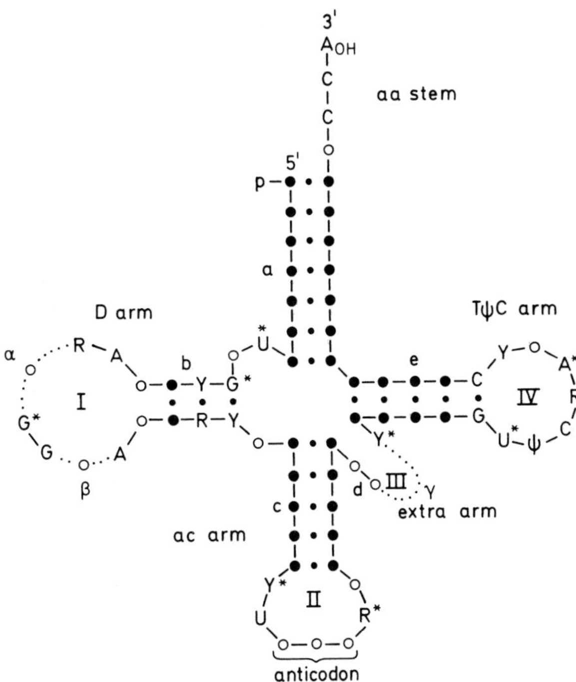

Fig. 5 Class 1 generalized cloverleaf. Solid circles ● indicate bases involved in helical stems containing Watson-Crick base pairs. Open circles o signify non-Watson-Crick base paired bases in cloverleaf arrangement. Starred nucleosides are positions where modifications of that nucleoside can possibly occur. This nomenclature is based on an original proposal by Clark, B.F.C., Jukes, T.H. and Cohn, W.E.

to decide whether these positions in the primary structure were conserved for metabolic or for structural reasons. Now that a three-dimensional structure has been determined (Robertus *et al.*, 1974a; Kim *et al.*, 1974; Ladner *et al.*, 1975; Quigley *et al.*, 1975) the primary and secondary structural information can be viewed in a new light and we can see a structural role for most of these special nucleosides. In this context it should be pointed out that a semi-invariant base pair in the D stem at position b2 (see Figure 5) was noted only after a knowledge of additional H-bonding interactions between bases and backbone in the tertiary structure (Ladner *et al.*, 1975). One invariant position which has not been properly explained in terms of structure is the constant Ψ in the TΨC loop.

As shown in Table II the class of small tRNAs (class 1) can be conveniently subdivided according to structural characteristics.

Class 1 (86 structures) contains 4 base pairs in the D stem (stem b) and a small extra loop (3-5 nucleotides). Class 2 (15 structures) has only 3 standard base pairs in the D stem and a large extra loop of 13 to 21 nucleotides. On the basis of the sequences in the extra loop and standard base pairs in the D stem, class 1 can be conveniently divided into subclasses A, B, C and D. Subclass 1A contains 5 nucleotides, one of which is m^7G, in the extra loop (III), 1B also contains 5 nucleotides in the extra loop, but now there is no m^7G, subclass 1C is irregular in that its extra loop contains only 4 nucleotides and subclass 1D contains only 3 Watson-Crick base pairs in the D stem. The tertiary structure has been determined for yeast $tRNA^{Phe}$, which belongs to the subclass 1A but the structure appears relevant to subclass 1B and will probably also accommodate subclass 1C and subclass 1D structures. For example a short extra arm, only 4 nucleotides in length, but making all the tertiary interactions, can be built with the excision of the residue U47, which is not involved in the tertiary bonding found for yeast $tRNA^{Phe}$.

The list of different structures shown in Table II is somewhat arbitrary since some closely similar structures such as the minor glutamic acid $tRNA_1$ (Ec Glu_1) and major glutamic acid $tRNA_2$ (Ec Glu_2) are both listed whilst uncertain sequences and spontaneous mutants showing evidence of gene duplication are not. In comparison with previous lists (Clark and Klug, 1975; Clark, 1977) Ec Arg_2 has been omitted since it is now agreed that the sequences determined for Ec Arg_1 and Ec Arg_2 are the same with Ec Arg_2 being correct but containing the modification reported for Ec Arg_1 (K. Chakraburtty and S. Nishimura, personal communication).

Biological Sources of Known Primary Structures

A simple way to review our state of knowledge regarding tRNA primary structures is given by Table III (see also Dirheimer *et al.*, 1976). Here I have listed known sequences chargeable with each of the 20 essential amino acids for the major sources so far studied. The table enumerates each of a number of published isoaccepting species including different strains and different cell types with +. Sequences not yet published are also indicated by ×. Plant and insect sources have been omitted. Currently there are known tRNA sequences corresponding to all 20 amino acids but this is not the case for one biological source. However, where bacteria is the source material tRNA sequences for all amino acids except for Pro are known. Nevertheless, as the table shows the sequence of bacteriophage $tRNA^{Pro}$ is known. Yeast has been the next most popular source material (28 sequences) after bacteria (including mycoplasma, 41 sequences) with mammalian cells still much less used (13 sequences) largely for reasons of availability and technology. It is generally accepted that there are of the order of 55 different tRNA gene types in a particular cell. So even for the case of the most popular bacterial sources, *Escherichia coli* (30 sequences) there is a long way to go before the complete set of primary structures will be known for one source material. Although there are considerable differences in sequences for different isoaccepting species, evidence so far for several different amino acids suggests that tRNA sequences are largely conserved in different mammalian cell types (see Table II and Dirheimer *et al.,* 1976).

TABLE III

List of tRNAs of known Primary Structure from different Sources

Charging Amino Acids	Bacterio-phage	Bacteria	Yeast	Mammals
Ala		+	++	
Arg	+	+	++	
Asn		+		+
Asp		+	+	
Cys		+	+	
Gln	+	++		
Glu		++	+	
Gly	+	++++	+	
His		+		
Ile		+	+	
Leu	+	++	++	
Lys		++	++	xxx
Met_f		+++++	+	+
Met_m		+	+	+
Phe		++++	++	+
Pro	+			x
Ser	+	++	++	++
Thr	+	++	++	
Trp		+	+	++
Tyr		+++	++	
Val		++++	++++	+

Recent technological developments in RNA sequencing has made tRNA sequencing easier especially so that small amounts of pure tRNA (about 10 μg) are enough to digest and radioactively label to gain information for a primary structure. DNA sequencing developments have shown a spin off in the RNA field by the recent reports by Simonscits *et al.*, (1977) and by Stanley and Vassilenko (1978). These developments should enable the sequence determination of many more mammalian tRNAs which are difficult to isolate.

Discussion of Exceptions to the Generalized Structure

Several exceptions to the generalized structure shown in Figure 5 are enumerated below and their locations in the generalized cloverleaf are shown in Figure 6. Apart from these, non-Watson-Crick base

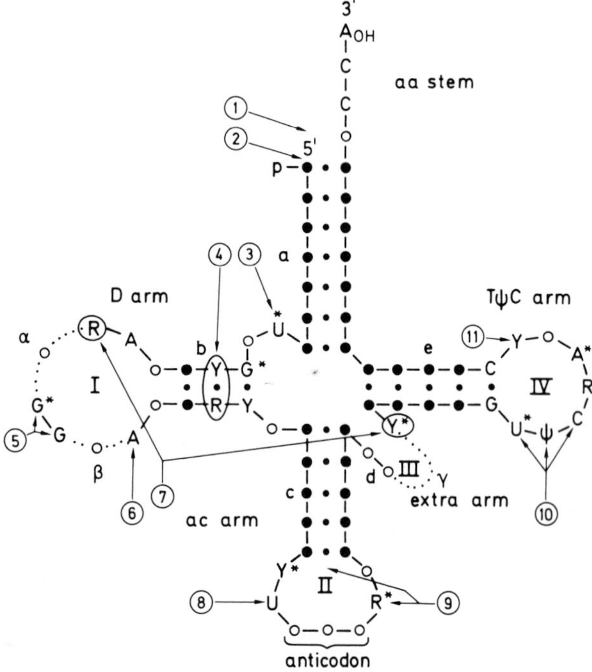

Fig 6 Exceptions to the generalized cloverleaf referred to in the text. The locations of the exceptions are given by the arrowheads. The locations with circled numbers are described accordingly in the text.

pairs sometimes occur in helical stems. The most frequent non-Watson-Crick base pairs occurring in stems are G·U, G·Ψ and A·Ψ.

The positions of G·U base pairs and Ψ in double helical stems will be discussed later. In these stems G·Ψ is considered equivalent to G·U and A·Ψ to A·U. When the simplified classification of tRNA structures into two classes is used (Figure 4) then the other non-

Watson-Crick base pairs which occur in stems are U·U, C·C, G·A, C·A, Ψ·U and Ψ·Ψ. No more than one of these unusual base pairs is found in a stem and there are only few instances altogether. Two interesting examples of such exceptions are found for position c5 in the ac stem: here the tRNAs for both yMet$_3$ and Mye + Rbl Met$_4$ contain Ψ·Ψ.

The exceptions cited in Rich and RajBhandary (1976) and Dirheimer et al., (1976) to the standardized cloverleaf are shown in Figure 6 and listed as follows:

1) Ec tRNA$_1^{His}$ contains 8 base pairs in the aa stem. It is thus longer by one nucleotide at the 5′-end and the single stranded region at the 3′-end of the aa stem contains only CpCpA.

2) Prokaryotic initiator tRNAs, tRNA$_f^{Met}$, do not contain a Watson-Crick base pair at position al of the aa stem, e.g. C·A in Ec tRNA$_f^{Met}$.

3) The modified U is s^{4}U in most *Escherichia coli* tRNAs but an unmodified U in yeast and mammalian tRNAs.

4) Prokaryotic initiator tRNAs contain an A·U instead of Y-R in the standard cloverleaf at position b2.

5) tRNAs which participate in protein biosynthesis have G-G or G*-G at location 5 (of Figure 6), but cell wall biosynthesis Sta tRNA^{Gly} species have this constant doublet replaced by U-U.

6) The second constant A in the D loop I is replaced by G in some class 2 tRNAs, e.g. Ec tRNA$_1^{Leu}$, and Ec tRNA$_2^{Leu}$ and by D in Ec tRNA^{Cys}.

7) Levitt (1969) noticed a common base change in tRNAs with different purines at position 15 (location 7 in Figure 6) co-ordinated with position 48 (numbering as in yeast tRNA^{Phe}). Usually there is a G15·C48 or A15·U48. Exceptions are A15·C48 in Ec tRNA$_1^{Gly}$ and y tRNA$_3^{Glu}$, and G15·G48 in Ec tRNA^{Cys}.

8) The constant U in the anticodon loop is replaced by C in mammalian tRNA$_f^{Met}$.

9) The modified R in the anticodon loop is replaced by a pyrimidine Y (usually C) in cell wall tRNA^{Gly} and there is a possible sixth Watson-Crick base pair in the ac stem.

10) The invariant sequence G-U*-Ψ-C where U* is T,U,Ψ or s^{2}T is changed to G-A-U-C in eukaryotic initiator tRNAs, to G-A-Ψ-C in the recently determined silk worm alanine tRNAs (Sw Ala$_1$ and Sw Ala$_2$, K.U. Sprague *et al.,* 1977) and to G-U-G-C in cell wall tRNA^{Gly} but also in part of Wg tRNA^{Gly}.

11) The semi-invariant Y in the TΨC loop is replaced by A in eukaryotic initiator tRNAs but also in Mye tRNA$_1^{Val}$, Sw tRNA$_1^{Ala}$ and Sw tRNA$_2^{Ala}$.

The most important exceptions to the standard cloverleaf concern tRNAs which have special functional roles such as initiator tRNAs and glycine tRNAs involved in cell wall metabolism. These special exceptions are considered to point out positions in the cloverleaf which could be concerned with these special functions. However, this kind of interpretation may be too simplistic. Nevertheless, the special role of prokaryotic initiator tRNAs may be related to their exceptional structures at locations (2) and (4) in Figure 6. Previously eukaryotic initiator tRNAs appeared to be characterized by the special sequence A-U instead of U* -Ψ in TΨC loop IV (location 10 in Figure 6). Now that Sw $tRNA^{Ala}_{1+2}$ contain A-Ψ at this position the special initiating role can be related only to U replacing the constant Ψ. In the case of mammalian initiator tRNAs, C before the anticodon (location 8 in Figure 6) may also play a special role.

Since the Sta $tRNA^{Gly}$ species do not play a role in protein biosynthesis their lack of constant features should point out parts of the generalized tRNA involved in protein biosynthesis. It is possible that the common sequence of the TΨC loop missing in Sta $tRNA^{Gly}$ is involved in ribosomal site binding of the tRNA and there is experimental evidence supporting this proposition (Erdmann *et al.*, 1973). The lack of the constant G*-G of the D loop in the cell wall tRNAs also implicates this constant feature in protein biosynthesis. However, rather than being a recognition point for some enzyme it may have a structural role because we know that it is concerned in holding the tertiary structure of yeast $tRNA^{Phe}$ together.

From the lack of a purine after the anticodon in the cell wall tRNAs we may infer that the purine is a structural requirement concerning proper fitting during the anticodon-codon decoding interaction. However, this feature may also be influenced by the possible sixth Watson-Crick base pair in the ac stem of cell wall tRNAs.

In addition to the pregoing exceptions to the generalized cloverleaf it should be pointed out that there is possibly one exception to what we have considered as possible sizes in the extra arm, loop III. Our knowledge of the tertiary structure permits the ready accommodation of a loop III containing as few as 4 bases. However, a shorter length such as 3 reported for Tu $tRNA^{Val}$ (see Barrell and Clark, 1974) does not appear possible based on the tertiary structural interactions found in y $tRNA^{Phe}$. Thus the sequence of Tu $tRNA^{Val}$ can be considered as a structural exception.

With respect to this discussion it should be pointed out that the short 3 nucleotide long extra loop III region previously reported for y $tRNA^{Gly}$ (see Barrell and Clark, 1974) has now on reinvestigation been shown to contain 4 nucleotides in reality (M. Yoshida, personal communication).

Non-Watson-Crick Base Pairs and Possible Significance of G·U

The wealth of tRNA primary structural information now available permits several interesting analyses of the secondary structure. In particular we have examined the occurrence of non-Watson-Crick base pairs such as G·U, U·U, C·A or G·A in stem regions. With the occasional exception of G·U there tends to be only one of these in a given molecule, and where there is more than one, no two are adjacent. It is therefore likely that the RNA double helix can accommodate a single non-Watson-Crick base pair without a serious distortion.

The most common non-Watson-Crick base pair occurring in tRNA is G·U. Figure 7 shows the positions and frequency of occurrence of this base pair in the known class 1 primary structures. The frequency of occurrence of the probably equivalent G·Ψ pair is shown in brackets. So far G·U base pairs have not been found in stem positions b3, c3 and c5 or in e4 and e5. Although G·U base pairs occur frequently and according to our knowledge of the tertiary structure (Ladner *et al.*, 1975; Quigley *et al.*, 1975) without gross helix distortion, there appears to be a particularly strong conservation of pure Watson-Crick base pairs in the loop ends of the anticodon stem (c) and the TΨC stem (e). A structural reason for this conservation is probable.

Why do G·U base pairs and other non-Watson-Crick base pairs occur at all in the stem regions? Clearly a G·U base pair in the middle of a piece of helix requires a certain accommodation of the helix backbone if two H-bonds are to be made. This is therefore a potential point of 'weakness' or rather 'irregularity' in such a stem or possibly a special point for enzyme recognition. For example this could apply to the G·U base pairs in the aa stem (a).

G·U base pairs at the end of a helix may play a different role (Clark and Klug, 1975). Thus it has been found in the 0.25 nm model of yeast $tRNA^{Phe}$ that the phosphate of nucleotide 49 in yeast $tRNA^{Phe}$ (i.e. at the break in the long double helix formed by the aa stem on the TΨC stem, (see later)) is moved from its regular double helical position in order to allow the adjacent nucleotide in the extra loop to make an almost right angle bend with it. A G·U base pair in this position could be very helpful in relieving this tight corner. It is perhaps significant that there is the highest frequency of G·U pairs observed in the base pair position e1, and moreover that in 13 of 14 cases, the G is always on one strand and U(Ψ) on the other. Similar

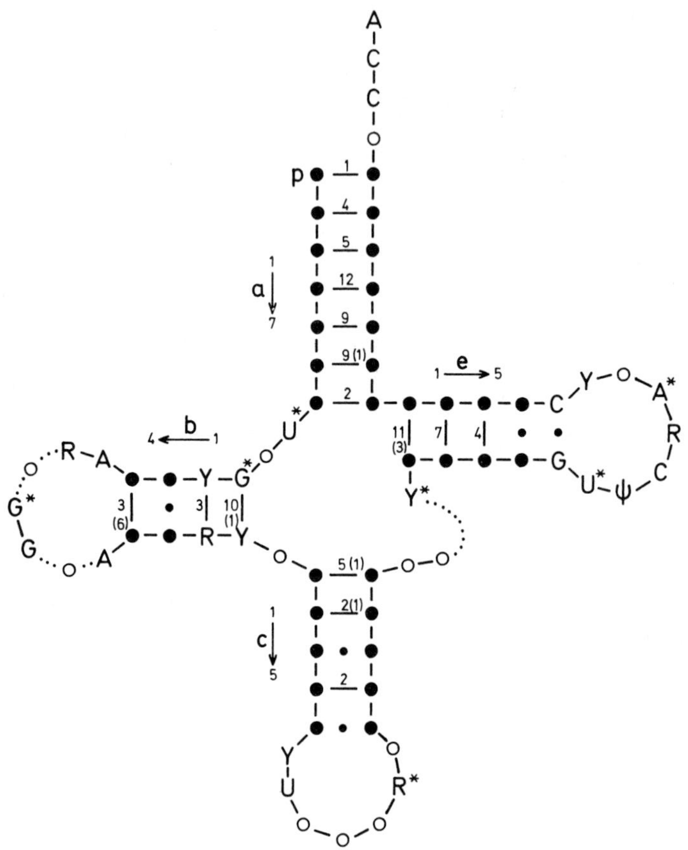

Fig. 7 The positions of G·U base pairs in the Class 1 cloverleaf, shown by ●—● The number of occurrences of each position are given by the numbers, with the number of G·Ψ occurrences in brackets. The arrows give the direction of the standard numbering of stem base positions.

remarks apply to G·U pairs in position b1 which is close to another sharp bend in the backbone (see Figure 11).

Position of Ψ and Possible Role of Modified Bases in tRNA

The positions of Ψ occurring in class 1 tRNA structures are shown in Figure 8 on a standard generalized cloverleaf. A circled nucleoside position indicates an invariant position and a nucleoside in parentheses indicates a semi-invariant position. Except for the possible structural involvement of the Ψ contained in the common loop IV sequence G-U*-Ψ-C, it seems unlikely that these Ψ positions have significant structural roles. More likely they are concerned with metabolic roles largely as yet unidentified with the exception that the Ψ in the position after R in the anticodon loop II appears to confer repressor activity on bacterial tRNA$_1^{His}$ in connection with regulation of the histidine biosynthetic pathway (Lewis and Ames, 1972).

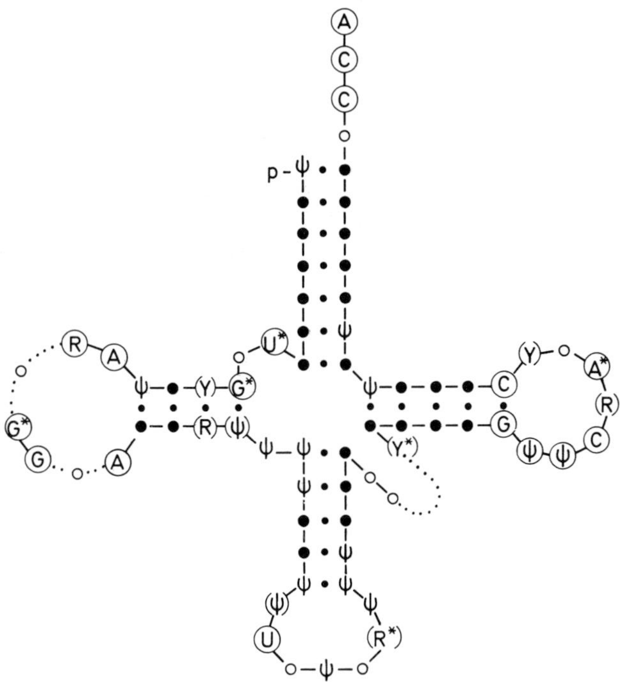

Fig. 8 Position of Ψ in class 1 tRNAs.

It remains difficult to generalize on the role of other modified bases in tRNA especially the many different methylated bases. For a general view of their structures and occurrence see McCloskey and Nishimura (1977).

The interesting observation on the modified purine nucleoside found following the anticodon tantalisingly still does not have a clearcut functional or structural interpretation. Nishimura (1972) pointed out an apparent connection between this particular modified base (location 9 in Figure 6) and the type of first base in the codon which the particular tRNA decodes. For example, tRNAs containing ms^{2}i^{6}A decode codons with U as the first base and tRNAs containing

t^6A decode codons with A as the first base. In other words this implies a connection between the modified base and the third position of the anticodon. Whether the connection is due to a structural specification or a metabolic pathway linkage is not known. The biosynthesis of the modified bases is discussed further in chapters by Abelson and Altman in this volume.

Analysis of Primary Structure and Significance for Function

Because of the extensive data on tRNA nucleotide sequences the question naturally arose as to whether a comparison of sequences showing common sequences (sequence homologies) can be interpreted in terms of tRNA functional sites. Therefore different tRNAs recognized by the same enzyme have been compared to determine the common parts of their primary structures. Of course, it is now known where these common parts lie in the three-dimensional structure so we can evaluate the results in a more sensible way than before the tertiary structure was known. The sequence homologies can be discussed in terms of several of the activities listed in Table I, for example initiator tRNAs binding to ribosomal sites, common charging of myeloma and *E. coli* methionine tRNAs by the appropriate synthetases, the common charging of different phenylalanine tRNAs and the transformylase recognition of initiator tRNAs.

Although our early studies comparing the sequences of bacterial methionine tRNAs (tRNAs chargeable with the same activating enzyme) did not permit the assignment of common parts of the cloverleaf structures (see Barrell and Clark, 1974) detailed studies of the different tRNAs which can be mischarged by yeast phenylalanine tRNA synthetase suggested obligatory common sequences for enzyme recognition (Williams *et al.*, 1974). In addition to invariant and semi-invariant regions and the fourth nucleotide from the 3′-end, the short extra loop III and a region of the D stem were implicated in the recognition studies.

It is difficult to judge whether this line of reasoning based on sequence homology by the criterion of mischarging is valid. For example we have data for two instances of normal charging by methionyl-tRNA synthetases where the sequence homology information leads to different conclusions (Clark and Klug, 1975). Furthermore, very recently another type of yeast $tRNA^{Phe}$ from *Schizosaccharomyces pombe*, chargeable by normal yeast Phe-tRNA synthetase has a different sequence from the normal yeast $tRNA^{Phe}$ in the D stem region (McCutchan *et al.*, 1978). Clearly such a proposition for recognition based on a consideration of similarities among sequences arranged in cloverleaf forms is not correct. This has also been pointed out earlier in section II in connection with the situation comparing protein amino acid sequences where similar conclusions have been drawn in regard to substrate binding sites for aminoacyl-tRNA synthetases (Bruton, this volume).

The Tertiary Structure of Yeast $tRNA^{Phe}$ and Functional Significance

After tRNA was first crystallized (Clark *et al.*, 1968) several laboratories sought a species which would give crystals suitable for an X-ray crystallographic analysis to high resolution (Brown *et al.*, 1972; Sigler, 1975; Rich and RajBhandary, 1976). A systematic study of many species from *E. coli* and yeast yielded suitable crystals of an orthorhombic form and a monoclinic form of yeast $tRNA^{Phe}$. The monoclinic form had a smaller unit cell (Ladner *et al.*, 1972; Ichikawa and Sundralingam, 1972) than the related orthorhombic form (Kim *et al.*, 1972) and was better ordered than a different orthorhombic form obtained earlier by Cramer and his colleagues (Cramer *et al.*, 1970).

The X-ray crystallographic determination of a structure generally goes through several stages in the following order: 1) the production of reproducible crystals of suitable size (about 0.2-0.5 mm in each dimension), 2) the introduction of heavy atoms into the crystals to provide isomorphous heavy atom derivatives, 3) the collection of X-ray data from the native and at least two derivatives, 4) the production of an electron density map and the fitting of an atomic model to the map, 5) the refinement of the model using computer techniques, and 6) further collection of higher resolution native data followed by further refinement of the atomic model. Naturally, there can be variations in the timing of the switch from the wire model to the computer and the timing of the inclusion of the model itself into the calculation of the electron density map.

The three-dimensional structure of the yeast $tRNA^{Phe}$ in the monoclinic form has been solved to 0.25 nm resolution in Cambridge by the method of isomorphous replacement (Ladner *et al.*, 1975), using 5 heavy atom derivatives. Independent structural determinations using the orthorhombic form (Quigley *et al.*, 1975) and also the monoclinic form (Stout *et al.*, 1976) have yielded the same structure in broad outline. A photograph of the Kendrew skeletal model built to fit the electron density is shown in Figure 9. The second phase of the X-ray crystallographic analysis to 0.25 nm resolution resolved some ambiguities in the structure. The chain tracing of the ribose-phosphate backbone was completed with certainty and many more

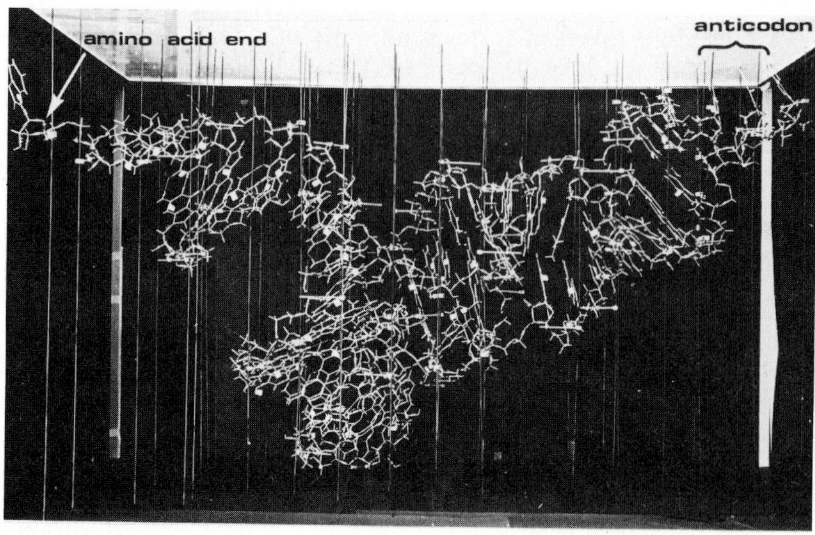

Fig. 9 Photograph of a 0.25 nm model of yeast tRNA^{Phe} built at Aarhus University under the guidance of Dr. J.E. Ladner. The two most significant functional parts are labelled. Actually on the model the distance from the anticodon to the terminal 3′ -end Å is about 85 Å (8.5nm).

features of the molecular stereochemistry have been revealed. In addition to the base pairs of the cloverleaf arrangement of yeast tRNA^{Phe} (Figure 10), many other interactions (see Figure 11) which fix the tertiary structure of yeast tRNA^{Phe} have been deduced (see Figure 11 showing how the cloverleaf of Figure 10 is folded). At present the work continuing on the structural elucidation of yeast tRNA^{Phe} is at stage 6.

From the X-ray studies we now have a picture of the ordered complexity of a folded RNA molecule, a complexity as great as that of a protein. An interesting by-product is a detailed picture of the stereochemistry of a G·U base pair in a double helical stem, where the pairing is that predicted by the 'wobble' hypothesis (Crick, 1966).

In the accounts of the 0.3 nm tertiary structure attention was focussed on the interactions between bases. Now there has been traced an extensive network of H-bonds including ribose and phosphate groups as well as bases, in which almost half the bonds involve 2′ -OH groups of riboses as acceptors, donors, or both (Ladner *et al.*, 1975; Quigley *et al.*, 1975). A prominent set of H-bonds stabilizes the T-joint made in the structure by the two long double helices which meet approximately at right angles. Another set reinforces the links between bases at the bends of the D and TΨC loops.

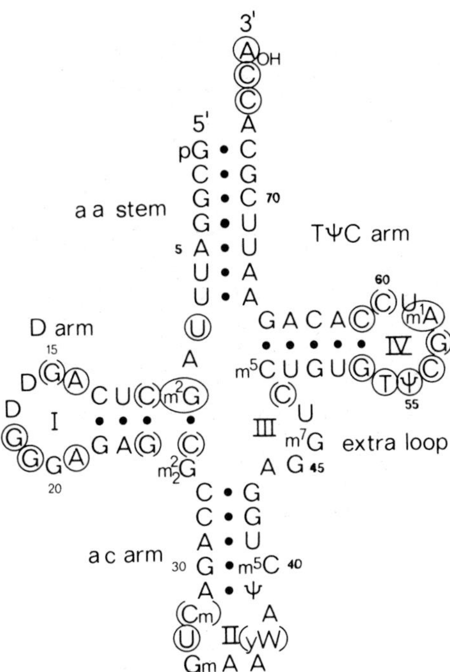

Fig. 10 Primary structure of y tRNA^{Phe} arranged in the cloverleaf form. Invariant residues are circled and semi-invariant ones are in parentheses.

In the 0.3 nm model all sugar conformations were assumed to be C3′-endo, as found in RNA double helices. However, in the more detailed model building occasioned by fitting the 0.25 nm map, the pucker of ten of the sugars was changed to the C2′-endo conformation (Ladner *et al.*, 1975). Although the resolution of the electron density map is still not sufficient to show the shape of a sugar unequivocally, the conformation can be deduced from the restrictions imposed on it by the relative dispositions of the two phosphate groups on either side and of the base that emanates from it. A good example is that of nucleotide m^1 A58 whose sugar is C2′-endo in the 0.25 nm model. In the 0.3 nm model the phosphate and base densities were well fitted, but the C3′-endo ribose was not. Fitting the ribose density moved the base of m^1 A58 too close to its base paired partner, T54, which was well fixed. After the change, the C2′-endo sugar resided in its proper density and the base pair was comfortably made.

It should also be pointed out that the topography of the model deduced from the crystallographic study agrees with a companion

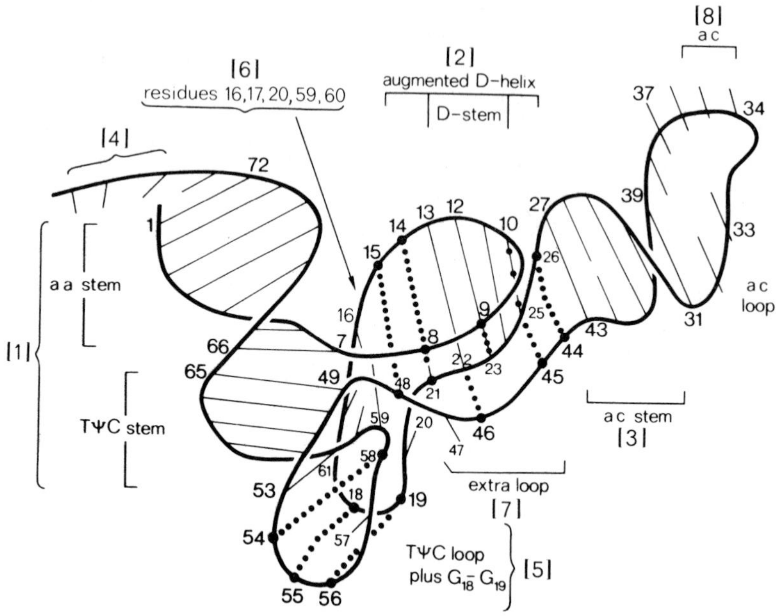

Fig. 11 A schematic diagram of the folding and the tertiary interactions between bases of y tRNA^{Phe} (adapted from Ladner *et al.*, 1975). The ribose-phosphate backbone is represented by a continuous line. Base pairs in the double helical stems are represented by long lightly weighted lines, and non-paired bases by shorter lines. Base interactions additional to those in the cloverleaf formula are indicated by dotted lines. Substructures referred to in the text are given by numbers in square brackets.

study of the chemical reactivity of yeast tRNA^{Phe} in solution (Robertus *et al.*, 1974b). There is thus no reason to presume that the structure in the crystal departs significantly from that in solution. This presumption is also confirmed by physical chemical studies, such as NMR of tRNA in solution.

In summary, some of the structural features which appear to have functional significance in the molecule depicted in the model (Figure 9 and schematically in Figure 11) can be described in terms of 3 major skeletal substructures: [1] the long double helix formed by the aa stem and TΨC stem stacking on top of each other; [2] the augmented D helix forming the central part, or 'core', or 'thorax' of the molecule, consisting of the D stem, augmented laterally by interactions with the short 'stretcher' region 8-9, with the extra loop III and with a part of the D loop; [3] the anticodon stem tilted off by about 20° to the D stem and apparently hinged to it by hydrogen bonds between nucleotides A44 and m_2^2G26 and between G45 and m^2G10. The molecule has been described as L-shaped (Kim *et al.*, 1974) but the two major long double helical stretches i.e. [1] and [2] + [3] above, are arranged as the letter T (Robertus *et al.*, 1974a). Thus the T-form of tRNA is a tertiary structural arrangement of the secondary structure represented by the cloverleaf form. The L-shape just reflects one side view of the molecule shown in Figure 9.

To these skeletal substructures described above are joined five regions with established, or potential, functional roles: [4] at one end of the long double helix there is the 3′-end CpCpA to which the amino acid is attached; [5] at the other end the invariant TΨC loop is tightly folded and interacts with a part (constant G*-G) of the D loop in a complex set of interactions such as those which have been interpreted in detail in the electron density map at 0.25 nm resolution (Ladner *et al.*, 1975). Briefly U59 stacks on to the pair G15·C48, T54 is base paired with m^1A58 in a reverse Hoogsten type, G57 intercalates between G18 and G19 giving rise to a stack of 4 purines 58, 18, 57 and 19, and interacts with the backbone of the D loop, C56 forms a Watson-Crick base pair with G19, and G18 appears to interact with Ψ55 and with the TΨC backbone on either side of it. The interaction of the TΨC loop with the invariant G18 and G19 probably masks the potential of the T-Ψ-C-G sequence for its proposed role of binding to the ribosomal A-site during peptide bond formation (Erdmann *et al.*, 1973). [6] the variable regions in the D loop perhaps form a discrimination site (or part of one) for enzymes to distinguish between tRNAs (not necessarily the amino acyl-tRNA synthetase recognition). The two sections α and β of the D loop (see Figure 5) are near each other and form a surface patch on the molecule; they each contain a variable number (1 to 3) of nucleotides and, either on their own or together with the nearby non-base paired residues 59 and 60, could well provide a general enzyme discrimination site. The variable extra loop (or, at least, a part of it, e.g. U47 in $tRNA^{Phe}$), which is numbered region [7] in Figure 11, may also play such a role in some cases. Finally the decoding function of tRNA is provided for by region [8], the anticodon loop, containing the anticodon. This loop contains a stack of five bases on the 3′-end side continuing from the anticodon stem but the two other bases Cm32 and U33 are also stacked separately on the end of the anticodon stem but with a sharp bend between U33 and Gm34 (Jack *et al.*, 1976; Quigley and Rich, 1976).

The structural integrity of the molecule as a whole depends upon the central part where elements of four chains come close together and a number of base triples or pairs are found. These additional base pairs or triples are stacked or intercalated so as to make their H-bonds inaccessible to water molecules. The tertiary structure is further stabilised by the stacking of non-paired bases on each other or on base pairs as for example in the anticodon loop. Indeed about 90% of the bases can be said to be internal as hydrophobic groups are internal in proteins.

The structural determination has shown that the hydrogen bonding potential is more varied than simple Watson-Crick type pairing, that the planar bases can form structural sheets which involve H-bonding of 3 nucleotides, and that the ribose residues can participate in forming structurally important H-bonds between themselves, with bases and with phosphate oxygens. In addition the presence of single stranded loops, a core region where four backbone chains run along together and regions of reasonably standard RNA-type double helix makes it of interest to determine in detail the torsional angles used and to identify what restrictions can be placed on particular angles when predicting an unknown structure.

Recent developments in refining the structures determined independently by the MRC group in Cambridge and by the MIT group have given evidence locating important metal ion binding sites and of particular biological importance, strong magnesium binding sites and spermine sites.

Both groups agree that the strong magnesium binding sites are found near 1) the phosphate oxygens of U8 and A9; 2) near phosphate oxygens of G20 and A21; and 3) at a site near the phosphate of G19, atoms N7 and 06 of G20 and 04 of U59 (Jack *et al.*, 1977; Quigley *et al.*, 1978). In addition a good case has been made for the location of a magnesium ion in the anticodon loop between Cm32 and A38 (Quigley *et al.*, 1978; Klug, personal communication). Earlier this density was identified as an H-bond (Quigley and Rich, 1976) and a water molecule or a weak magnesium binding site (Jack *et al.*, 1977). Two spermine molecule positions have been located by Quigley *et al.*, (1978) in the major groove at one end of the anticodon stem and near phosphate 10 where the polynucleotide chain makes a sharp turn. The positions of these cations may have important biological implications. Certainly spermine and magnesium ions often appear to be important in maintaining the overall folding of tRNA. There is experimental evidence (Bolton and Kearns, 1977) that varying combinations of polyamine and magnesium in the tRNA structure can induce structural changes.

Recently, new data collection and refinement has resulted in a slight modification of the structure in the D loop by the MRC group (B. Hingerty and A. Klug, personal communication) to the earlier 0.25nm model (Ladner *et al.*, 1975). Now the base D16 is turned inwards rather than pointing out into solution so that it points towards C60, i.e. to what Klug has called the variable pocket (Ladner *et al.*, 1975). So far the MIT has not commented on this alteration.

Generality of Yeast $tRNA^{Phe}$ Tertiary Structure

The elucidation of the tertiary structure of yeast $tRNA^{Phe}$ naturally led one to ask whether other tRNA primary structures could be folded in the same way. In particular, could other bases be substituted into the base pair or base triple positions of yeast $tRNA^{Phe}$ so as to give equivalent interactions? Many changes of sequence in class 1 can indeed be compensated by concomitant changes elsewhere, so that local geometry in the 3-dimensional structure remains essentially unchanged (Klug *et al.*, 1974). Such changes are listed in Table IV where the dots indicate H-bonds.

Levitt (1969) first noticed a co-ordinated base change at positions 15 and 48 ($tRNA^{Phe}$ numbering, see Figure 10) by examining tRNA primary structures. However, the pair was found to be not of the Watson-Crick type (Robertus *et al.*, 1974a) and moreover is asymmetric since only a purine R is found in position 15 and a pyrimidine Y at position 48 and not vice versa. Structural considerations require such an asymmetry in forming a reverse Watson-Crick base pair. Thus G·C can be replaced by A·U to give the same disposition of H-bonds in this non-standard pair. Both y $tRNA_3^{Glu}$ and Ec $tRNA_1^{Gly}$ provide interesting exceptions to this structural correlation; they contain A15-C48 instead of the usual semi-invariant A15·U48 or G15·C48. Furthermore, the recently determined Ec $tRNA^{Cys}$ contains G15·G48.

Subdivisions of class 1 into 1A and 1B in Table II are based on the two nucleoside positions 46 and 9 (in the numbering for $tRNA^{Phe}$). Position 46 has either m^7G or not (G or A) and this is the basis for the subdivision into the two groups 1A and 1B. However, the base triple 46-22-13 can be made equivalently in these subclasses, with nearly the same disposition or glycosyl bonds (Klug *et al.*, 1974). Nucleoside 9 is G, m^1G or A and here again a base triple 9-23-12 can be made for all members of subclasses 1A and 1B, though the number of hydrogen bonds is not the same in all cases. The coordinated base changes are summarized in Table IV.

The model shown in Figure 9 is perhaps also directly relevant to sequences of class 1C. In the model it is possible to remove the exposed nucleotide U47 and still bridge the gap between 46 and 48 by

TABLE IV

Correlated Base Changes in Class 1 tRNAs

base pair at positions	Directions of chains
15 - 48	
G15 : C48	Parallel
A15 : U48	Parallel
base triple at positions	
13 - 22 - 46	
C13 : G22 : m^7G46	Antiparallel/antiparallel
U13 : A22 : A46	Antiparallel/antiparallel
U13 : G22 : G46	Antiparallel/antiparallel
base triple at positions	
12 - 23 - 9	
U12 : A23 . A9	Antiparallel/parallel
G12 : C23 . G9	Antiparallel/parallel
U12 : A23 . m^1G9	Antiparallel/parallel
base pair at	
26 - 44	
A26 . G44	Antiparallel
G*26 : A44	Antiparallel

the phosphate-sugar linkage, maintaining all the tertiary interactions. Class 1C could have this type of structure: any base triples equivalent to 13-22-46 would now have to be made in a somewhat different way, but still preserving the relative disposition of the pieces of backbone being bridged by the base-base interactions. By an extension of the argument, the model is also relevant for class 1D. If, as discussed earlier, a base pair, albeit non-standard, can be fitted into position b4, an equivalent augmented D helix might be made without disruption of the pattern of the four chains which come together to make it. Hence it is supposed that the yeast $tRNA^{Phe}$ structure is common to all class 1 structures. We have no firm tertiary structural information about class 2 structures although there have been recent speculations (Brennan and Sundaralingam, 1976; Brown and Rubin, personal communication; Altman, personal communication) about such structures. These speculations are rather too premature to be discussed here.

Solution Structure of tRNA

The bases of yeast $tRNA^{Phe}$ which react with a carbodiimide reagent and with methoxyamine in solution have been identified (Rhodes, 1975). This information has been combined with other results from the literature to produce a composite picture (Clark and Klug, 1975; Robertus *et al.*, 1974b) of base accessibility in yeast $tRNA^{Phe}$ shown in Figure 12 in secondary structure and tertiary structure forms. For the chemistry of the reactions indicated in Figure 12 and for additional reactions of other tRNA species the reader is referred to a reasonably comprehensive review by Brown (1974). The bases which react chemically can be correlated with exposed positions in the three dimensional structure of tRNA (see Figure 12B). Those which do not react are either in the double helical regions or are involved in maintaining the tertiary structure. These results confirm the usual structural assumptions made about reactivity in chemical studies and hence give confidence for extending such work to other tRNA structures. Furthermore, since the chemical studies are carried out on tRNAs in solution rather than in the crystalline state, they also support the contention that there is no significant change in the conformation of yeast $tRNA^{Phe}$ upon crystallization.

In addition, physical chemical studies especially using the NMR technique have identified the H-bonds involved in the tertiary interactions between bases (Robillard *et al.*, 1976; Reid and Hurd, 1977). These studies are now also interpreted in terms of the same structure existing in solution as in the crystal form. Some of the difficulties encountered by workers attempting to interpret spectra in terms of three-dimensional structure have recently been reviewed by Ladner (1978). At any rate the yeast $tRNA^{Phe}$ conformation identified with the crystal form appears also to be the most stable form of lowest potential energy in solution so that if the conformation does become changed during function once it is released by its reacting macromolecule then it reverts to the crystal form's conformation.

Structural Changes During Function

Although we have strong evidence now confirming that the tRNA conformation is the same in solution as in the crystal form, it is unclear how the structure changes its conformation for its role in protein biosynthesis. Indeed it is likely that the conformation does change and is effected by the many different interactions which the tRNA undergoes with other protein and nucleic acid components. However, to gain more insight into putative conformational changes,

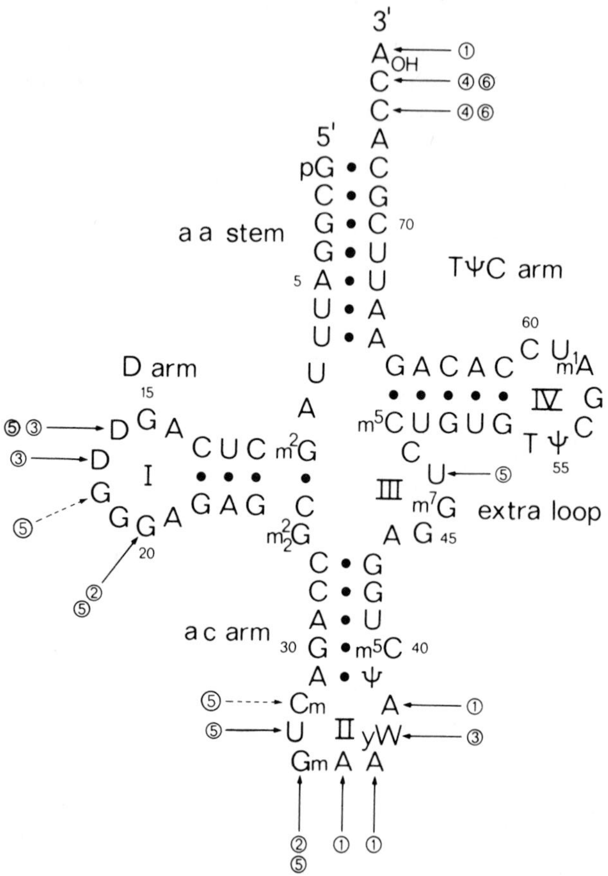

Fig. 12A Summary of the chemical reactivity of y tRNA^{Phe} arranged in the cloverleaf form. The arrows indicate points of chemical reaction. Dotted arrows indicate only a minor reaction (Rhodes, 1975). Reagents used were: 1) perphthalic acid, specific for A residues, 2) kethoxal for exposed G residues, 3) $NaBH_4$ reduction, 4) methoxyamine, 5) carbodiimide and 6) $I_2/T1C1_3$ (for refs. see Clark and Klug, 1975).

we need information on the different tRNA-protein complexes formed during the steps of protein biosynthesis from chemical, X-ray diffraction and other physical chemical studies.

Charging

There is a surprising lack of hard facts concerning the possible difference in conformations between uncharged and charged tRNA. The literature contains evidence both in favour and against a change

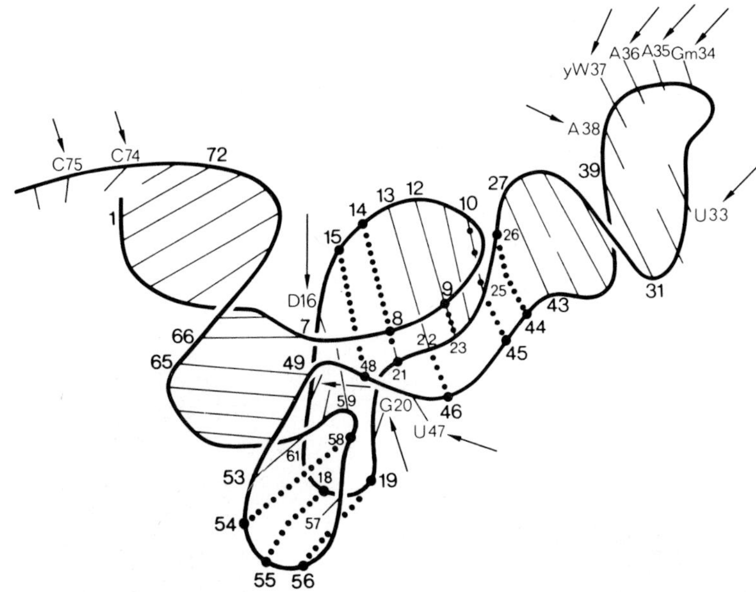

Fig. 12B Chemically reactive positions shown on the surface of the folded y tRNA^{Phe} structure of Figure 11 by arrows.

in conformation. Those in favour suggest that the methods used where negative results have been obtained are not capable of measuring the possibly subtle changes which occur. It does seem that large structural rearrangements which would be readily detectable do not occur. One method which does appear to show differences in charged and uncharged tRNA conformations is the use of laser light scattering (Potts *et al.*, 1977) but the general feeling is that corroboration is needed from other methods.

Binding to EF-Tu

So far it has not been possible to determine the chemical reactivity of an aminoacyl-tRNA complexed with EF-Tu and GTP because the chemical reagents inactivate the protein too rapidly. However, there is some evidence that the nuclease sensitive regions of aminoacyl-tRNA are not different in the ternary complex with EF-Tu and GTP (Miller, 1978). Again it is reasonable to conclude that no large changes of conformation occur on binding to EF-Tu.

Quite often mistaken ideas of how EF-Tu might recognize aa-tRNA can arise without keeping in mind the relative sizes of the macromolecules concerned. For information purposes Figure 13 shows a comparison of 0.5nm Balsa wood models of both EF-Tu:GDP and uncharged yeast $tRNA^{Phe}$. The data are taken from the high resolution X-ray analyses by Morikawa *et al.*, (1978) and Ladner *et al.*, (1975), respectively. Although the protein and the tRNA have very different molecular weight ratios of about 45,000 and 25,000 the spatial sizes are very similar.

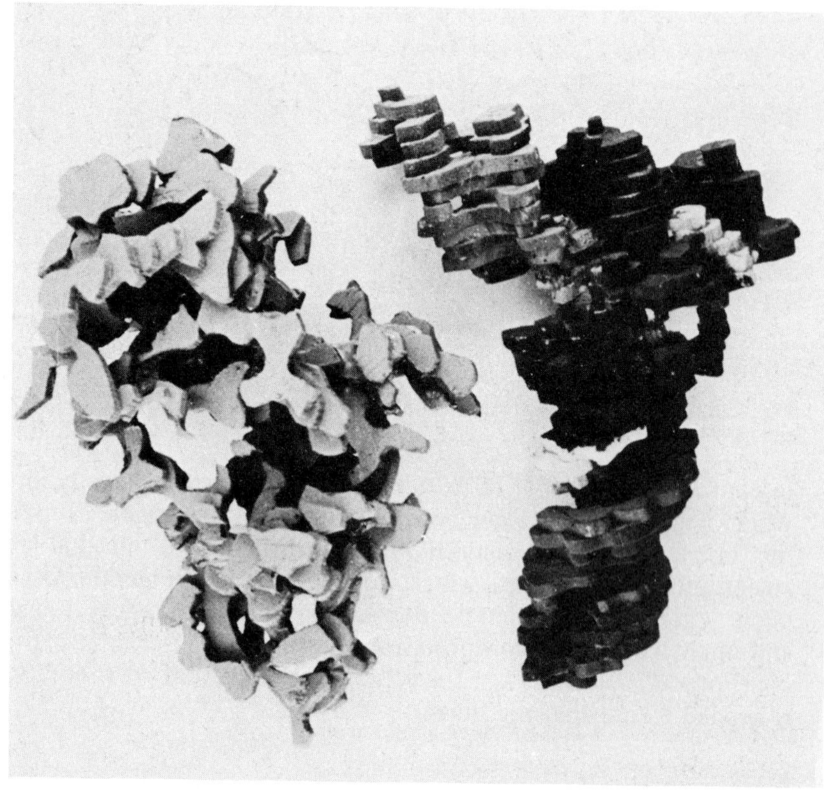

Fig. 13 A photograph of 0.5nm Balsa wood models of EF-Tu:GDP containing essentially all of the EF-Tu molecule (Morikawa *et al.*, 1978) on the left and y $tRNA^{Phe}$ (with the ac arm pointing downwards) on the right.

Binding to the Ribosome

Although much evidence is being collected about the ribosomal environment for the bound aa-tRNA we have little idea so far about the conformation of the aa-tRNA on the ribosome. Indeed recent work has done little to clarify the situation. Based on oligonucleotide inhibition studies there is evidence that the binding of the aminoacyl-tRNA to the ribosome requires the exposure of sequence T-Ψ-C-G for complementary binding to the ribosomal 5S RNA (Erdmann *et al.*, 1973). Should this be really the case a change in conformation of the tRNA is necessary from the structure represented in Figure 11 where the common sequence is unavailable for further base pairing because of its involvement in the tRNA tertiary structural design. Further evidence for the tightness of the tertiary structure in the regions including the T-Ψ-C-G which interacts with the D loop comes from monitoring various intermediate conformations of tRNA seen on heating the tRNA until it melts completely, i.e. with all of its tertiary and secondary structure destroyed. Results from both proton magnetic resonance studies (Kastrup and Schmidt, 1978) and from chemical reactivity studies (Rhodes, 1977) at different temperatures confirm that the interaction between the TΨC and D loops is the stablest part of the tertiary structure. Of course if T-Ψ-C-G must become exposed for binding to the ribosome the change in conformation could be effected by the elongation factor EF-Tu or ribosomal proteins.

Although the ribosomal 5S RNA contains C43-G-A-A46 (Brownlee *et al.*, 1968) which is complementary with T-Ψ-C-G, it has recently been shown that G44 is inaccessible in the ribosome structure to chemical modification by kethoxal (H. Noller and R. Garrett, personal communication). Thus if the aa-tRNA binds to 5S rRNA in a complementary way then there must also be a rearrangement of ribosome structure as well as the tRNA's.

The Anticodon Loop Conformation

The anticodon loop is another place where conformational changes possibly occur during protein synthesis. The hinge-like region of the structure where the anticodon arm is attached to the rest of the structure (see Figure 11 around residues 26 and 44) could provide a point of directing conformational change in the anticodon loop or for conformational change concomitant with movement of the tRNA from the A-site to the P-site (Figure 2) during protein synthesis. As can be seen in the diagrammatic Figure 11 the two bases following the anticodon to the 3′-side are stacked on the anticodon. This conformation found in the crystal form can be the called 3′-S conformation (see also Figure 14). There is a break in the regular stacking to the 5′-side of the anticodon but the two preceding pyrimidines are stacked on each other and on to the anticodon stem. This structure is very similar to that originally proposed by Fuller and Hodgson (1967) from model building studies. In the crystal form

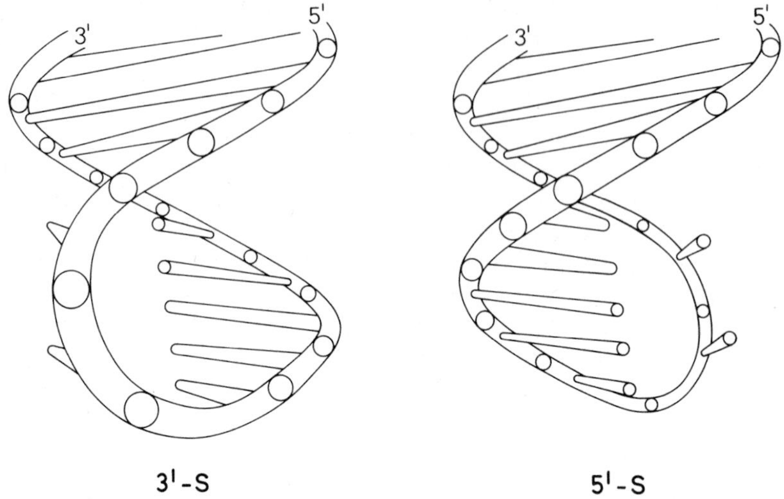

Fig. 14 Possible conformations of the anticodon loop dependent on stacking of bases on either end of the anticodon.

stacking of the anticodon, with somewhat more overlapping of bases than in a regular RNA double helix, appropriately exposes the anticodon bases for interaction with a mRNA codon. An alternative conformation for the anticodon loop where the 2 bases to the 5′ -side of the anticodon are stacked on it (the 5′ -S form) is shown in Figure 14. This has also been called the anti-Fuller Hodgson conformation and used by Woese (1970) to define the P-site of the ribosome in a model for protein biosynthesis. Actually, there is recent evidence (Kuechler, personal communication) from fluorescent studies of tRNA crosslinked to mRNA that the anticodon conformation does not change when the tRNA is translocated from A-site to P-site. Nevertheless, another model proposed by Lake (1977) to involve tRNA conformational changes during protein synthesis can incorporate these two conformations of the anticodon loop. In this model the 5′ -S conformation exists in a third ribosomal binding site, the recognition or R site where codon recognition (checking) is proposed to occur. The anticodon loop conformation flips to the normal 3′ -S conformation for the A-site binding of the aa-tRNA and this also brings the amino acid into the correct environment for peptide bond formation. Although there is no what we might call conclusive evidence on these proposals very recently Urbanke and Maas (1978) by investigating the fluorescence of the yW base in y $tRNA^{Phe}$ by a temperature jump method have concluded that they have evidence for a transition between two such conformations illustrated in Figure 14. Reassuringly their low temperature state of the tRNA is the 3′ -S or crystal form conformation.

Another interesting hypothesis which assumes a conformational change with a transition between such forms as in Figure 14 has recently been made by Crick *et al.*, (1976). This hypothesis concerns a model for primitive protein biosynthesis in the absence of ribosomes. In this model the aminoacyl-tRNA binds to the mRNA in a pentanucleotide base pairing interaction for a stable interaction. Adjustment of the tRNA is then facilitated by a change in the base stacking of the anticodon loop defining conformational changes from 5′ -S to 3′ -S.

Decoding of mRNA

Before the crystal structure was known results from experiments testing the strength of binding of specific complementary oligonucleotides to regions of the anticodon loop of y $tRNA^{Phe}$ were interpreted to mean that at least the next base (the invariant U33) was stacked on the 5′ -side of the anticodon (Pongs and Reinwald, 1973). Similar results were obtained with Ec $tRNA_f^{Met}$ (Uhlenbeck, 1972). Further physical chemical studies using oligonucleotides complementary to the y $tRNA^{Phe}$ anticodon loop led Yoon *et al.*, (1975) to postulate that the anticodon loop changes conformation from the 3′ -S (crystal form) when pairing with the codon to allow the formation of fourth base pair on the 5′ -side of anticodon (i.e. into a 5′ -S form). However, Bald and Pongs (personal communication quoted by Ladner, 1978) contradicted this interpretation after measuring similarly high binding constants to y $tRNA^{Phe}$ for U-U-C-R and for U-U-C-m^1A since the latter can form only 3 Watson-Crick base pairs with the anticodon loop. In addition Grosjean *et al.*, (1976) have indications that the strong binding of two tRNAs with complementary anticodons does not involve any slow reorganization of the tRNA structure.

In summary these results suggest that the 3′ -S form of the anticodon loop is not changed on binding to a complementary oligonucleotide and that the increased binding constant of the tetranucleotide ending in 3′ -A over the triplet is due to stacking forces rather than additional Watson-Crick base pairing increasing stability. Clearly with respect to oligonucleotides binding to other parts of the tRNA structure one does not need to interpret strong binding in terms of full Watson-Crick base complementarity.

An interesting extension of oligonucleotide binding studies has recently been reported by Möller *et al.*, (1978). These workers have shown that in some cases anticodon-complementary tetranucleotides terminating in 3′ -A show efficient presumptive binding to the complementary tRNA but are inactive as codons for binding tRNA to the ribosome. They have concluded that there is no straightforward relationship between effective binding of a complementary oligonucleotide to the tRNA and the coding activity of the respective oligonucleotide.

At present there are technical difficulties to measuring anticodon-mRNA interactions under natural conditions. As the next best thing Grosjean *et al.*, (1978) have measured the binding constants for the interactions of a large number of tRNA pairs (more than 60) which have complementary anticodons. Because the rates of formation of these complementary pair complexes are all roughly similar the binding constants are reflected in their rates of dissociation and hence in their mean lifetimes which have been measured by temperature-jump relaxation technique. The results indicate that interactions involving base pairs allowed by the Wobble Hypothesis (Crick, 1966) have, by and large, lifetimes in excess of 10 msec at 9°C, whereas those involving forbidden pairs are more unstable. Grosjean *et al.*, (1978) have made the striking conclusion that the stability of a tRNA pair in complex does not increase regularly according to the number of G·C base pairs involved (i.e. not in order of increasing H-bonds as might have been the case). Presumably there must be some effect that stabilises A·U pairs. Furthermore, Grosjean *et al.*, (1978 and H.J. Grosjean, personal communication) conclude that stacking effects conditional on the sequence of bases are probably important especially as there is an effect due to the nature of the hypermodified base next to the 3′ -end of the anticodon (see also McCloskey and Nishimura, 1977). In other words there is a context effect in the decoding interaction. Although these studies by Grosjean *et al.*, (1978) have given valuable information also on stability of non-Watson-Crick interactions especially in support of Wobble-base interactions, one still is uncertain how far the system of interacting tRNAs can reflect the native tRNA decoding of mRNA.

Another view of the decoding interaction has recently been proposed by Lagerkvist (1978). The latter bases his proposal on a reconsideration of tRNA decoding patterns and contradicts the findings of Grosjean et al., (1978) as too artefactual by suggesting that more G·C base pairs in codon-anticodon interactions lead to an enhanced stability as in double helical DNA. Lagerkvist's proposal arose out of some earlier work (Mitra et al., 1977) where it was found that under the conditions of *in vitro* protein biosynthesis only one valine tRNA could decode all four valine codons, GUN (N = any nucleotide). The third position of the codon did not appear to have any discriminatory function. Thus Lagerkvist (1978) has introduced the idea of codon families where a family of codons is a set of 4 codons differing only in the third base and all coding for the same amino acid. Furthermore, this decoding called 'two out of three' by Lagerkvist has been proposed as an alternative method of reading of the genetic code to the Wobble Hypothesis (Crick, 1966) and to occur under certain conditions *in vivo.* Clearly in two out of three reading only one tRNA is needed to decode a family of codons whereas the Wobble Hypothes requires at least two tRNAs to decode such a family. Because the genetic code is not made up only in such families Lagerkvist has introduced mixed families and the idea that the need for more discrimination arises only with more A·U base pairs being involved. Something to distinguish non-family codons is needed because, if two out of three reading was general, translational errors would occur on a large scale. It is difficult to decide whether Lagerkvist is on the right lines because his evidence is also taken from an *in vitro*, i.e. an unnatural, system. Whether it is more native than the system of Grosjean *et al.,* (1978) is unclear. One drawback of Lagerkvist's proposal is that the role of the hypermodified base is not considered.

Nevertheless, under *in vitro* conditions of protein synthesis and the triplet binding assay one tRNA can decode all four codons differing only in the third base. This has been demonstrated with pure *E. coli* tRNAs specific for tryptophan (decoding UGN, Buckingham and Kurland, 1977) and for arginine (decoding CGN, Kruse *et al.*, 1978). Actually, the non specific decoding effect by $tRNA^{Trp}$ is inhibited by the presence of other tRNAs such as $tRNA^{Cys}$ when $tRNA^{Trp}$ is used to decode the Cys codon. Clearly *in vitro* events are not necessarily reflecting normal *in vivo* events.

In the case of the non-specific decoding by $tRNA^{Trp}$(Buckingham and Kurland, 1977) very interesting information arose about the requirement for tRNA fitting on the ribosome for proper decoding. Interestingly the UGA suppressor $tRNA^{Trp}$ (Hirsh, 1971) gives more non-specific two out of three decoding of the Cys codons than does the wild type (su^-) $tRNA^{Trp}$ *in vitro.* Both the UGA suppressor $tRNA^{Trp}$ and the su^- $tRNA^{Trp}$ have the same anticodon (CCA) but the suppressor tRNA contains U11·A24 instead of U11·G24 in the su^-tRNA at position b2 of the D stem (see Figures 5 and 6). It is difficult to speculate on how this base change could affect the decoding activity of the anticodon through the tertiary structure to the anticodon which is contained in an arm which does not interact with the rest of the tRNA (see Figure 11). Most likely the structural environment of base pair b2 is in some way recognized by the ribosomal interaction which then influences the decoding. This implicates ribosomal components in the anticodon's decoding interaction, perhaps another way of increasing the specificity of translation. The increased non-specific decoding activity of the UGA suppressor tRNA can be explained by a transmitted effect through the ribosome starting from an altered fit about base pair b2 to relieve the specificity of decoding designed by the anticodon's ribosomal environment. It should be remembered here also that position b2 is one of the two outstanding differences of prokaryotic initiator tRNAs from other tRNAs (see Figure 6).

For some time there have been attempts to prove that the interaction of a tRNA with an oligonucleotide complementary to its anticodon causes a rearrangement of the tRNA's tertiary structure which in particular allows tight binding of C-G-A-A which was assumed to be in complementarity with the constant T-Ψ-C-G region (Schwarz and Gassen, 1977). The binding of the C-G-A-A was facilitated by the presence of 30S ribosomal subunits. There is also further physical chemical evidence for a structural change in the D loop (and thus implicating the interacting TΨC loop) from fluorescence measurements of a probe in the D loop of y $tRNA^{Phe}$ with and without the presence of oligonucleotides complementary to the anticodon (Robertson *et al.*, 1977). The interpretation of effects involving the binding of an oligonucleotide is difficult because the oligonucleotide may change the tRNA structure (although as discussed earlier, it is unlikely in the case of the anticodon loop) and the location of the oligonucleotide may not be predictable because binding need not depend on Watson-Crick base pairing (also discussed earlier).

Consideration of tRNA's tertiary structural interactions makes it unlikely that the interaction of tRNA and an oligonucleotide complementary to the anticodon in the absence of ribosomal components could signal the large rearrangement needed to unhitch the TΨC loop to allow complementary base pairing with C-G-A-A as suggested earlier by Schwarz *et al.*, (1974) and again by Schwarz and Gassen (1977). Nevertheless it is possible, just as in the case of the anticodon oligonucleotide interactions discussed earlier, that a concoction of stereochemical fitting, stacking and some base pairing could permit C-G-A-A to bind to the TΨC loop region without significant structural rearrangement. In support of this idea are some recent NMR studies of the methyl group of T54. No significant structural changes are observed around T54 when a complementary oligonucleotide is bound to the anticodon of y $tRNA^{Phe}$ (M. Sprinzl, personal communication). However, very recently the method of chemical modification which is more direct than the earlier oligonucleotide binding studies have given evidence that G57 in Ec $tRNA^{Lys}$ becomes chemically accessible to kethoxal when the complementary oligonucleotide A-A-A-A is present (Wagner and Garrett, 1978, and personal communication). No such accessibility is seen when a non-complementary oligonucleotide is used. Although this result cannot at present be completely uncontroversial in the light of the NMR studies, the current yield of the modified G57 is of the order of 10%, high enough to make the result good evidence that the binding of an oligonucleotide to tRNA can cause a transmitted structural effect changing the local structure at a location far from the binding. What all this now means for protein biosynthesis in the light of the finding that the C-G-A-A in 5S rRNA is buried is rather unclear. Perhaps these unnatural isolated systems at best give us a hint of structural changes that occur in the presence of ribosomal components which facilitate and direct the process.

Role of Metal Ions and Spermine

As described earlier, magnesium and spermine binding sites have recently been located in the crystal structure of y $tRNA^{Phe}$ (Jack *et al.*, 1977; Quigley *et al.*, 1978). These cations appear to play a role in holding the folded tRNA structure together. Indeed, many physical chemical studies involving melting the tRNA structure under various ionic conditions confirm this suggestion. Thus the sensitivity of the tRNA structure to Mg^{2+} conditions (Stein and Crothers, 1976; Bolton and Kearns, 1977) could well be an indication that selective ribosome-mediated removal of structurally crucial cations such as Mg^{2+} or spermine will trigger unfolding of the aa-tRNA which could return to the crystal form upon restoration of these special cations (Bina-Stein and Stein, 1976; Ladner, 1978).

Other Biological Activities of tRNA

In addition to their role in protein synthesis, tRNAs have activities concerned with a number of reactions conveniently classified as in Table V, as being concerned with RNA metabolism and aminoacyl-tRNA transferases (Clark and Klug, 1975; Rich and RajBhandary, 1976). The tRNA is trimmed to size and matured from the precursor molecule by a series of as yet poorly characterized enzymes which

TABLE V

Other Biological Activities of tRNA

RNA Metabolism

1. As precursor recognized by cleavase and maturation enzymes
2. Enzymes for making modified bases
3. CCA repair enzyme
4. Pep-tRNA degraded by hydrolase
5. Nuclease degradation
6. Reverse transcriptase primer
7. Selection during viral encapsulation
8. Correlation with 3′-end of viral RNA
9. Alteration of *E. coli* ENDO I specificity

Aminoacyl-tRNA Transferases

10. Cell Wall Biosynthesis
11. Membrane components
12. Protein modification

presumably are linked to other metabolic roles (activities 1-3, Table V). Recent work in this field is reviewed by S. Altman in this volume. The CAA repair enzyme (activity 3) also called tRNA nucleotidyl transferase certainly repairs cellular tRNAs with incomplete 3′-ends and most likely is concerned with maturation as well. If peptidyl-tRNA should fall off the ribosome, there is a peptidyl-tRNA hydrolase (activity 4) which can hydrolyse off the peptide thus permitting the tRNA to be recycled through protein biosynthesis. Little is known about tRNA turnover but specific nucleases (activity 5) must be involved.

Recently some interesting properties of eukaryotic tRNAs have been identified (activities 6-8) with regard to virus metabolism. Reverse transcriptases from RNA tumour viruses use a specific $tRNA^{Trp}$ or $tRNA^{Pro}$ as a primer during synthesis of virally coded DNA (see Rich and RajBhandary, 1976). Furthermore, a certain number of selected tRNA species (perhaps 10-15) are incorporated non-covalently into RNA tumour particles during encapsulation from the cell membrane (Waters and Mullin, 1977). It is also well established that many viruses, especially plant viruses, have elements of tRNA structure so that their 3′-ends can be charged specifically with an amino acid (activity 8), e.g. turnip yellow mosaic viral RNA with valine (Haenni *et al.*, 1975). Activity 9 is not easily interpreted but the specificity of bacterial endonuclease I for double stranded cuts in DNA is altered to a nicking property when it binds tRNA.

Finally there is a special class of enzymes called aminoacyl-tRNA transferases (Rich and RajBhandary, 1976) which transfer amino acids from charged tRNAs to a variety of acceptor molecules without the involvement of a decoding mechanism and ribosomes. At least in the case of activity 10 where the amino acids are incorporated into cell walls a special type of tRNA best characterized in a series of Staphylococcal glycine tRNA species is involved (Roberts, 1972). These special cell wall tRNAs do not contain all the constant features of the general cloverleaf structure, presumably excluding those for ribosomal binding. However, their primary structures can be arranged in normal cloverleaf structures (Holley et al., 1965). The acceptor molecule can also be a phospholipid (Gould and Lennarz, 1970) so that the transferred amino acid becomes incorporated into a cell membrane referred to in activity 11. Activity 12 refers to the presently last known acceptor molecule being a finished protein. Thus for unexplained reasons these enzymes transfer amino acids to the N-terminal ends of certain proteins (Scarpulla et al., 1976) including membrane proteins (Kaji and Rao, 1976) from specific aminoacyl-tRNAs.

Clearly we are only at the beginning in trying to explain the variety of biological functions of tRNA in terms of three-dimensional structure. I have listed these other biological activities to illustrate that there is a lot more to tRNA than just its important role in protein biosynthesis. The near future will certainly provide a wealth of new information about the biochemical mechanisms involved. Now that we know a three-dimensional structure for this fascinating type of molecule we have the bare bones for progressing in our understanding of function in molecular terms.

References

Allende, J.E. (1975). *PAABS* **4**, 343.

Barrell, B.G. and Clark, B.F.C. (1974). Handbook of Nucleic Acid Sequences, Joynson Bruvvers Ltd., Oxford.

Bell, J.B., Gelugne, J.P. and Jacobson, K.B. (1976). *Biochim. Biophys. Acta* **435**, 21.

Bina-Stein, M. and Stein, A. (1976). *Biochemistry* **15**, 3912.

Bolton, P.H. and Kearns, D.R. (1977). *Biochim. Biophys. Acta* **477**, 10.

Brennan, T. and Sundaralingam, M. (1976). *Nucleic Acids Res.* **3**, 3235.

Brown, D.M. (1974): *In:* Basic Principles in Nucleic Acid Chemistry, (ed. T' so P.O.P) Vol. 2, Chap. 1, Academic Press, London.

Brown, R.S., Clark, B.F.C., Coulson, R.R., Finch, J.T., Klug, A. and Rhodes, D. (1972). *Eur. J. Biochem.* **32**, 130.

Brownlee, G.G., Sanger, F. and Barrell, B.G. (1968). *J. Mol. Biol.* **34**, 379.

Buckingham, R.H. and Kurland, C.G. (1977) *Proc. Nat. Acad. Sci. USA*, **74**, 5496.
Chen, E.Y. and Roe, B.A. (1978). *Biochem. Biophys. Res. Commun.* **82**, 235.
Clark, B.F.C. (1977). *In:* Progress in Nucleic Acid Research and Molecular Biology (ed. Cohn, W.E.) Vol. 20, 1. Academic Press, London and New York.
Clark, B.F.C., Doctor, B.P., Holmes, K.C., Klug, A., Marcker, K.A., Morris, S.J. and Paradies, H.H. (1968). *Nature*, **219**, 1222.
Clark, B.F.C. and Klug, A. (1975). *In:*'Proceedings of the Tenth FEBS Meeting' (eds. Chapeville, F. and Grunberg-Manago, M.) Vol. 39, 183, North Holland/American Elsevier.
Cramer, F., von der Haar, F., Holmes, K.C., Saenger, W., Schlimme, E. and Schulz, G.E. (1970). *J. Mol. Biol.* **51**, 523.
Crick, F.H.C. (1957). *Biochem. Soc. Symp.* **14**, 25.
Crick, F.H.C. (1966). *J. Mol. Biol.* **19**, 548.
Crick, F.H.C., Brenner, S., Klug, A. and Pieczenick, G. (1976). *Origins of Life* **7**, 389.
Dirheimer, G., Keith, G., Martin, R. and Weissenbach, J. (1976). *In:* 'Synthesis, Structure and Chemistry of Transfer Ribonucleic Acids and their Components, Proceedings of the International Conference, Dymaczewo', Sept. 1976. (ed. Wiewiórowski, M.), 273. Polish Academy of Sciences, Poznan.
Erdmann, V.A., Sprinzl, M. and Pongs, O. (1973). *Biochem. Biophys, Res. Commun.* **54**, 942.
Fuller, W. and Hodgson, A. (1967). *Nature* **215**, 817.
Gould, R.M. and Lennarz, W.J. (1970). *J. Bacteriol.* **104**, 1135.
Grosjean, H.J., de Henau, S., and Crothers, D.M. (1978). *Proc. Nat. Acad. Sci. USA* **75**, 610.
Grosjean, H.J., Söll, D.G. and Crothers, D.M. (1976). *J. Mol. Biol.* **103**, 499.
Guthrie, C., Scholler, C.A., Yesian, H. and Abelson, J. (1978). *Nucleic Acids Res.* **5**, 1833.
Haenni, A.L., Bénicourt, C., Teixeira, S., Prochiantz, A. and Chapeville, F. (1975). *In:* 'Proc. Tenth FEBS Meeting', (eds. Chapeville, F. and Grunberg-Manago, M.). Vol. 39, 121. North Holland/American Elsevier.
Hirsh, D. (1971). *J. Mol. Biol.* **58**, 439.
Hoagland, M.B., Zamecnik, P.C. and Stephenson, M.L. (1959). *In:* 'A Symposium on Molecular Biology' (ed. Zirkle, R.E.), 105. University of Chicago Press.
Holley, R.W., Apgar, J., Everett, G.A., Madison, J.T., Marquisee, M., Merrill, S.H., Penswick, J.R. and Zamir, A. (1965). *Science* **147**, 1462.
Ichikawa, T. and Sundaralingam, M. (1972). *Nature New Biol.* **236**, 174.
Jack, A., Ladner, J.E. and Klug, A. (1976). *J. Mol. Biol.* **108**, 619.
Jack, A., Ladner, J.E., Rhodes, D., Brown, R.S. and Klug, A. (1977). *J. Mol. Biol.* **111**, 315.
Kabsch, W., Gast, W.H., Schulz, G.E. and Lebermann, R. (1977). *J. Mol. Biol.* **117**, 999.
Kaji, H. and Rao, P. (1976). *FEBS Letters* **66**, 194.
Kastrup, V. and Schmidt, P.G. (1978). *Nucleic Acids Res.* **5**, 257.
Kim, S.H., Quigley, G., Suddath, F.L., McPherson, A., Sneden, D., Kim, J.J., Weinzierl, J., Blattmann, P. and Rich, A. (1972). *Proc. Nat. Acad. Sci. USA*

69, 3746.

Kim, S.H., Suddath, F.L., Quigley, G.J., McPherson, A., Sussman, J.L., Wang, A.H.J., Seeman, N.C. and Rich, A. (1974). *Science,* **185**, 435.

Klug, A., Ladner, J.E. and Robertus, J.D. (1974). *J. Mol. Biol.* **89**, 511.

Kruse, T.A., Clark, B.F.C. and Sprinzl, M. (1978). *Nucleic Acids Res.* **5**, 879.

Ladner, J.E. (1978). *In:* 'Biochemistry of Nucleic Acids' (ed. Clark, B.F.C.), 1. University Park Press, Baltimore.

Ladner, J.E., Finch, J.T., Klug, A. and Clark, B.F.C. (1972). *J. Mol. Biol.* **72**, 99.

Ladner, J.E., Jack, A., Robertus, J.D., Brown, R.S., Rhodes, D., Clark, B.F.C., and Klug, A. (1975). *Proc. Nat. Acad. Sci. USA* **72**, 4414.

Lagerkvist, U. (1978). *Proc. Nat. Acad. Sci. USA,* **75**, 1759.

Lake, J.A. (1977). *Proc. Nat. Acad. Sci. USA,* **74**, 1903.

Levitt, M. (1969). *Nature* **224**, 759.

Lewis, J.A. and Ames, B. (1972). *J. Mol. Biol.* **66**, 131.

Mazzara, G.P. and McClain, W.H. (1977). *J. Mol. Biol.* **117**, 1061.

Mazzara, G.P., Seidman, J.G., McClain, W.H., Yesian, H., Abelson, J. and Guthrie, C. (1977). *J. Biol. Chem.* **252**, 8245.

McCloskey, J.A. and Nishimura, S. (1977). *Accounts Chemical Res.* **10**, 403.

McCutchan, T.M., Silverman, S., Kohli, J. and Söll, D. (1978). *Biochemistry* **17**, 1622.

Miller, D.L. (1978). *In:* 'Gene Expression Symposium A2 of the 11th FEBS Meeting, Copenhagen 1977' (eds. Clark, B.F.C., Klenow, H. and Zeuthen, J.) Vol. 43, 59, Pergamon Press.

Miller, D.L. and Weissbach, H. (1977). *In:* 'Molecular Mechanisms of Protein Biosynthesis' (eds. Weissbach, H., Pestka, S.), 323. Academic Press, London and New York.

Mitra, S.K., Lustig, F., Akesson, B. and Lagerkvist, U. (1977). *J. Biol. Chem.* **252**, 471.

Möller, A., Schwarz, U., Lipecky, R. and Gassen, H.G. (1978). *FEBS Letters* **89**, 263.

Morikawa, K., la Cour, T. F. M., Nyborg, J., Rasmussen, K.M., Miller, D.L. and Clark, B.F.C. (1978). *J. Mol. Biol.* **125**, 327.

Nishimura, S. (1972). *In:*'Progress in Nucleic Acids Res. and Mol. Biol.' (eds. Davidson, J.N. and Cohn, W.E.), Vol. 12, 49. Academic Press, New York and London.

Pedersen, F.S., Lund, E. and Kjeldgaard, N.O. (1973). *Nature New Biol.* **243**, 13.

Pongs, O. and Reinwald, W. (1973). *Biochem. Biophys. Res. Commun.* **50**, 357.

Potts, R., Fournier, M.J. and Ford, N.C. (1977). *Nature* **268**, 563.

Quigley, G.J. and Rich, A. (1976). *Science,* **194**, 796.

Quigley, G.J., Teeter, M.M. and Rich, A. (1978). *Proc. Nat. Acad. Sci. USA* **75**, 64.

Quigley, G.J., Wang, A., Seeman, N.C., Suddath, F.L., Rich, A., Sussman, J.L. and Kim, S.H. (1975). *Proc. Nat. Acad. Sci. USA* **71**, 3711.

Reid, B.R. and Hurd, R.E. (1977). *Accounts Chemical Res.* **10**, 396.

Rhodes, D. (1975). *J. Mol. Biol.* **94**, 449.
Rhodes, D. (1977). *Eur. J. Biochem.* **81**, 91.
Rich, A. and RajBhandary, U.L. (1976). *Ann. Rev. Biochem.* **45**, 805.
Robillard, G.T., Tarr, C.E., Vosman, F. and Berendsen, H.J.C. (1976). *Nature* **262**, 363.
Roberts, R.J. (1972). *Nature New Biol.* **237**, 44.
Robertson, J.M., Kahan, M., Wintermeyer, W. and Zachau, H.G. (1977). *Eur. J. Biochem.* **72**, 117.
Robertus, J.D., Ladner, J.E., Finch, J.T., Rhodes, D., Brown, R.S., Clark, B.F.C. and Klug, A. (1974a). *Nature* **250**, 546.
Robertus, J.D., Ladner, J.E., Finch, J.T., Rhodes, D., Brown, R.S., Clark, B.F.C. and Klug, A. (1974b). *Nucleic Acids Res.* **1**, 927.
Scarpulla, R.C., Deutch, C.H. and Soffer, R.L. (1976). *Biochem. Biophys. Res. Commun.* **71**, 584.
Schwarz, U. and Gassen, H.G. (1977). *FEBS Letters,* **78**, 267.
Schwarz, U. Lührmann, R. and Gassen, H.G. (1974). *Biochem. Biophys, Res. Commun.* **56**, 807.
Sharma, O.K., Beezley, D.W. and Roberts, W.K. (1976). *Biochemistry,* **15**, 4313.
Sigler, P.B. (1975). *Ann. Rev. Biophys. Bioeng.* **4**, 477.
Simoncsits, A., Brownlee, G.G., Brown, R.S., Rubin, J.R. and Guilley, H. (1977). *Nature,* **269**, 833.
Smith, J.D. (1972). *Ann. Rev. Genetics,* **6**, 235.
Smith, D.W. (1975). *Science,* **190**, 529.
Sprague, K.U., Hagenbüchle, O. and Zuniga, M.C. (1977). *Cell,* **11**, 561.
Sprinzl, M. and Cramer, F. (1978). *In:* Progress in Nucleic Acids Res. Mol. Biol. Vol. 22, 2 (ed. Cohn, W.E.), Academic Press.
Sprinzl, M., Grüter, F. and Gauss, D.H. (1978). *Nucleic Acids Res.* **5**, r15.
Stanley, J. and Vassilenko, S. (1978). *Nature*, **274**, 87.
Stein, A. and Crothers, D.M. (1976). *Biochemistry,* **15**, 160.
Stout, C.D., Mizuno, H., Rubin, J., Brenner, T., Rao, S.T. and Sundaralingam, S. (1976). *Nucleic Acids Res.* **3**, 1111.
Uhlenbeck, O.C. (1972). *J. Mol. Biol.* **65**, 25.
Urbanke, C. and Maas, G. (1978). *Nucleic Acids Res.* **5**, 1551.
Wagner, R. and Garrett, R.A. (1978). *FEBS Letters,* **85**, 291.
Waters, L.C. and Mullin, B.C. (1977). *Progr. Nucleic Acids Res. Mol. Biol.* (ed. Cohn, W.E.), **20,** 131 Academic Press.
Williams, R.J., Nagel, W., Roe, B. and Dudock, B. (1974). *Biochem. Biophys. Res. Commun.* **60**, 1215.
Woese, C R. (1970). *Nature,* **226**, 817.
Yoon, K., Turner, D.H. and Tinoco, I. (1975). *J. Mol. Biol.* **99**, 507.

PROBING THE SUB-STRUCTURE, EVOLUTION AND INTERACTIONS OF AMINOACYL-tRNA SYNTHETASES

C.J. BRUTON

Department of Biochemistry, Imperial College, London, England.

Introduction

Significant quantities of homogeneous aminoacyl-tRNA synthetases first became available about twelve years ago (Baldwin and Berg, 1966). It was immediately apparent that they offered one of the best systems in which to study one of the fundamental problems of molecular biology - the molecular interactions involved in protein/nucleic acid recognition. The existence of isoaccepting tRNA species (Clark and Marcker, 1966) aminoacylated by a single aminoacyl-tRNA synthetase (Bruton and Hartley, 1968) sparked the false hope that the basic topology of the recognition problem might be elucidated very rapidly by primary structure studies. Nothing could have been further from the truth. To take one example, the two methionine accepting tRNAs have 41 nucleotides in similar positions many of which are in regions common to all tRNA molecules (Dube *et al.*, 1968; Cory *et al.*, 1968) whereas $tRNA_m^{Met}$ and $tRNA_1^{Val}$ have 51 nucleotides in similar positions (Yaniv and Barrell, 1969), yet the MTS from *E. coli* aminoacylates both methionine accepting species and not the valine one.

Structural work on the aminoacyl-tRNA synthetases has lagged behind such studies on the tRNA molecules for a variety of reasons including the problems of supply of material, the relative instability of the purified proteins, the sheer size of the molecules, the comparative inferiority of protein sequencing techniques to nucleic acid sequencing techniques and the 'pot luck' associated with growing

---

*Abbreviations: Aminoacyl-tRNA synthetases are abbreviated TS preceded by the amino acid letter from the one letter code. Thus: MTS = methionyl-tRNA synthetase; YTS = tyrosyl tRNA synthetase etc.

TABLE I

Quaternary Structure and Protomeric Molecular Weights of Purified Aminoacyl-tRNA Synthetases

Enzyme	Source	Quaternary Structure	Reference
α_1 Class			
Arg	*B. stearothermophilus*	1 x 78,000	Parfait and Grosjean, 1972
Arg	*E. coli*	1 x 75,000	Marshall and Zamecnik, 1969
Arg	*N. crassa*	1 x 85,000	Naxario and Evans, 1974
Cys	*B. stearothermophilus*	1 x 54,000	This paper
Gln	*E. coli*	1 x 70,000	Folk, 1971
Ile	*E. coli*	1 x 110,000	Arndt and Berg, 1970
Ile	*B. stearothermophilus*	1 x 110,000	Charlier and Grosjean, 1972
Leu	*E. coli*	1 x 100,000	Waterson and Konigsberg, 1974
Leu	*B. stearothermophilus*	1 x 110,000	Koch *et al.* 1974
Val	*E. coli*	1 x 110,000	Berthelot and Yaniv, 1970
Val	*B. stearothermophilus*	1 x 110,000	Koch *et al.*, 1974
Val	Yeast	1 x 113,000	Lagerkvist and Waldenstrom, 1967
α_2 Class			
His	*E. coli*	2 x 43,000	Kalousek and Konigsberg, 1974
Lys	*E. coli*	2 x 52,000	Rymo *et al.*, 1972
Lys	Yeast	2 x 72,000	Rymo *et al.*, 1972
Met	*E. coli*	2 x 85,000	Bruton and Koch, 1974
Met	*B. stearothermophilus*	2 x 85,000	Mulvey and Fersht, 1976

TABLE 1 *contd.*

Pro	*E. coli*	2 x 47,000	Lee and Muench, 1969
Ser	*E. coli*	2 x 50,000	Waterson and Konigsberg, 1974
Ser	Yeast	2 x 60,000	Hertz and Zachau, 1973
Trp	*E. coli*	2 x 37,000	Joseph and Muench, 1971
Trp	*B. stearothermophilus*	2 x 37,000	Koch *et al.*, 1974
Trp	Yeast	2 x 50,000	Hossein and Kallenbach, 1974
Trp	Bovine pancreas	2 x 54,000	Iborra *et al.*, 1973
Tyr	*E. coli*	2 x 45,000	Chousterman and Chapeville, 1973
Tyr	*B. stearothermophilus*	2 x 45,000	Koch *et al.*, 1974
$\alpha\beta$ Class			
Glu	*E. coli*	1 x 56,000 1 x 46,000	Lapointe and Soll, 1972
$\alpha_2\beta_2$ Class			
Gly	*E. coli*	2 x 33,000 2 x 80,000	Ostrem and Berg, 1974
Phe	*E. coli*	2 x 37,000 2 x 98,000	Fayat *et al.*, 1974
Phe	Yeast	2 x 62,000 2 x 72,000	Fasiolo *et al.*, 1970

protein crystals suitable for single crystal X-ray diffraction analysis. Although Francis Crick has said that, given the structure of an aminoacyl-tRNA synthetase and its cognate tRNA, he will tell us how they interact, I remain sceptical and believe that the final answer will only emerge when a three-dimensional structure of an enzyme/tRNA complex is elucidated.

That goal remains a long way off. However the full three-dimensional structure of one tRNA molecule ($tRNA^{Phe}$) is now known (Kim *et al.*, 1974; Robertus *et al.*, 1974) and at least two aminoacyl-tRNA synthetases - MTS from *E. coli* (Zelwer *et al.*, 1976) and YTS from *B. stearothermophilus* (Irwin *et al.*, 1976) are well advanced. The amino acid sequences of these proteins is also nearing completion (Bruton, unpublished; Winter and Koch, unpublished). Unfortunately FTS is a particularly large and complex molecule (see Table I) and one hopes that $tRNA^{Met}$ and $tRNA^{Tyr}$ structures will also be determined.

Besides these full structure determinations, chemical studies are in progress on a number of other aminoacyl-tRNA synthetases including ITS (Tsai and Kula, 1976), WTS (Kuehl, Lee and Meunch, 1976) and YTS (Barker and Bruton, unpublished), all from *E. coli* and the complete amino acid sequence of WTS from *B. stearothermophilus* is known (Winter and Hartley, 1977). In this paper, I propose to review the results of structural studies on aminoacyl-tRNA synthetases and discuss some of the questions which these results pose. The general theme which emerges is a lack of generality: results in one system are frequently not applicable in another. However as our knowledge becomes more detailed some order is found. Light travels at 186,000 miles per second; enlightenment somewhat more slowly.

Production of Gramme Quantities of Pure Enzymes

Large quantities of homogeneous protein are a prerequisite of all these studies. Most aminoacyl-tRNA synthetases represent about 0.1% of the cell's soluble protein and hence the production of material has always presented a logistic problem. Biomass is most easily obtained in prokaryotic systems and these also offer the possibility of manipulation to produce high levels of enzyme protein. However the critical role played by the aminoacyl-tRNA synthetases in the translation process makes it difficult to devise methods to put an organism under sufficient pressure to select the required high synthesis mutants. The semi-industrial scale-up of laboratory procedures first reported by Baldwin and Berg (1966) remains the chief method of production.

Cassio, Lawrence and Lawrence (1970) observed that *E. coli* EM2 EM20031 (McFall, 1967) contained elevated levels of MTS as it carried the structural gene for that enzyme on a stable episome and this strain (a K12) has been used to purify not only MTS, but also YTS, QTS and several other proteins (Bruton, Jakes and Atkinson, 1975). Working on a 50 kg scale 3 g of pure MTS was obtained.

The long term storage of aminoacyl-tRNA synthetases in active form has always been a problem. The enzymes are highly susceptible to mild proteolysis (Cassio and Waller, 1968) sometimes with loss of and sometimes retention of enzymic activity, but always producing a mixture of components requiring further purification. Moreover enzymes from *E. coli* tend to exhibit a slow progressive loss of activity on storage.

Enzymes from thermophiles are, however, considerably more stable than those from mesophiles and not merely in respect of their thermal stability - e.g. Aminopeptidase I from *B. stearothermophilus* is stable in 8 *M* urea and also in 0.3% sodium dodecyl sulphate (Roncari and Zuber, 1969). Accordingly a procedure has been devised for the simultaneous isolation of many enzymes in large quantities from 70 kg batches of *B. stearothermophilus* in the expectation that the enzymes would be more stable, hence more easily stored and more tractable. This multi-enzyme preparation (see Figure 1) performed annually at Imperial College, London provides aminoacyl-tRNA synthetases for research in many laboratories throughout the world. The main objective has also been achieved with the successful crystallisation of YTS from *B. stearothermophilus* (Reid *et al.*, 1973).

The Diversity of Aminoacyl-tRNA Synthetase Structures

A summary of the sub-unit structures and protomeric molecular weights of aminoacyl-tRNA synthetases purified from a variety of organisms is shown in Table I. These data alone immediately expose the naivety of the early assumptions about the degree of homology expected within this family of enzymes. In marked contrast to other extensively investigated families, for example the serine proteases (Hartley *et al.*, 1965) and the tRNA substrate molecules themselves, there is a total lack of conservation of structural features. The native molecular weights vary from 54,000 to 260,000 and the protomers from 30,000 to 110,000: the quaternary structure from α_1 to α_2 and $\alpha_2\ \beta_2$. The presumed evolution of such a family of enzymes all catalysing an identical reaction with one identical substrate and two very similar ones from one ancestral gene must have taken a tortuous path and the end result as we see it gives very few clues about that

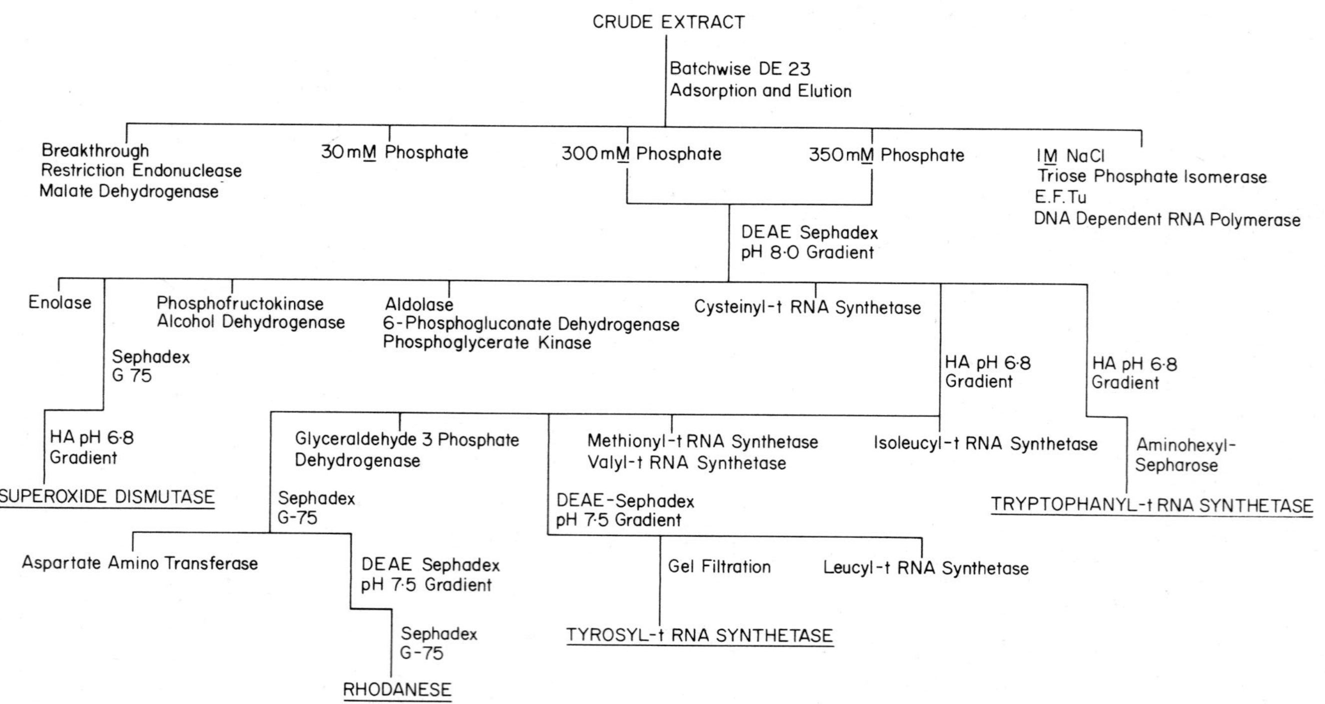
CRUDE EXTRACT
Batchwise DE 23
Adsorption and Elution
Breakthrough
Restriction Endonuclease
Malate Dehydrogenase
30 mM Phosphate
300 mM Phosphate
350 mM Phosphate
1M NaCl
Triose Phosphate Isomerase
E.F.Tu
DNA Dependent RNA Polymerase
DEAE Sephadex
pH 8·0 Gradient
Enolase
Phosphofructokinase
Alcohol Dehydrogenase
Aldolase
6-Phosphogluconate Dehydrogenase
Phosphoglycerate Kinase
Cysteinyl-t RNA Synthetase
HA pH 6·8
Gradient
HA pH 6·8
Gradient
Sephadex
G 75
HA pH 6·8
Gradient
SUPEROXIDE DISMUTASE
Glyceraldehyde 3 Phosphate
Dehydrogenase
Methionyl-t RNA Synthetase
Valyl-t RNA Synthetase
Isoleucyl-t RNA Synthetase
Aminohexyl-
Sepharose
TRYPTOPHANYL-t RNA SYNTHETASE
Sephadex
G-75
DEAE-Sephadex
pH 7·5 Gradient
Aspartate Amino Transferase
DEAE Sephadex
pH 7·5 Gradient
Gel Filtration
Leucyl-t RNA Synthetase
Sephadex
G-75
TYROSYL-t RNA SYNTHETASE
RHODANESE

path's nature.

Some of the apparent diversity may be explained by the observation that the larger protomeric units contain repeating sequences (Kula, 1973; Bruton *et al.*, 1974; Koch *et al.*, 1974; Waterson and Konisberg, 1974; Bruton, 1975). In every case which has been studied where the polypeptide chain is longer than 700 residues a significant proportion of the chain is repeated. This has presumably arisen by gene duplication and fusion. Furthermore in the two cases which have been studied in detail - MTS from *E. coli* (Bruton, 1976 and unpublished) and ITS from *E. coli* (Kula, 1973) - there is a very high degree of conservation, up to 100% within the repeats. Table II shows a sequence of 47 amino acid residues which is repeated exactly in MTS.

In contrast α_2 type enzymes with smaller polypeptide chains have unique amino acid sequences. This was originally surmised from fingerprinting studies but subsequently proved by the determination of the complete amino acid sequence of WTS (molecular structure $2 \times 37{,}000$) from *B. stearothermophilus* (Winter and Hartley, 1977).

It was not possible to detect, by eye or by computer (McLachlan, 1971) any homologous internal repetition. Hence the putative ancestral aminoacyl-tRNA synthetase gene would have coded for a polypeptide chain of about 300 amino acid residues.

All of the available amino acid sequence information on WTS from both *E. coli* and *B. stearothermophilus,* MTS and ITS from *E. coli* and YTS from *B. stearothermophilus* has been subjected to a systematic analysis for homology by computer (McLachlan, 1971). As expected for the same enzyme from two different bacteria, WTS *E. coli* and WTS *B. stearothermophilus* showed a considerable degree of homology (Winter *et al.*, 1977). Residues totalling over two thirds of the enzymes were compared and tentative alignments were made corresponding to one third of the molecule. However none of the other pairs of enzymes tested revealed any detectable homology. Although none of the other sequences is complete, large fragments over one hundred residues in length, including for example the N-terminal portions, were scanned 11 residues at a time. The success of the WTS *E. coli*/WTS *B. stearothermophilus* search was achieved with considerably less data. It seems, therefore, that we can safely conclude that, at the level of primary sequence, there is no homology

Fig. 1 See opposite. A multi-enzyme preparation. The flow diagram shows the main steps in the multi-enzyme preparation from *B. stearothermophilus* (Atkinson *et al.*, 1978). All the enzymes listed have been purified to homogeneity from one or more preparations and the four in capitals are regularly purified.

between different aminoacyl-tRNA synthetases even from the same organism.

TABLE II

*A Repeat in MTS from E. coli**

-Gly-Met-Phe-Leu-Pro-Asp-Arg-Phe-Val-Lys-Cys-Gly-Ser-Pro-Asx-Gly-Tyr-(Cys$_2$, Asx$_2$, Thr, Gly$_2$, Ala, Val, Lys)-Tyr-Ser-Pro-Thr-Glu-Leu-Ile-Gly-Pro-Lys-Ser-Val-Val-Ser-Gly-Ala-Thr-Pro-Met-Arg.

*A 47 amino acid sequence which occurs twice within the stable tryptic fragment of MTS from *E. coli.* The sequence was determined mainly by combined dansyl-Edman sequencing of peptides isolated from tryptic, chymotryptic and cyanogen bromide digests of the MTS fragment. The sequence in brackets is not unambiguously overlapped but no variations within it have been found. Two different leader sequences have been identified.

The functional homology persists and the challenge now is to accelerate the elucidation of its structural basis. Active site directed inhibitors, such as *p*-nitrophenyl-carbamylaminoacyl-tRNA (Bruton and Hartley, 1970), and photoaffinity labelling with analogues of ATP and aminoalkyl-adenylates (Wetzel and Soll, 1977; Wetzel, Bruton and Winter, unpublished) may reveal small regions of similarity but conservation of three dimensional features remains the best prospect.

The double domain structure of bovine rhodanese is an obvious example (Ploegman, 1977). This molecule contains two domains each of 109 amino acids and the tertiary structure of these domains are very similar. Indeed the polypeptide chains can be essentially superimposed and the root mean square distance of the 109 pairs of α-carbon atoms (Rao and Rossmann, 1973) is only 1.72 Å. Yet there is virtually no similarity between the amino acid sequences of the two domains. Only 14 out of the 109 residues are identical and never more than two of these are in tandem. The aligned sequences have been tested for homology both by the amino acid alternative method (McLachlan, 1971) and by the minimum base change per codon method (Dickerson, 1971). Both reveal no significant homology and the value of 1.28 mutations per codon required to change from one domain to the other is very close to the calculated figure (1.24) for sequences which are similar because of structural constraints, but otherwise random. This striking example allows us to retain the expectation of tertiary structure homology within the aminoacyl-tRNA synthetases; particularly as there is considerable evidence of domain structures within the molecules.

MTS from *E. coli* has a stable and active fragment formed by mild proteolysis which consists of three quarters of the native polypeptide chain and has an intact N-terminus (Bruton and Koch, 1974). It is this fragment which has been crystallised and this three-dimensional structure which is being determined (Waller *et al.*, 1971). Extensive protection studies using methionine, ATP, methioninyl-adenylate and $tRNA^{Met}$ species have all failed to release the predicted remaining 20,000 dalton fragment intact. That this does not form a stable domain of its own is indicated by two observations. Sodium dodecyl sulphate polyacrylamide gels of extremely mild digests of MTS during the preparation of the stable fragment always reveal intermediates larger than 65,000 but smaller than the native sub-unit. A predominate species around 75,000 is often seen and an intact 20,000 unit is never observed. Furthermore approximately molar yields of small peptides are released concurrently with the 65,000 stable fragment. Mild proteolysis of WTS from *B. stearothermophilus* and bovine pancreas also produces some very stable intermediates (Winter *et al.*, 1977; Epely *et al.*, 1976). With these molecules the more stable portion is the C-terminus of the chain and the relationship between the mammalian enzyme and the bacterial one is best explained by assuming an N-terminal appendage to the former. However the conservation of sequence discussed above between WTS from *E. coli* and *B. stearothermophilus* is markedly greater in the N-terminal portion of the (bacterial) chain. The mammalian enzyme is considerably bigger but an active stable fragment of very comparable size is formed by mild proteolysis. Further digestion of this entity removing only about twenty more amino acids renders the molecule inactive. This reinforces the view that this region of the molecule may be of catalytic importance.

Perhaps a more impressive case is YTS from *B. stearothermophilus.* From the electron density map it is possible to trace about 270 residues of the polypeptide chain (Irwin *et al.*, 1976) and the remainder of the molecule (approximately 150 residues) may be less rigid even in the crystal. This conclusion is supported by the results of partial proteolysis experiments which again reveal a more stable C-terminal domain. Tyrosine, ATP and tyrosinyl-adenylate all bind to the crystal and have been located within the traced portion of the chain (Monteilhet and Blow, 1978). Insufficient data exists on other aminoacyl-tRNA synthetases to conclusively establish any general pattern, but the existence of these domains and the lessons of the bovine rhodanese structure strengthen our fundamental conviction that there will be a structural foundation to the functional homology.

Are Multiple Sites Necessary?

Most aminoacyl-tRNA synthetases have molecular weights in the range 75,000 to 120,000 daltons and contain within the native molecule substantial portions of the polypeptide chain twice by virtue of either two identical sub-units or repeated sequences. That the presence of repeated sequences within a polypeptide chain could have similar functional significance to the existence of two sub-units was demonstrated by Fersht (1975). The monomeric VTS from *B. stearothermophilus* was shown to have two active sites in an analagous manner to the dimeric YTS from the same organism. Both enzymes were found to bind a second molecule of ATP and amino acid after one aminoacyl-adenylate had been formed on the enzyme. In that paper a general mechanism for the enhancement of catalysis through interacting sites and the catalytic advantages in the facility of using the binding energies of the substrates more than once during the reaction are described. However it is stressed that the existence of two sites does not solve the problem of specificity and a verification mechanism is required to account for the fidelity of aminoacyl-tRNA synthetases in translation.

In view of the emerging pattern of aminoacyl-tRNA synthetase infrastructure and these potential catalytic advantages, it has been suggested that the existence of two or more sets of binding sites would be a general feature of this family. The smallest reported molecule is CTS from *E. coli* (Fayat *et al.*, 1974). (I discount the structure of 1 × 46,000 proposed for YTS from yeast by Beikirch *et al.*, (1972) as inadequate steps were taken to inhibit proteolytic activity known to be prevalent in yeast and by analogy with the known dimeric structures of YTS from *E. coli* and *B. stearothermophilus.*) Although the report of the molecular weight of CTS deals only with the elution position of the enzymic activity from gel filtration of a crude extract, in this case considerable precautions were taken to inhibit proteolytic activity - MTS remained in its native form - and hence it seemed to me likely that this did represent the true size of CTS in *E. coli.*

We have purified CTS from *B. stearothermophilus* on a large scale during the multi-enzyme preparation referred to above. The homogeneous enzyme has a native molecular weight as judged by its elution position from a calibrated Sephacryl column of 55,000 daltons. Sodium dodecyl sulphate polyacrylamide gel electrophoresis (SDS-

PAGE) (Laemmli, 1970) on 10% and 15% gels indicate a sub-unit molecular weight of 53,000 and 55,000 respectively (see Figure 2).

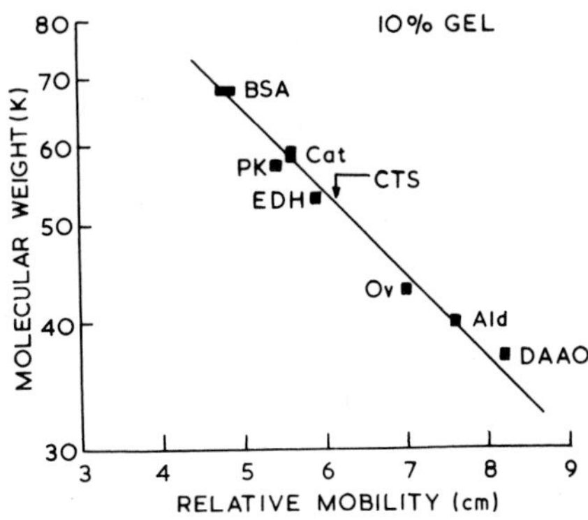

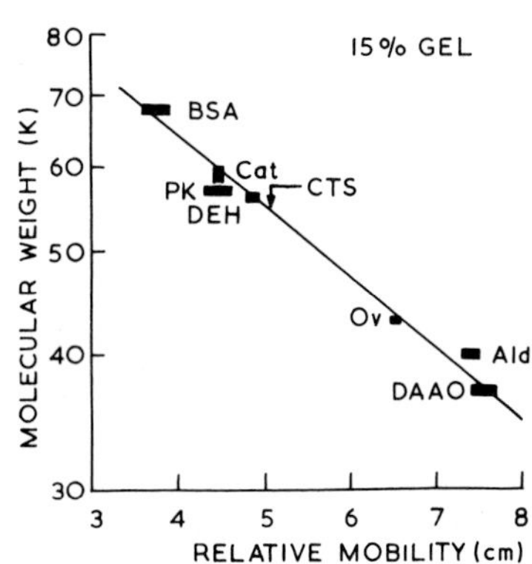

Fig. 2 Semi-logarithmic plots of sodium dodecyl sulphate gels of CTS and standard proteins. 10% and 15% acrylamide gels were run as described by Laemmli (1970). BSA - Bovine serum albumin; PK - pyruvate kinase; EDH - glutamate dehydrogenase; Cat - catalase; Ov - ovalbumin; Ald - aldolase; DAAO - D-amino acid oxidase.

These results show that the bacterial CTS is considerably smaller than the other aminoacyl-tRNA synthetases and consists of a single polypeptide chain which is of a comparable size to the sub-unit or the 'intramer' of most of the other enzymes.

It is infinitely simpler to demonstrate unequivocally the existence of repeated sequences (Bruton *et al.*, 1974) than to conclusively prove their absence. The final proof of the latter really requires a complete amino acid sequence determination which could hardly be justified to establish this one point. Fortunately in favourable cases fingerprinting and peptide counting (Bruton, 1975) can give a very strong indication. Such an experiment on the pure CTS from *stearothermophilus* (see Figure 3) reveals the presence of about 60 tryptic peptides of which 37 contain arginine compared to figures of 61 and 30 respectively predicted by the amino acid analysis and assuming a unique sequence. Furthermore preparative enzymic digests of the (^{14}C)-carboxymethylated protein contain all the cysteine residues in different sequences. These results are entirely con-

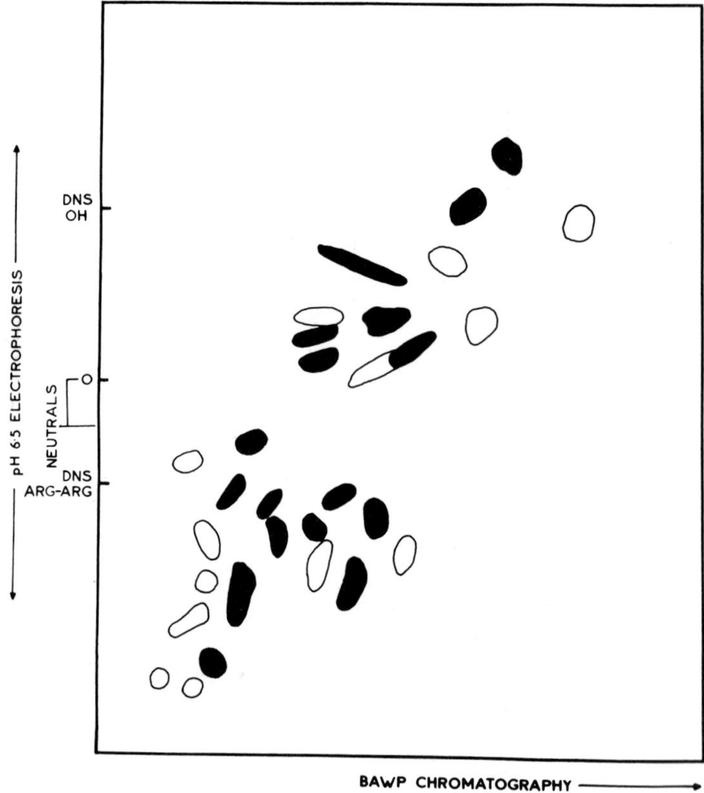

Fig. 3a (legend opposite).

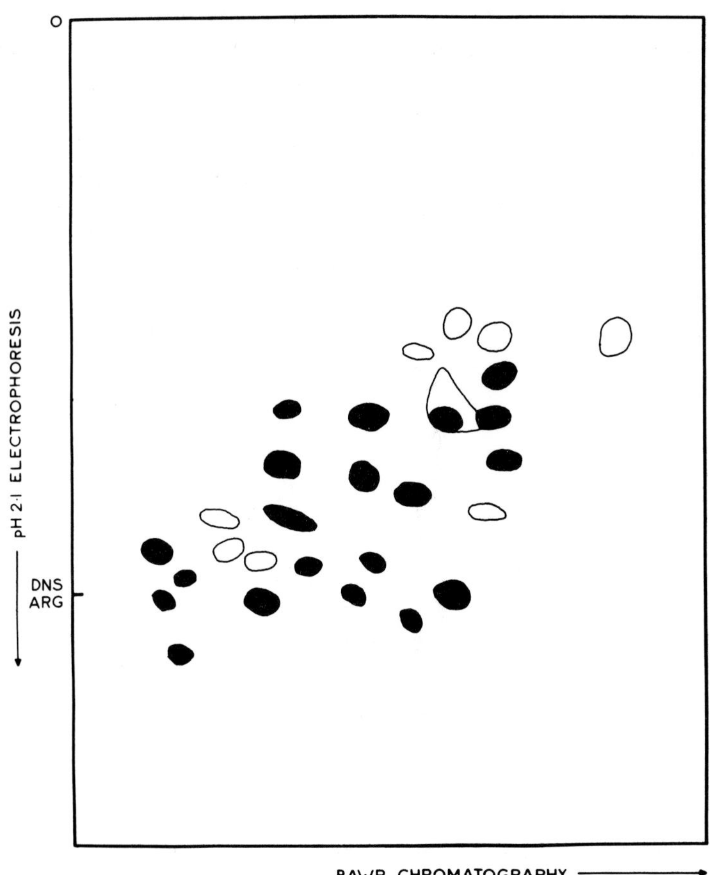

Fig. 3 Tryptic fingerprint of CTS. 2 mg of performic acid oxidised (Hirs, 1956) CTS from *B. stearothermophilus* was digested with 20 μg of trypsin. Peptides were separated as described by Bruton (1975). ⬭ Peptides visualised with fluorescamine (Udenfriend *et al.*, 1972). ⬬ Peptides visualised by the phenanthroquinone reagent (Yamada and Itano, 1966).

sistent with a non-repeating primary structure for CTS, but, of course, do not exclude the possibility of a very small region of duplicated sequence.

The active site titration (Fersht *et al.*, 1975) of CTS shows that one (0.72-1.02) cysteinyl-adenylate is bound per molecule of enzyme

(M.W. 54,000). Equilibrium dialysis using the small scale apparatus designed and kindly loaned by Prof. Fersht reveals that 0.9 moles of L-cysteine are bound per mole of enzyme with a K_s of 50 μM. This binding stoichiometry is not increased by the presence of ATP and pyrophosphatase nor is the slope of the linear portion of the active site titration affected by excess substrates. These were the techniques used to demonstrate the existence of the second sites of VTS and YTS from *B. stearothermophilus* (Fersht, 1975).

The possibility exists, of course, that in the presence of its substrates the CTS monomer aggregates. This was explored by running a series of parallel gel filtration columns with the buffers containing saturating concentrations of L-cysteine or L-cysteine + ATP + $MgCl_2$. These experiments are summarised in Table III. There is no indication of any major change in molecular size or shape in the presence of these substrates.

TABLE III

*Gel Filtration of CTS**

Additions	Temperature °C	Void Volume (ml)	CTS Elution (ml)
None	4	54	80
None	25	52	79
None	40	52	79
None	60	51	78
50 mM L-cysteine	4	54	80
50 mM L-cysteine + 10 mM ATP + 20 mM $MgCl_2$	4	54	80

*2 mg of pure CTS, dissolved in 100 mM tris-HCl, pH 7.8 with the additions indicated, was applied to a column (1.8 x 60 cm) of Sephacryl S-200 Superfine equilibrated with the same buffer. The column was eluted with the same buffer at 20 ml/hr. The elution position of the protein was determined spectrophotometrically at 280 nm or, in the case of the buffer containing ATP, by the cysteine-dependent ATP-PPi exchange assay (Bruton and Hartley, 1968).

No change in native molecular weight is observed from 4° to 60° as judged by the elution position from gel filtration columns (Table III). Furthermore the enzyme aminoacylates $tRNA^{Cys}$ from 4° to 65°. This is in marked contrast to PTS from *E. coli* which is monomeric at 4°, dimerises at higher temperatures and is only active in the aminoacylation assay at the higher temperature (Lee and Muench, 1969).

CTS is absolutely specific for L-cysteine. Even in the ATP-pyrophosphate exchange reaction which examines the less specific first partial reaction leading to the formation of cysteinyl-adenylate, no activation has been observed with any other naturally occuring amino acid. Moreover although, like YTS, CTS aminoacylates both the 2′ and 3′-OH of the terminal adenosine (Sprinzl and Cramer, 1975), unlike YTS, CTS neither activates nor aminoacylates the D-isomer. The turnover number of the pure enzyme is 20 sec^{-1}. Thus the monomeric CTS achieves both discrimination and catalytic efficiency comparable to the dimeric α_2 or the duplicated large α_1 enzymes.

The correspondence between the native molecular weight estimated from the elution positions from gel filtration columns calibrated with globular proteins and the molecular weight of the polypeptide chain estimated by SDS-PAGE suggests that the CTS molecule is itself globular and not particularly elongated. A globular protein of 54,000 daltons is barely large enough to encompass a tRNA molecule of the dimensions of the structure determined for crystals of $tRNA^{Phe}$. The tRNA may well adopt a different conformation when bound to the enzyme, but models of the interaction in which the tRNA molecule spans both sub-units of the small dimeric aminoacyl-tRNA synthetases (Bosshard *et al.,* 1978) are unlikely to be appropriate for the CTS/$tRNA^{Cys}$ interaction.

Mapping the tRNA Binding Site

When the full three dimensional structures of both an aminoacyl-tRNA synthetase and its cognate tRNA are determined, it is possible that the details of the interactions at the molecular level will be immediately apparent. It is also possible that Francis Crick will, as he forecast, unravel any remaining mysteries. We believe, however, that it is far more likely that a number of possible interactions will suggest themselves, but that it will be extremely difficult, if not impossible, to distinguish between these options by simple model building. Further experimental evidence concerning the details of the interactions will be critical.

There have been relatively few attempts to attack this problem from the aminoacyl-tRNA synthetase standpoint. These have usually been based on introducing a reactive group into the tRNA molecule and identifying the site of reaction with the protein (Bruton and Hartley, 1970; Santi and Cunnion, 1974). The design of this type of experiment predetermines the residues in the protein which can be identified and, even when successful, the experiment reveals only one or two side chains which are in close proximity to the region of the tRNA molecule into which the reactive group was placed. It can not give an overall picture of the multitude of interactions made over the entire surface of the two macromolecules. Although photochemical cross-linking (Schoemaker and Schimmel, 1974) has revealed some information about the regions of the tRNA molecule which could be involved, the restricted information available on the primary structure of the aminoacyl-tRNA synthetases and the limitations of the technique make it inappropriate and uninformative as far as the protein is concerned.

The approach of choice from the protein angle is a modification of the competitive labelling technique of Kaplan *et al.*, (1971). This method involves comparing the reactivities of the ϵ-amino groups of the lysine side chains of the protein towards an acylating reagent in the free enzyme and the enzyme/tRNA complex. Trace-labelling is performed with extremely low concentrations of high specific activity (^{3}H)acetic anhydride. Under these conditions a single enzyme molecule will react at only one position and hence the possibility of the modification of neighbouring groups affecting the reactivity of any particular group is negligible. After the trace-labelling the protein is denatured and fully acetylated with excess unlabelled acetic anhydride. A control sample of denatured enzyme completely acetylated with (^{14}C) acetic anhydride is also prepared. Peptides are isolated from a digest of a mixture of chemically homogeneous but radioactively heterogeneous protein and the (^{3}H)/(^{14}C) ratio determined. This ratio is a measure of the reactivity of the amino group in the protein. A further trace-labelling is carried out in the presence of tRNA under conditions where all the protein will be in the form of enzyme/tRNA complex and the (^{3}H)/(^{14}C) ratios for the same peptides determined again. Phenylalanine is used as an internal standard in both trace-labelling experiments. The reactivities of particular lysine residues in the free enzyme and the enzyme/tRNA complex can then be compared. This technique has been applied to YTS from *B. stearothermophilus* (Bosshard *et al.*, 1978) and three lysine side chains, all with reduced reactivity in the enzyme/tRNA complex, were identified.

The structure of MTS from *E. coli* is also well advanced and this enzyme interacts with two tRNA species of considerably different primary structure and different biological roles. Their interactions are indistinguishable by steady state kinetic methods (Bruton and Hartley, 1968). Unlike tRNA^{Tyr}, both tRNA^{Met} molecules have small extra arms and hence can be expected to adopt a configuration similar to that determined for $tRNA^{Phe}$. To assist the elucidation of the MTS/$tRNA^{Met}$ interactions and in an attempt to unmask any detailed difference in the structures of the initiator and internal tRNA species, we have performed competitive labelling studies on native MTS, MTS/$tRNA_f^{Met}$ complex and MTS/$tRNA_m^{Met}$ complex.

Five peptides were isolated with changed reactivities ((^{3}H)/(^{14}C) ratio) in one or both of the complexes (Table IV).

TABLE IV

*Peptides with Altered Reactivities in MTS/tRNA^{Met} Complexes**

Peptide	Sequence	Ratio of (^{3}H)/(^{14}C) Ratios	
		MTS/$tRNA_f^{Met}$ / MTS	MTS/$tRNA_m^{Met}$ / MTS
IB9	DAGFIN$\underline{K}$R	0.4	0.58
IA20	$\overline{A}$QV	0.32	0.94
IIIA5	$\overline{L}$TLB-G-$\underline{K}$-R	0.84	3.44
VA9	FGISZGMVMAAGPGG$\underline{K}$	0.38	0.50
VIIIA25	IPD-G$\underline{K}$L	1.81	1.98

*Competitive labelling was performed as described by Kaplan *et al.*, (1971). The acetylated MTS was digested thoroughly with trypsin (1/25, w/w) and the peptides purified by ion exchange chromatography, high voltage paper electrophoresis and paper chromatography. The vigorous digestion was deliberately used to give some additional non-tryptic cleavages and hence a soluble digest. Over sixty radioactive peptides were purified and their (^{3}H)/(^{14}C) ratios determined in native MTS, MTS/$tRNA_f^{Met}$ and MTS/$tRNA_m^{Met}$. The majority of peptides showed no major differences in ratios in all three enzyme samples (ratio of ratios 0.8-1.2). The peptides listed showed greater deviations.

The amino acid sequences of the peptides have been determined independently from the main sequence determination of MTS. These sequences are shown. Peptides 1B9, 1A20 and VA9 are unambiguously overlapped within the MTS tryptic fragment. The peptide nomenclature refers to the purification used.

The underlined residue in the sequence is the one acetylated.

These changes could be due to a change in pK_a of the amino group or to a change in its nucleophilicity. To distinguish between these possibilities a series of experiments at different pH values in the range from 8.0 to 11.0 (around the pK'_a of the groups involved) must be done (Kaplan *et al.*, 1971), and from these data it is possible to determine both the pK_a and the reactivity relative to an unhindered nucleophile for each group. The affinity of MTS for $tRNA^{Met}$ varies considerably in this pH range and hence different interactions may be made. Furthermore each experiment at each pH point would require several mg of homogeneous enzyme and tRNA. Hence all the experiments were performed at one pH only (8.5) which is a compromise designed to give a fairly high affinity for the MTS/$tRNA^{Met}$ interaction without reducing the amino group reactivity too markedly. It is impossible to assign the contribution of pK_a shifts and nucleophilicity changes and in individual cases these contributions could have opposite effects. An unchanged (^{3}H)/(^{14}C) ratio could be the result of a decreased pK_a accompanied by a decreased nucleophilicity or vice versa so we may be under-estimating the number of amino groups involved. However the changes observed must be due to some interaction in the complexes. The three which are unambiguously overlapped in the sequence of the MTS tryptic fragment are widely separated, reflecting the three dimensional folding. They are *not* repeating sequences. The other two peptides have not been completely sequenced and may be in the C-terminal portion of the MTS polypeptide which is cleaved off by mild proteolysis.

In any attempt to provide a basis for quantitative interpretation, it is necessary to make some assumptions concerning the contribution of the various factors in the reactivity and we have calculated the shift in pK_a required to bring about the observed changes in reactivities in the complexes assuming no change in nucleophilicity. These pK_as and shifts are presented in Table V.

These interactions will be of most value when the three-dimensional structure of MTS is known. Several features stand out now. In contrast to the observations with YTS from *B. stearothermophilus* both increased and decreased reactivities are found. More surprisingly, despite the failure to detect any differences in the interactions of the two $tRNA^{Met}$ species with the enzyme kinetically, this technique reveals major differences. Two of the five groups interact with only one of the tRNA species and a third shows a different degree of shielding. These results illustrate the power of the competitive labelling approach, which can be extended further, if necessary, by performing a full pH scan. Perhaps the most striking observation is that $tRNA_f^{Met}$ but not $tRNA_m^{Met}$, causes a reduction in the reactivity of the N-terminal amion group (Peptide IA20) equivalent to a pK_a shift of +0.5. This may lend support to the ideas that the N–terminal region of aminoacyl-tRNA synthetases are critically involved in the enzyme action and also are more likely to be conserved, discussed above.

The use of acetic anhydride restricts reaction to amino groups, but other side chains can be readily studied using different reagents. Arginine residues are obvious candidates to interact with the exposed tRNA phosphates and cyclohexane-1,2-dione can label them although the relatively high pH of the reaction would lead to lower enzyme/tRNA affinities and perhaps non-physiological interactions. Tyrosine and tryptophan side chains could make hydrophobic stacking-like interactions with the nucleotide bases and iodination and *o*-nitrophenylsulphenyl chloride respectively could be employed to investigate these possibilities.

TABLE V

A Quantitative Interpretation of Altered Reactivities

Peptide	pK_a in native MTS	pK_a Shift in	
		$MTS/tRNA_f^{Met}$ Complex	$MTS/tRNA_m^{Met}$ Complex
IB9	9.13	+ 0.5	+ 0.3
IA20	9.7	+ 0.5	+ 0.1
IIIA5	10.2	+ 0.2	– 0.5
VA9	9.5	+ 0.4	+ 0.2
VIIIA25	10.2	– 0.2	– 0.3

*The pK_a of each peptide described in Table IV in the native enzyme and both complexes was calculated from the $(^3H)/(^{14}C)$ ratio of the peptides and the phenylalanine internal standard assuming that each amino group behaves as un hindered nucleophile. pK_a of L-phenylalanine was taken as 9.13 and of the ϵ-amino group of lysine as 10.4.r values were taken from the Bronsted plot of Kaplan *et al.*, (1971). The isolation of the peptides is described in Table IV.

I am, therefore, optimistic that we can probe most of the surface of the enzyme molecule in this way by suitable choice of reagent and confidently expect that the further experiments will add considerably to our detailed understanding of the $MTS/tRNA^{Met}$ interactions.

Conclusions

Within the next two years three-dimensional structures for two aminoacyl-tRNA synthetases (YTS from *B. stearothermophilus* and MTS from *E. coli*) should emerge from the single crystal X-ray diffraction analyses. Until then, as more and more primary structure data accumulates and as increasing evidence for the existence of domains is found, we can merely speculate and hope that the lessons learned from other enzyme families will not prove totally inappropriate in this case. The partial rationalisation of the diversity of quaternary structures brought about by the discovery of repeated sequences in the longer polypeptide chains and the bizarre parallel of the double domain structure of bovine rhodanese give me some confidence that this conviction will not prove totally unfounded.

However the elucidation of these protein structures actually only defines the fundamental question which is: 'How does an aminoacyl-tRNA synthetase select the right tRNA and reject the wrong one?' We have developed an approach which allows us to study this problem in fine detail from the protein standpoint and I have discussed above how it can be extended to probe most of the surface of the enzyme molecule. As work concentrating on the tRNA is already well ahead, even if not in such fine detail, I live in hope that the final answer will not have to await the X-ray diffraction analysis of a crystal of the aminoacyl-tRNA synthetase/tRNA complex. As such a crystal has yet to be grown, this could be a very long wait.

References

Arndt, D.J. and Berg, P. (1970). *J. Biol. Chem.* **245**, 665.

Atkinson, A., Banks, G.T., Bruton, C.J., Comer, M.J., Jakes, R., Kamalagharan, A., Whitaker, A.R. and Winter, G.P. (1978). *Biochem. J.* (in press).

Baldwin, A.N. and Berg, P. (1966). *J. Biol. Chem.* **241**, 831.

Beikirch, H., von der Haar, F. and Cramer, F. (1972). *Eur. J. Biochem.* **26**, 182.

Berthelot, F. and Yaniv, M. (1970). *Eur. J. Biochem.* **16**, 123.

Bosshard, H.R., Koch, G.L.E. and Hartley, B.S. (1978). *J. Mol. Biol.* **119**, 377.

Bruton, C.J. (1975). *Biochem. J.* **147**, 191.

Bruton, C.J. (1976). *In:* Abstr. EMBO Workshop on tRNA. Sandjberg.

Bruton, C.J. and Hartley, B.S. (1968). *Biochem. J.* **108**, 281.

Bruton, C.J. and Hartley, B.S. (1970). *J. Mol. Biol.* **52**, 165.

Bruton, C.J., Jakes, R. and Atkinson, A. (1975). *Eur. J. Biochem.* **59**, 327.

Bruton, C.J., Jakes, R. and Koch, G.L.E. (1974). *FEBS Lett.* **45**, 26.

Bruton, C.J. and Koch, G.L.E. (1974). *FEBS Lett.* **40**, 180.

Cassio, D., Lawrence, F. and Lawrence, D.A. (1970). *Eur. J. Biochem.* **15**, 331.

Cassio, D. and Waller, J-P. (1968). *Eur. J. Biochem.* **5**, 33.

Charlier, J. and Grosjean, H. (1972). *Eur. J. Biochem.* **25**, 163.

Chousterman, S. and Chapeville, F. (1973). *Eur. J. Biochem.* **35**, 46.

Clark, B.F.C. and Marcker, K.A. (1966). *J. Mol. Biol.* **17**, 394.

Cory, S., Marcker, K.A., Dube, S.K. and Clark, B.F.C. (1968). *Nature (London)*

220, 1039.
Dickerson, R.E. (1971). *J. Mol. Biol.* **57**, 1.
Dube, S.K., Marcker, K.A., Clark, B.F.C. and Cory, S. (1968). *Nature (London)* **218**, 252.
Epely, S., Gros, C., Labouesse, J. and Lemaire, G. (1976). *Eur. J. Biochem.* **61**, 139.
Fasiolo, N., Befort, F., Bolloch, C. and Ebel, J-P. (1970). *Biochim. Biophys. Acta* **217**, 305.
Fayat, G., Blanquet, S., Dessen, P., Batelier, G. and Waller, J-P. (1974). *Biochimie* **56**, 35.
Fersht, A.R. (1975). *Biochemistry* **14**, 5.
Fersht, A.R., Ashford, J.S., Bruton, C.J., Jakes, R., Koch, G.L.E. and Hartley, B.S. (1975). *Biochemistry* **14**, 1.
Folk, W.R. (1971). *Biochemistry* **10**, 1728.
Hartley, B.S., Brown, J.R., Kauffman, D.L. and Smillie, L.B. (1965). *Nature (London)* **207**, 1157.
Hertz, H.S. and Zachau, H.G. (1973). *Eur. J. Biochem.* **37**, 203.
Hirs, C.H.W. (1956). *J. Biol. Chem.* **219**, 611.
Hossein, A. and Kallenbach, N.R. (1974). *FEBS Lett.* **45**, 202.
Iborra, F., Donizzi, M. and Labouesse, J. (1973). *Eur. J. Biochem.* **39**, 275.
Irwin, M.J., Nyborg, J., Reid, B.R. and Blow, D.M. (1976). *J. Mol. Biol.* **105** 577.
Joseph, D.R. and Meunch, K.H. (1971). *J. Biol. Chem.* **246**, 7602.
Kalousek, F. and Konigsberg, W.H. (1974). *Biochemistry* **13**, 999.
Kaplan, H., Stevenson, K.J. and Hartley, B.S. (1971). *Biochem. J.* **124**, 289.
Kim, S.H., Suddath, F.L., Quigley, G.J., McPherson, A., Sussman, J.l., Wang, A., Seeman, N.C. and Reid, A. (1974). *Science* **185**, 435.
Koch, G.L.E., Boulanger, Y. and Hartley, B.S. (1974). *Nature (London)* **249**, 316.
Kuehl, G.V., Lee, M. and Muench, K.H. (1976). *J. Biol. Chem.* **251**, 3254.
Kula, M-L. (1973). *FEBS Lett.* **35**, 299.
Laemmli, U. (1970). *Nature (London)* **227**, 680.
Lagerkvist, U. and Waldenstrom, J. (1967). *J. Biol. Chem.* **242**, 3021.
Lapointe, J. and Soll, D. (1972). *J. Biol. Chem.* **247**, 4966.
Lee, M. and Muench, K.H. (1969). *J. Biol. Chem.* **244**, 223.
Marshall, R.D. and Zamecnik, P.C. (1969). *Biochim. Biophys. Acta* **181**, 454.
McFall, E. (1967). *J. Bacteriol.* **94**, 1982.
McLachlan, A.D. (1971). *J. Mol. Biol.* **61**, 409.
Monteilhet, C. and Blow, D.M. (1978). *J. Mol. Biol.* (in press).
Mulvey, R.S. and Fersht, A.R. (1976). *Biochemistry*, **15**, 243.
Naxario, M. and Evans, J.A. (1974). *J. Biol. Chem.* **239**, 4934.
Ostrem, D.L. and Berg, P. (1974). *Biochemistry* **13**, 1338.
Parfait, R. and Grosjean, H. (1972). *Eur. J. Biochem.* **30**, 242.
Ploegman, J.H. (1977). Ph.D. Dissertation, University of Groningen.
Reid, B.R., Koch, G.L.E., Boulanger, Y., Hartley, B.S. and Blow, D.M. (1973). *J. Mol. Biol.* **80**, 199.

Robertus, J.D., Ladner, J.E., Finch, J.T., Rhodes, D., Brown, R.S., Clark, B.F.C. and Klug, A. (1974). *Nature (London)* **250**, 546.
Roncari, G. and Zuber, H. (1969). *Int. J. Protein Res.* **1**, 45.
Rymo, L., Lundvik, L. and Lagerkvist, U. (1972). *J. Biol. Chem.* **247**, 3888.
Santi, D.V. and Cunnion, S.O. (1974). *Biochemistry* **13**, 481.
Schoemaker, H.J.P. and Schimmel, P.R. (1974). *J. Mol. Biol.* **84**, 503.
Sprinzl, M. and Cramer, F. (1975). *Proc. Nat. Acad. Sci., U.S.A.* **72**, 3049.
Tsai, H. and Kula, M-R. (1976). *In:* Abstr. 10th International Congress Biochem. Hamburg, 104.
Udenfriend, S., Stern, S., Bohlen, P., Dairman, W., Leimgruber, W. and Weigele, M. (1972). *Science* **178**, 871.
Waller, J-P., Risler, J-L., Montheilhet, C. and Zelwer, C. (1971). *FEBS Lett.* **16**, 186.
Waterson, R.M. and Konigsberg, W.H. (1974). *Proc. Nat. Acad. Sci. U.S.A.* **71**, 376.
Wetzel, R. and Soll, D. (1977). *Nucleic Acids Res.* **4**, 1681.
Winter, G.P. and Hartley, B.S. (1977). *FEBS Lett.* **80**, 340.
Winter, G.P., Hartley, B.S., McLachlan, A.D., Lee, M. and Muench, K.H. (1977). *FEBS Lett.* **82**, 348.
Yamada, S. and Itano, H.A. (1966). *Biochim. Biophys, Acta* **130**, 538.
Yaniv, M. and Barrell, B.G. (1969). *Nature (London)* **222**, 278.
Zelwer, C., Risler, J-L. and Montheilhet, C. (1976). *J. Mol. Biol.* **102**, 93.

RIBOSOMAL SELECTION OF INITIATOR CODONS ON mRNA

K.A. MARCKER

Department of Molecular Biology,
University of Aarhus, Denmark.

Introduction

Initiation of protein biosynthesis consists of a series of events insuring the recognition by the ribosome of the initiator regions present in mRNA. During this process the initiator tRNA is placed opposite the initiator triplet present within the initiation signal so that correct translation ensures. The formation of the initiation complex is thus the critical event in phasing the mRNA on the ribosome. Initiation which is probably the rate limiting step in protein synthesis also represents an obvious control point in translation. Any alteration of any one of the components involved in formation of the initiation complex could in principle eliminate recognition of a particular initiation signal or alter the relative efficiency of ribosome binding to various initiation regions. In view of the critical importance for the cell of correct initiation it is therefore not surprising that both prokaryotes and eukaryotes have evolved a rather complex machinery for making certain that mRNAs are read correctly. In the case of prokaryotes the components are now well characterized (Grunberg-Manago *et al.*, 1978) while eukaryotic initiation is as yet less well understood.

Ribosomal mRNA Recognition in Prokaryotic Cells

Initiation of protein synthesis involved the specific binding of the small ribosomal subunit -fMet $tRNA_f$ complex to a region of mRNA containing the initiator codon, AUG or GUG. Polynucleotides having an AUG or GUG at the 5′ end initiate polypeptide synthesis with formyl methionine and in the beginning it was, therefore, thought possible that natural mRNAs simply started with an AUG or GUG codon at the 5′ terminal end. However, as sequences of natural mRNAs became available it was evident that the cistron usually begins some distance from the 5′ end of the mRNA. Thus the MS2 genome (Fiers *et al.,* 1976) starts with a 129 nucleotide long untranslated sequence followed by the 'A' protein gene, an intracistronic region of 26 nucleotides, the coat protein gene, an intercistronic region of 36 nucleotides, the replicase gene and finally finishes with a 74 nucleotide long untranslated region at the 3′ terminus. Prior to the complete elucidation of the primary structure of MS2 RNA the ribosomal binding sites of several mRNAs (which are the only ones protected by RNase digestion after initiation complex formation) had been sequenced (Steitz, 1969, 1977). Analysis of such sequences showed that in all cases an initiator triplet precedes the first translated codon and that the initiator triplet is always preceded by a non-translated sequence. It is noteworthy that a common feature in all ribosome protected sequences is a polypurine stretch of 4-8 nucleotides located about 10 bases 5′ to the initiator codon. Taking these facts into consideration it soon became obvious that the small ribosomal subunit somehow must be able to discriminate against the many internal AUG (or GUG) codons present in mRNA only selecting the AUG (or GUG) triplet at the beginning of a cistron. Consequently, some feature other than the presence of the initiator codon on mRNA must be necessary for proper ribosomal recognition.

In 1974 Shine and Dalgarno (1974, 1975) presented a hypothesis to explain the selection of initiator regions on mRNA. They suggested that the 3′ terminus of *E. coli* 16S rRNA directly participates in the selection of initiation sites by forming base pairs with the polypurine sequence located in the vicinity of the initiator codon. In the case of R17 RNA as shown in Figure 1, 7 base pairs are formed in the interaction between 16S rRNA and the initiation site for the 'A' protein cistron, 4 for the replicase and 3 for the coat protein, the latter being the minimum for a stable interaction. The lengths of the complementary regions in different mRNAs vary between 3 and 9 nucleotides, the average being 4-5. The number of nucleotides separating the complementary regions from the initiator codon varies with an average of 10 nucleotides separating the middle of each. When two or more initiator triplets occur in a single sequence the triplet preceded by the strongest and most appropriately positioned complementary region functions in polypeptide initiation. Examination of the available collection of about 40 different initiator regions from phage and *E. coli* mRNAs reveals that the rRNA sequence -CCUCC- is most consistently utilized in this interaction (Hägenbüchle *et al.,* 1978). In all cases at least two or usually three or more bases

5′-CUAGGAGGUUU-3′	'A' Protein
-ACAUGAGGAUU-	Replicase
-ACCGGGGUUUG-	Coat Protein
$_{OH}$AUUCCUCCACpy-	16S rRNA

Fig. 1 R17 ribosome binding site sequences and proposed pairing with the 3′ terminus of 16S rRNA. The sequences shown are from the 'A' protein, replicase and coat protein initiator regions of R17 RNA. The initiator codon is located eight to nine bases from the middle of this sequence. Underlined sequences can form base pairs with the 3′ terminus of 16S rRNA.

in this sequence is involved in the proposed complementarity with mRNA. In addition all eubacteria examined so far seem to retain at least four bases of this sequence in a comparable position near the 3′ end of the respective 16S rRNAs.

In addition to the presence of the postulated sequence complementarity to 16S rRNA in each known ribosome binding site other data support the Shine Dalgarno hypothesis. An intact 3′ terminus of 16S rRNA is necessary for initiation as indicated by the inhibitory action of colicin E3 which removes about 50 nucleotides from the 3′ terminus of 16S rRNA by a single endonucleolytic cleavage (Boon, 1971; Dahlberg *et al.*, 1973). Several inhibitors of initiation such as streptomycin (Crépin *et al.*, 1973) and kasugamycin (Melser *et al.*, 1971) act close to the 3′ terminus of 16S rRNA. Cross linking experiments indicate that the 3′ terminus of 16S rRNA is close to the binding site for initiation factors and to ribosomal proteins known to be involved in initiation.

Direct evidence for the Shine Dalgarno interaction has been obtained by Steitz and Jakes (1975) who treated initiation complexes formed between *E. coli* ribosomes and a fragment containing the initiation region for the 'A' protein of R17 RNA with colicin E3. After treatment of the initiation complex with colicin E3 a mRNA-rRNA hybrid could be isolated which only appeared in the presence of all the components necessary for initiation complex formation. Furthermore, it does not appear if the colicin treatment is omitted. Analysis revealed that the hybrid consisted of approximately equimolar amounts of a 30 nucleotide mRNA fragment and a 49-50 nucleotide fragment originating from 3′ terminus of 16S rRNA. Detailed sequence analysis further showed that the structure of the isolated RNA:RNA hybrid was in complete accordance with that predicted by the Shine Dalgarno hypothesis.

Recently, genetic analysis of mutations in the initiator region of

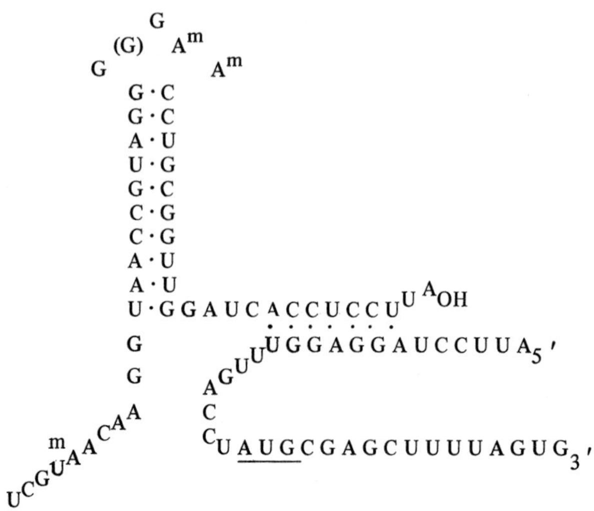

Fig. 2 Postulated structure of the colicin fragment of 16S rRNA and the R17 'A' protein initiator region. From Steitz and Jakes (1975).

the T7 0.3 protein strongly argues in favour of the postulated interaction between mRNA and 16S rRNA (Dunn *et al.*, 1978). In a particular mutant a decrease of about ten fold in the efficiency of the synthesis of the gene 0.3 protein in T7 infected cells was observed. Detailed analysis revealed that the mutation affects the initiation region by reducing the complementarity of the Shine Dalgarno interaction from a 5 base interaction GAGGU to a less stable interaction. A suppressor mutation that restores the rate of 0.3 protein synthesis to essentially wild type levels was subsequently found. This mutation is adjacent to the original mutation creating a new four base complementarity (GGAG) with 16S rRNA. Thus all the evidence so far obtained strongly supports the idea that a base pairing interaction between 16S rRNA and mRNA is involved in prokaryotic initiation of protein synthesis and that this interaction is important in selecting the site in mRNA at which ribosomes bind.

An important aspect is now whether the Shine Dalgarno hypothesis is able to explain species specificity in protein biosynthesis. *E. coli* ribosomes recognize the beginning of the three cistrons of phage f2 RNA, protein 'A', coat and replicase, coat being the most efficient initiation site. Bacterium stereathermophilus, however, translates only cistron 'A' either at 65°, its physiological temperature, or at 47° where *E. coli* translates all three cistrons (Lodish, 1970). The 3′ terminal

```
                    II I
                    GA
 15       20        ↑↑          30
-ACUGCACGAGGUAACACAAGAUG-
         • • • • •
  HOAUUCCUCCACUAG-
```

Fig. 3 Mutations in the initiator region of the 0.3 protein of T7. Mutation I reduces the efficiency of initiation to about 1/10 of normal level. This mutation is suppressed in mutation II. It is evident that in mutation II a new stable base pair interaction (GGAG) can be formed with the 3′ terminus of 16S rRNA. (Dunn *et al.*, 1978).

of both types of 16S rRNA are identical except that the 3′ terminal adenosine of *E. coli* rRNA is replaced by $UCUA_{OH}$ in the bacillus (Sprague *et al.*, 1977). Thus B. stereathermophilus ribosomes in principle have the same potential form binding to f2 RNA initiator regions as have E. coli ribosomes unless one postulates that the four extra bases somehow prevent Shine Dalgarno base pairing from taking place, which is not very likely. The specificity of stereathermophilus ribosomes is not altered when initiation factors from *E. coli* are used instead of homologous factors. Subsequent experiments showed that cistron specificity resides in the 30S particle. Reconstitution of 30S subunits using the technique developed by Nomura (Held *et al.*, 1974) indicated that the ability to discriminate among the initiation sequences present in f2 RNA resides in 16s rRNA and the ribosomal protein S12. Other experiments (Isono and Isono, 1975) indicate that also protein S1 is involved in this process. Thus not surprisingly proteins are also implicated as specificity determinants. The possibility remains, however, that proteins can create other interactions between mRNA and 16S rRNA than the Shine Dalgarno base pairing and further experiments are needed before definite conclusions can be drawn concerning the mechanism of prokaryotic translational specificity.

The requirement in matural initiation for the initiation factor F3 is more drastic than with synthetic polynucleotides. With natural mRNAs the requirement varies according to the nature of the different cistrons. For the three R17 cistron ribosomal recognition of the coat and replicase initiator regions is several fold more dependent on IF3 than is binding to the 'A' protein initiator region (Steitz *et al.*, 1977). It is noteworthy that this reflects an inverse relationship between the amount of IF3 required for initiation and the number of potential base pairs formed during the Shine Dalgarno interaction. Although the function of IF3 in this context is not fully understood it is not unreasonable to suppose that one role of IF3 is to stabilize base pairing in the Shine Dalgarno interaction. Thus, when only three base pairs can be formed stabilization is needed and the requirement for IF3 will be high. On the other hand if 7 base pairs are possible requirement for IF3 wll be less pronounced. There may even be cases where recognition will occur without any need for IF3.

Examination of the sequence of MS2 RNA reveal that several regions fullfil the criteria for a Shine Dalgarno interaction and yet do not bind to ribosomes. A segment in the replicase gene has the sequence -UAAGGAGCCUGAUAUG- and accordingly should interact with *E. coli* ribosomes. However, this is not so and consequently its accessibility must therefore be prevented by secondary and tertiary structures. Inspection of the MS2 sequence also suggests that the sequences surrounding this potential initiator triplet is sesquestered in RNA secondary structure. Thus the secondary structure of mRNA plays an active negative role in limiting initiation to the proper regions. Whether secondary structure is involved in positive control is not clear. There are hints that such may be the case, but the evidence is scanty and no definite statement concerning this problem is possible at this stage. In conclusion the binding of initiator regions of mRNA to 30S subunits is stabilized by two independent RNA:RNA interactions: 1) the anticodon of fMet-tRNA with the initiator codon AUG or GUG; and 2) the 3′ terminus of 16S rRNA with a region on mRNA according to Shine and Dalgarno. The first interaction depends on initiation factor IF2 while the second reaction is influenced by the ribosomal proteins S12 and S1 in addition to initiation factor IF3. S12 and S1 are probably also specificity determinants while IF3 may be differentially required according to the number of potential base pairs formed, thus probably having a stabilizing function. RNA secondary and tertiary structures have an active negative control function in that potential initiation sites are not exposed for interaction with the ribosome, thus preventing faulty initiation.

The Structure of Eukaryotic mRNA

Before discussing selection of initiator codons in eukaryotic cells it is appropriate to consider some aspects of the structure of eukaryotic mRNAs. This is because it seems probable that some of the significant differences between prokaryotic and eukaryotic initiation are a direct consequence of the peculiar structure of eukaryotic mRNAs. In contrast to prokaryotic mRNAs most eukaryotic mRNAs have a polyadenylate sequence at the 3′ terminal end. This segment can be up to 200 nucleotides long and is added post transcriptionally. The evidence so far indicate that the poly A segment has no role in polypeptide initiation, but rather that it is of importance for the half life of the mRNA. An important feature of the structure of eukaryotic mRNA has recently been discovered. It appears that most eukaryotic mRNAs have a m^7G residue (the cap) added post transcriptionally at the 5′ end linked through a 5′ -5′ pyrophosphate bond to the next residue in the mRNA. First it was claimed that the presence of the m^7G residue on the mRNA was absolutely required for its translation. Later experiments have shown that this is not so. In the absence of the cap, initiation still proceeds but with a much reduced efficiency *(Progress in Nucleic Acid and Molecular Biology, 1976).* Thus the effect of the cap is quantitative not qualitative. Eukaryotic ribosomes therefore somehow recognize the presence of the cap on mRNA but how this is done is at present unknown except that one of the many eukaryotic initiation factors seems to be specifically involved in this process.

Initiation of protein synthesis in the eukaryotic cell proceeds from a single site on the mRNA molecule whereas the corresponding process in prokaryotes can take place from multiple initiation sites. Eukaryotic proteins consisting of more than one peptide chain (e.g. α and β chains of haemoglobin) are synthesized singly from monocistronic mRNAs. Some viral proteins are generated from cleavage of a large polyprotein precursor which is synthesized from one initiation site on a polycistronic mRNA (e.g. picorna virus protein). However, many viral mRNAs (e.g. semliki forest virus) do have more than one initiation site, but only the site close to the 5′ end is active (Smith, 1977). The other site is not functional. The reason for the curious phenomenon that eukaryotic mRNA only have one functional initiation site is not known. It is unclear whether this restriction is a property of eukaryotic mRNA or whether there is a fundamental restriction in the eukaryotic translational machinery which prevents active initiation at internal sites.

Ribosomal mRNA Recognition in Eukaryotic Cells

Since most molecular mechanisms in translation have been conserved during the evolution of eukaryotes from prokaryotes a Shine Dalgarno interaction might be expected to contribute to mRNA recognition by eukaryotic ribosomes. The structure of the 3′ terminus of several 18S rRNAs from widely different eukaryotic cells have recently been determined. The sequences of the six 3′ termini are highly conserved. In addition they also show a strong homology to the 3′ end of *E. coli* 16S rRNA. The major difference is that eukaryotic 18S rRNAs do not contain the -CCUCC- sequence found in prokaryotic 16S rRNAs. Thus the 3′ terminus of eukaryotic 18S rRNA lacks the very sequence

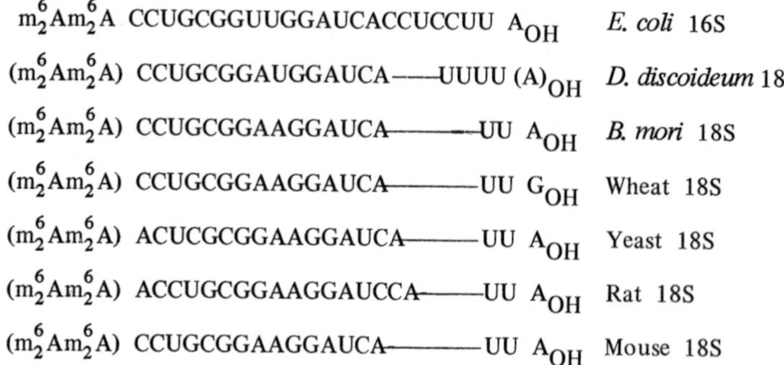

Fig. 4 3 terminal sequences of 16S and 18S rRNAs. The data are from Alberty *et al.*, (1978), Hagenbüchle *et al.*, (1978) and Jonge *et al.*, (1977). The 18S rRNA sequences are continuous. The lines are drawn to maximise homology.

implicated in binding prokaryotic mRNAs to the ribosome during translational initiation. On the other hand eukaryotic 18S rRNA do contain a purine rich region (-GCGAAGG-) which in eukaryotes may have the same function as -CCUCC- in prokaryotes. The initiator regions of about 25 different eukaryotic mRNAs are now known. Examination of the structures show that the distance from the initiator codon to the cap varies considerably (9 nucleotides in BMV to 50 in globin mRNA) and in certain viral mRNAs the 5′ end does not carry a cap. A disappointing feature is that no universally conserved sequence other than AUG is apparent. It has therefore been proposed (Baralle and Brownlee, 1978) that AUG itself functions in ribosomal initiator selection in a modified Shine Dalgarno interaction.

In this model the AUG of the mRNA first binds to the 3′ end of 18S rRNA which has the methionine anticodon 3′UAC5′ thus positioning the mRNA correctly on the 40S subunit. Subsequently Met-$tRNA_f$ itself binds to the small subunit thus displacing the 18S rRNA and translation can then proceed. Experiments, however, have shown that removal of either the 5′ non-coding region including the cap of sequences on the 3′ site of AUG reduces the efficiency of binding to the ribosomes (Kozak and Shatkin, 1978). An alternative model (Hagenbüchle *et al.*, 1978) suggests that the purine rich sequence on 18S rRNA is involved in base pairing with mRNA in a way similar to the interaction in prokaryotes. The 5′ non-coding regions of many eukaryotic mRNAs are indeed complementary to at least a part of the purine rich sequence present in 18S rRNA. However, the regions of complementarity are at widely variable distances from the AUG initiator codon (8-44 nucleotides) and several mRNAs do not show

β rabbit globin mRNA	m^7 GpppACACUUG<u>CUUUUG</u>ACACAACUGUGUUUACUUGCAAUCCCCCAAAACAGACAGA<u>AUG</u> –
BMV coat protein	m^7GpppGUAUUAAUA<u>AUG</u> –

Fig. 5 Sequence of the 5′ ends of β rabbit globin mRNA and BMV coat protein. The globin mRNA contains the sequence -CUUpyUG- present in many mRNAs, which can base pair with the 3′ terminus of 18S rRNA. The coat protein RNA does not contain this sequence. Data from Baralle and Brownlee (1978).

any complementarity at all. It can be argued that for mRNAs having short leader sequences, interaction with the initiation codon only is sufficient whereas with mRNAs having long non-coding regions an additional mRNA-rRNA interaction is required. In addition different mRNAs might have differential requirements for initiation factors. For example the sequence -CUUpyUG- present in many mRNAs and which is complementary to the purine stretch in 18S rRNA might be specific for an initiation factor whereas other less specific signals might be used in other mRNAs. Investigations relating to these possibilities have unfortunately not been performed so far. Finally it should be stressed that the models mentioned above are purely speculative since there is at present no direct evidence involving 18S rRNA in the binding of mRNA to eukaryotic ribosomes. Thus we must conclude that selection of initiator regions on mRNA by eukaryotic ribosomes is still very much an unknown process.

Whatever the mechanism of eukaryotic initiator codon selection it is evident that it must be different from the corresponding process in prokaryotes. Even if a Shine Dalgarno type of interaction is postulated for eukaryotes the base pairs involved will be different from those of prokaryotes because the 3′ terminus of 18S rRNA does not contain the functional sequence -CCUCC- present in the appropriate position in 16S rRNAs. Why has this particular sequence been deleted in eukaryotes? A possibility is that the eukaryotic ribosome could thereby avoid interacting strongly with mRNAs designed for efficient utilization by prokaryotic ribosomes. Such a translational barrier might have been important at some stage in the evolution of eukaryotes from prokaryotes. Is it possible to cross this barrier? Several prokaryotic mRNAs can be correctly translated in eukaryotic in vitro protein synthesizing systems (Hagenbüchle *et al.*, 1978). However, their efficiency relative to eukaryotic mRNAs is much reduced. Conversely very few if any eukaryotic mRNAs investigated so far are translated with fidelity in a prokaryotic cell-free system. In the few cases where correct translation has been claimed the efficiency of translation is very low, indeed. Insertion of genes from yeast (Struhl *et al.*, 1976; Ratzkin and Carbon, 1977) and neurospora (Vapnek *et al.*, 1977) into *E. coli* by recombinant DNA techniques do in fact in some cases lead to correct translation, since introduction into *E. coli* of plasmids carrying genes from either of the two eukaryotic organisms can lead to complementation of a nutritional requirement. Intervening sequences do exist in the yeast genome (Knapp *et al.*, 1978), but the yeast genes which can be translated in *E. coli* cannot contain such sequences since *E. coli* does no possess a mechanism for processing genes with interrupting non-coding sequences. Complementation of *E. coli* with higher eukaryotic DNA similar to these performed with yeast DNA have to my knowledge not been performed so far. The presence of genes in higher eukaryotes without intervening sequences do, however, raise the possibility that in some cases complementation will occur. This of course further requires that the metabolic pathway in question is more or less the same in the two cells and that the properties of the particular eukaryotic enzyme are such that it can function in the prokaryotic cell. This further restricts the number of cases where complementation is possible. Some genes from higher eukaryotes that fulfill these criteria and do not contain intervening sequences might in fact be translated with reasonable efficiency if they are present in several copies and the corresponding mRNAs have short leader sequence in front of the initiator codon. As argued previously it may be that in eukaryotic mRNAs having short leader sequences the only initiation signal required is the AUG codon itself, and that in such cases several initiation factors may not be required for translation. From our knowledge of prokaryotic systems such mRNAs will probably have a low but definite chance of being correctly translated in *E. coli* and if they are present in several copies which can be achieved in certain host/vector systems they may overcome their low affinity for ribosomal interaction so that a reasonable translation efficiency ensues. It is not unreasonable to suppose that the yeast genes which can complement nutritional requirements in *E. coli* have short leader sequences and for this reason can be translated. Yeast genes containing long leader sequences in the corresponding mRNA will probably not be translated even if they do not have any intervening sequences. The number of translated yeast genes in *E. coli* therefore, only reflects a minimum number of the genes without intervening sequences present within the yeast genome. Similar arguments also apply for genes in higher eukaryotes. Whatever the number of genes from higher eukaryotes which eventually will be found to be translable in *E. coli* it will only represent a minimum of the genes without intervening sequences present in the genome. On the basis of complementation studies alone there is therefore no compelling reason for believing that intervening sequences are more common in higher eukaryotes than they are in primitive eukaryotes.

References

Alberty, H., Raba, M. and Gross, H.J. (1978). *Nucleic Acids Res.* **5**, 425.
Baralle, F.E. and Brownlee, G.G. (1978). *Nature (London)* **274**, 84.
Boon, T. (1971). *Proc. Nat. Acad. Sci.* **68**, 2421.
Cohn, W. E., and Volkin E. (1976). *In:* Progress in Nucleic Acid Research and

Molecular Biology 19. Academic Press.
Crépin, M., Lelong, J.C. and Gros, F. (1973). *In:* Research Methods in Reproductive Endovirology 6, Karolinska Symposium.
Dahlberg, A.E., Lund, E., Kjeldgaard, N.O., Bowman, C.M. and Nomura, M. (1973). *Biochemistry* **12**, 948.
Dunn, J.J., Buzash-Pollert, E. and Studier, W. (1978). *Proc. Nat. Acad. Sci.* **75**, 2741.
Fiers, W., Contreras, R., Duerinck, F., Haegeman, G., Aserentant, D., Merregaert, J., Min Jou, W., Molemans, F., Raeymaekers, A., Van den Berghe, A., Volckaert, G. and Ysebaert, M. (1976). *Nature (London)* **260**, 500.
Grunberg-Manago, M., Buckingham, R.H., Cooperman, B.S. and Hershey, M. (1978). Symposium 28. Relations between Structure and Function in the Prokaryotic Cell, Cambridge University Press.
Hagenbüchle, O., Santer, M., Steitz, J.A. and Mans, R.J. (1978). *Cell* **13**, 551.
Held, W.A., Gette, W.R. and Nomura, M. (1974). *Biochemistry* **13**, 2115.
Isono, S. and Isono, K. (1975). *Europ. Journ. of Biochemistry* **56**, 15.
Jonge, P. de Klootwijk, J. and Planta, R.J. (1978). *Nucleic Acid Res.* **4**, 3655.
Knapp, J.S., Beckmann, P.E., Johnson, S., Fuhrman, A. and Abelson, J. (1978). *Cell* **14**, 221.
Kozak, M. and Shatkin, A.J. (1978). *Cell* **13**, 201.
Lodish, H.F. (1970). *Nature (London)* **226**, 705.
Melzer, T.L., Davies, J.E. and Dahlberg, J.E. (1971). *Nature New Biology* **232**, 12.
Ratzkin, G. and Carbon, J. (1977). *Proc. Nat. Acad. Sci.* **74**, 487.
Shine, J. and Dalgarno, L. (1974). *Proc. Nat. Acad. Sci.* **71**, 1342.
Shine, J. and Dalgarno, L. (1975). *Nature (London)* **254**, 34.
Smith, A.E. (1977). *In:* Gene Expression. FEBS Meeting, Copenhagen. (eds. B.F.C. Clark, H. Klenow and J. Zeuthen), 37. Pergamon Press.
Sprague, K.U., Steitz, J.A. Grenley, R.M. and Stocking, C.E. (1977). *Nature (London)* **267**, 462.
Steitz, J.A. (1969). *Nature (London)* **224**, 957.
Steitz, J.A. (1977). *In:* Biological Regulations and Development (ed. R. Goldberger) Plenum, New York.
Steitz, J.A. and Jakes, K. (1975). *Proc. Nat. Acad. Sci. U.S.A.* **72**, 4734.
Steitz, J.A., Wahba, A.J., Langrea, M. and Moore, P.B. (1977). *Nucleic Acids Res.* **4**, 1.
Struhl, K., Cameron, J.R. and Davis, R.W. (1976). *Proc. Nat. Acad. Sci.* **73**, 1471.
Vapnek, D. Hautala, J.A., Jacobsen, J.W., Gites, N.H. and Kushner, S.R. (1977). *Proc. Nat. Acad. Sci.* **74**, 3508.

PEPTIDE CHAIN TERMINATION

C.T. CASKEY AND J.M. CAMPBELL

The Howard Hughes Laboratory for the Study of Genetic Disorders, Departments of Medicine and Biochemistry, Baylor College of Medicine, Houston, Texas, USA.

Introduction

Peptide chain termination results in the release of the completed peptide from its ultimate ribosomal-bound tRNA. Peptide chain termination occurs on the ribosome and requires soluble protein factor(s) in both prokaryotic and eukaryotic extracts. Two proteins (release factor, RF) identified in *Escherichia coli* cells differ in their codon specificity; RF-1, UAA or UAG; RF-2, UAA or UGA. A single rabbit reticulocyte RF is active with UAA, UAG, and UGA. Binding and release of bacterial RF-1 and RF-2 from ribosomes involves a third protein factor, RF-3, which interacts with GDP and GTP. Reticulocyte RF activity is stimulated by GTP and expresses a ribosomal dependent GTPase activity. The hydrolysis of peptidyl-tRNA at chain termination also requires the ribosomal enzyme peptidyl transferase.

Soluble Release Factors

Prokaryotes

The recognition of peptide chain termination codons requires the participation of special protein factors whose existence was initially suggested by Ganoza (1966). Using randomly ordered poly (A_3,U) to direct the synthesis and release of peptides from ribosomes, Ganoza observed dissociation of protein synthesis and protein release from tRNA when reaction mixtures contained purified sources of transfer enzymes. These studies suggested that the supernatant, not the tRNA or ribosomal fraction of *E. coli* extracts, contained a factor(s) essential for peptide chain termination which differed from transfer factors. A protein factor was later identified in the supernatant fraction of bacterial extracts by Capecchi (1967). In these studies Capecchi used a mutant of bacteriophage R17 carrying an amber nonsense mutation in the seventh codon of the coat protein gene.

The requirement for a protein factor(s) participation in peptide chain termination has been confirmed by release of fMet from an (fMet-tRNA·AUG·ribosome) intermediate using terminator trinucleotides. The expression of all terminator codons was found to be dependent upon RF, and the rate of fMet release was proportional to its concentration (Caskey *et al.*, 1968). This crude *E. coli* RF preparation was fractionated by DEAE-Sephadex column chromatography (Scolnick *et al.*, 1968) and was found to contain two codon-specific RF fractions. The early-eluting fraction, RF-1, is active with UAA or UAG but not UGA. The later-eluting fraction, RF-2 is active with UAA or UGA, but not UAG. Each RF has been purified to homogeneity without any change in this codon specificity (Klein and Capecchi, 1971; Milman *et al.*, 1969). Studies which utilize phage mRNA as *in vitro* templates to direct protein synthesis have confirmed the codon specificity of RF-1 and RF-2 (Capecchi and Klein, 1970; Beaudet and Caskey, 1970). Klein and Capecchi (1971) reported a molecular weight of 44,000 for RF-1 and 47,000 for RF-2. We have recently re-examined the physical properties of the release factors. The molecular weight estimates by SDS PAGE are 47-49,000 for RF-1 and 48-50,000 for RF-2 and thus in agreement with earlier results. A remarkable difference from this value was observed for RF-2 using other methods. By Sepharose 6B-100 chromatography (in 6*M* guanidine-HC1) and sedimentation equilibrium under native and denaturing conditions the molecular weight estimates were 34,300-35,000 (Prakash *et al.*, 1978). We have examined RF-2 for carbohydrate content and found none (<2%). Both RF's are insensitive to treatment with ribonuclease A or T1. Analysis of RF-1 has revealed <1.0 atom of phosphorus/molecule (Capecchi and Klein, 1969). Analysis of RF-2 for nucleic acid by Smrt *et al.*, (1970) did not detect Up or Ap (<0.1 residue/molecule). Presently the most accurate molecular weight estimate would appear to be 34,500. The possibility that nucleic acid or carbohydrate contribute to the SDS PAGE behaviour has been excluded, and the molecular basis for the phenomenon is yet unresolved.

The detailed molecular comparison of RF-1 and RF-2 has been initiated and certain features are now apparent. The RF-2 combined mole % of glutamate and aspartate is 30.5% while arginine and lysine is 12.4%. This composition accounts for the pI of 5.1 for RF-2. The pI of RF-1 is 5.5-5.8. Antibodies have been prepared to each RF and immune determinants examined. The major immune determinants of RF-1 and RF-2 differ as measured by RF inactivation and immunoprecipitation in Ouchterlony plates. Using antibody preparations developed against both RF molecules minor common immune determinants were identified for RF-1 and RF-2. These studies indicate that RF-1 and RF-2 have some similar physical characteristics (Ratliff and Caskey, 1977). In an effort to more fully characterize these similarities the immune determinants for each RF are under study. RF-2 is cleaved by trypsin in the presence of SDS in a limited manner resulting in two peptides (TP1, 30,00-33,700 and TP2, 12,300-14,000). Immunological study of each has identified a minimum of 3 antigenic sites for RF-2 (Campbell, 1978).

The relationship between structure of the terminator codons and activity as mRNA templates for release factors has been explored in a limited way by Smrt *et al.*, (1970). Using modified trinucleotides (3me-UAG, 3me-UAA, 5me-UAG, 5me-UAA, Br-UAG, h-UAG, and UAI) as template in the formylmethionine release assay, they showed that the RF specificity for codon recognition closely resembles the specificity of Watson-Crick and wobble base pairing.

Release factors can be bound to E. coli ribosomes with codon specificity in the absence of peptidyl-tRNA hydrolysis, as shown by Scolnick and Caskey (1969). The binding of RF to ribosomes is determined by Millipore filter retention of radioactive terminator trinucleotides. Ribosomes, RF, and trinucleotide codon form a stable intermediate in the presence of ethanol. RF-1 binds to ribosomes with $[^3H]$UAA or $[^3H]$UAG and radioactivity is subject to competition by nonradioactive UAA or UAG, but not UGA. Capecchi and Klein (1969) performed equilibrium dialysis studies by which they demonstrated ribosome-independent RF binding to oligonucleotides containing terminator codons. However, the specificity of this recognition was not absolute in the absence of ribosomes.

Since suppressor aa-tRNA species are known to recognize terminator codons, RF and su^+ tRNA might be expected to compete for the translation of terminator codons in mRNA. Studies which examine such postulated competition have been performed, earlier, by addition of su^+ aa-tRNA to crude bacterial extracts; and later, by varying the concentration of RF (Beaudet and Caskey, 1970; Ganoza and Tompkins, 1970). By varying the concentration of RF or su^+ aa-tRNA, the level of chain termination (RF-mediated) or suppression (su^+ aa-tRNA-mediated) is varied. The results of these studies indicate that RF and suppressor aa-tRNA compete, and therefore support the concept that RF-1 and RF-2 are codon recognition molecules.

These conclusions have been challenged by Dalgarno and Shine (1973) and Shine and Dalgarno (1975). These investigators have demonstrated homology of the 3′-terminal sequence of ribosomal RNA isolated from the smaller subunit of *Saccharomyces cerevisiae,*

Drosophila melanogaster, rabbit reticulocytes, avian myeloblasts, and *E. coli*. All terminate in the sequence UUA_{OH} and also contain the trinucleotide UCA within eight residues of the terminus. The Watson-Crick base pair of UAA, therefore, is found at a conserved 3′-terminus of these molecules. On the basis of this observation, it was suggested that UUA_{OH} could base pair with UAA and UAG and that UCA could pair with UGA.

The 3′ terminus of the 16S rRNA is important to recognition of termination codons as recently demonstrated by studies employing the specific nuclease cloacin DF13 (Caskey *et al.*, 1977). Cloacin disrupts the ribosome by site specific cleavage 49 nucleotides from the 3′ terminus of the 16S rRNA. Such cleavage leads to inactivation of ribosomes for recognition of termination codons but not RF-mediated peptidyl-tRNA hydrolysis. Thus it is clear that 16S rRNA integrity plays an important role in codon recognition. It remains unresolved if this is due to specific nucleic acid interaction or ribosomal structure. The recognition of *E. coli* RF molecules on *B. subtilis* ribosomes was not qualitatively altered although the 3′ terminus of 16S rRNA of *B. subtilis* does not contain complementary sequences to termination codons. The preponderence of experimental data still favour RF as the recognition molecule.

During the development of procedures for purification of *E. coli* release factors, an additional peptide chain termination protein factor was discovered (Milman *et al.*, 1969). Factor RF-3 has no release activity, but stimulates fMet release mediated by RF-1 or RF-2 and terminator codons in the formylmethionine release assay. Since peptide chain termination involves at least two events (binding of RF to ribosomes upon recognition of terminator codon, and hydrolysis of ribosomal bound peptidyl-tRNA), RF-3 could stimulate release by acting at either or both events. The available data indicate that RF-3 affects terminator codon recognition. In the absence of RF-3, the K_m for UAA and UGA for RF-2 is $8.3 \times 10^{-5} M$ and $5.6 \times 10^{-5} M$, respectively. The K_m for both codons is lowered to $1.3 \times 10^{-5} M$ when RF-3 is added without affecting the maximal rate of hydrolysis (V_{max}), which suggests that RF-3 acts at codon recognition, and not during peptidyl-tRNA hydrolysis (Goldstein *et al.*, 1970a).

Protein RF-3 has been shown to stimulate the binding and release of RF-1 or RF-2 from ribosomes (Goldstein and Caskey, 1970) using radioactive terminator codons to quantitate the formation of intermediates. RF-3 stimulates the formation of (RF-1 or RF-2·[^{3}H] UAA·ribosome) intermediates which are actively dissociated by the addition of GTP, GDP, and less well by GDPCP, but not by GMP. RF-3 becomes incorporated into the (RF-1 or RF-2·UAA·ribosome) complex during the binding reaction and dissociates upon the addition of GTP or GDP. These studies suggest that RF-3 acts in two ways in peptide chain termination: (1) binding of RF-1 or RF-2 to ribosomes, and/or (2) dissociation of (RF-1 or RF-2·UAA·ribosome) intermediates.

The exact in vivo role of RF-3 is uncertain because of apparent *in vitro* conditional effects. At low trinucleotide concentration RF-3 stimulates fMet release, and this stimulation is eliminated by GTP or GDP. At 20-fold higher trinucleotide concentration, RF-3 alone reduces the rate of release while RF-3 plus GTP or GDP stimulates release. This stimulation is probably due to the more rapid dissociation of RF-1 or RF-2 from ribosomes. The lack of guanine nucleotide specificity may suggest that other factors (e.g. additional proteins, intact mRNA) will be required for clarification of the *in vivo* role of RF-3. Our precise understanding of RF-3 is further complicated by the report by Capecchi and Klein (1969) of purification of a protein factor which stimulates fMet release at low trinucleotide concentration and is inhibited by GTP. They suggest that RF-3 and the transfer factor EF-Tu are equivalent since this protein preparation contains both EF-Tu and stimulatory activity, and appears homogeneous by analytical disc gel analysis. Milman *et al.*, (1969) have contrasting observations since their most purified RF-3 preparations contain no detectable EF-Tu activity, and their purified EF-Tu preparations contain no RF-3 activity. They therefore have concluded that RF-3 is a protein factor involved specifically in peptide chain termination, not chain elongation.

Eukaryotes

Goldstein *et al.*, (1970b) first reported evidence for a release factor from rabbit reticulocyte extracts using a modification of the formylmethionine release assay. Formylmethionine release required a protein release factor; was stimulated by GTP, but not GDP; was inhibited by GDPCP; and required tetranucleotides UAAA, UAGA, UGAA, or UAGG (Beaudet and Caskey, 1971). Release factor purified from guinea pig liver, Chinese hamster liver, and rabbit reticulocytes functioned similarly. Using similar techniques an insect soluble factor (Ilan, 1973) and a rat liver factor (Innanen and Nicholls, 1973) have been identified with characteristics compatible with RF. The molecular weight of reticulocyte RF, as determined by Sephadex G-200 chromatography and equilibrium sedimentation suggests that it is large (105,000). Molecular weight determinations by sucrose gradient analysis and SDS polyacrylamide gel analysis indicate that the RF subunit molecular weight is smaller (56,500). Thus rabbit reticulocyte RF can exist as a dimer (Konecki *et al.*, 1977). It is unclear presently if the dimer or monomer interact at the ribosomal surface.

Reticulocyte RF-ribosomal binding can be assessed with reticulocyte RF, ribosomes, and [^{3}H] UAAA in a manner analogous to that discussed for prokaryotic RF-1 or RF-2 (Tate *et al.*, 1973). This method has been used to give functional confirmation that a single reticulocyte RF recognizes all terminator codons. Reticulocyte RF·ribosome·[^{3}H] UAAA complex was measured alone and in the presence of increasing amounts of nonradioactive UAAA, UAGA, and UGAA. The complete competition by all templates suggests that any RF molecule which binds with [^{3}H] UAAA also recognizes UAGA and UGAA. The results differ if a similar experiment is performed with a mixture of *E. coli* RF-1 and RF-2. UAGA competes out only that portion of complex formation due to RF-1 and UGAA that portion due to RF-2. The tetranucleotide UAAA, which recognizes RF-1 and RF-2, competes out all radioactive complex formation. These functional studies support the results of purification of prokaryotic and eukaryotic RF which have indicated that prokaryotes have two RF molecules (RF-1 and RF-2), while eukaryotic reticulocytes possess a single RF molecule capable of recognizing all terminator codons.

Mammalian peptide chain termination is stimulated by GTP and inhibited by GDPCP. These results suggested that γ-phosphate hydrolysis by GTP is requisite for the event. A ribosomal dependent GTPase activity has been demonstrated to be associated with highly purified reticulocyte RF. This hydrolysis does not occur with RF or ribosomes alone and is stimulated by the terminator oligonucleotide UAAA but not AAAA, suggesting that GTP hydrolysis is somehow related to terminator codon recognition. Since GTP hydrolysis occurs on ribosomes not carrying nascent peptidyl-tRNA, and furthermore, is not inhibited by antibiotics which inhibit peptidyl-tRNA hydrolysis, it appears that GTP hydrolysis is not directly linked to peptidyl-tRNA hydrolysis. The binding of reticulocyte RF to ribosomes is, however, stimulated by the guanine nucleotides GTP and GDPCP, but not GDP. This nucleotide specificity for RF-ribosomal binding is analogous to the protein factor-dependent ribosomal binding of initiator tRNA and aminoacyl-tRNA. In all three cases, initiation, elongation, and termination, protein factor-dependent binding utilizes GTP or GDPCP but completion of all three protein synthetic events requires phosphate hydrolysis of the specific nucleotide, GTP.

A summary of the eukaryotic and prokaryotic RF's is given in

Table I.

TABLE I

Soluble Chain Termination Factors

Designation	Molecular weight	Codon Specificity	GTP recognition
Prokaryotic			
RF-1	47,000[a]	UAA and UAG	None
RF-2	48,000[a], 35,000[b]	UAA and UGA	None
RF-3	46,000[a]	None	Yes
Eukaryotic			
RF	56,500[a], 105,000[b]	UAAA, UAGA, and UGAA	Yes

[a]SDS PAGE; [b]Equilibrium sedimentation under non-denaturing conditions

The Ribosomal Role

The necessity of ribosomal participation in peptide chain termination can be shown from a variety of studies. Ribosomal inactivation by ionic strength variation, antibiotic inhibition, pH variation, and disaggregation indicate that a ribosome must be 'active' (i.e., able to form peptide bonds) to function in peptide chain termination. Vogel *et al.*, (1969) showed a requirement for 'active' 50S ribosomal particles. They found that ribosomes prepared in buffers devoid of K^+ or NH_4^+ could not form peptide bonds or participate in peptide chain termination. The peptide bond-forming and peptide chain termination activity of these 'inactive' ribosomes or ribosomal subunits could be restored by addition of K^+ or NH_4^+ - containing buffers. Initially, fMet-tRNA was bound to 'active' 30S ribosomal subunits, 'active' or 'inactive' 50S ribosomal subparticles were added, and such intermediates examined for the capacity to form peptide bonds and participate in peptide chain termination. These studies indicated requirement of the 50S ribosomal subparticles for peptidyl-tRNA hydrolysis, and suggested that peptidyl transferase may catalyze this hydrolysis. Erdmann *et al.*, (1971) have reconstituted 5S RNA-deficient 50S ribosomal subunits. These deficient particles will not bind RF (measured by [^{3}H]UAA binding), nor will they form peptide bonds.

Antibiotic Inhibitors

The antibiotics tetracycline, streptomycin, sparsomycin, chloramphenicol, gougerotin, amicetin, and lincocin are all inhibitors of codon-directed peptide release in *E. coli* (Scolnick *et al.*, 1968; Vogel *et al.*, 1969; Caskey et al., 1971). With the development of methods for evaluating terminator codon recognition and peptidyl-tRNA hydrolysis independently, the site of action of these inhibitors was determined. Antibiotics usually affect the codon recognition event or the peptidyl-tRNA hydrolysis event, but not both. Tetracycline and streptomycin inhibit terminator codon recognition. Amicetin, lincocin, chloramphenicol, sparsomycin and gougerotin inhibit release of fMet without significant effect on terminator codon recognition. These latter antibiotics have been shown to be inhibitors of the peptide bond-forming enzyme, peptidyl transferase (Monro *et al.*, 1969). The relative inhibition of the codon-directed formylmethionine release and peptide bond formation has been compared at a variety of concentrations for these four antibiotics. Amicetin, lincocin, chloramphenicol, and sparsomycin inhibit each reaction in parallel (Vogel *et al.*, 1969). Menninger (1971) has obtained similar antibiotic effects with a different assay for peptide chain termination. He finds, however, that poly (U,A)-stimulated release of oligolysine from oligolysyl-tRNA is more sensitive to sparsomycin, gougerotin, and erythromycin than is release of oligolysine by puromycin. The significance of this difference is uncertain. Goldstein *et al.*, (1970b) examined antibiotic sensitivity of release of formylmethionine from mammalian ribosomes by puromycin and reticulocyte RF. Sparsomycin, gougerotin, anisomycin, and, less well, amicetin, inhibited both mammalian peptidyl transferase and RF-mediated fMet release. Thus, peptidyl transferase and RF-mediated peptidyl-tRNA hydrolysis activity are inhibited by the same antibiotics for two types of ribosomes which differ in their antibiotic sensitivity.

The putative role of peptidyl transferase in peptide chain termination is enhanced by the reports of Scolnick *et al.*, (1970) and Fahnestock *et al.*, (1970) that ribosomes (peptidyl transferase) can form ester bonds. In addition, Caskey *et al.*, (1971) have found that ribosomes can catalyze RF-independent peptidyl-tRNA hydrolysis. A useful substrate is f[^{3}H]Met-tRNA bound to *E. coli* or reticulocyte ribosomes. The addition of puromycin yields f[^{3}H]Met-puromycin, i.e. peptide bond formation. In the presence of ethanol, the peptidyl transferase can catalyze ester formation yielding f[^{3}H]Met-ethyl ester dependent on the presence of deacylated tRNA or its 3′-terminal sequence CCA. If acetone is substituted for ethanol, peptidyl-tRNA hydrolysis results, yielding f[^{3}H]Met, again dependent on deacylated tRNA or CCA. Results are similar with *E. coli* and reticulocyte ribosomes. With *E. coli* components each reaction will occur with 50S ribosomal subunits, is inhibited by the same antibiotics, has identical cation requirements, and has identical pH optima, thus suggesting mediation by a common enzyme, peptidyl transferase. The peptidyl transferase appears capable of peptide bond formation with an amino group as the nucleophilic agent, ester formation with an alcohol as the nucleophilic agent, and hydrolysis with water as the nucleophilic agent. The ability of peptidyl transferase to hydrolyze peptidyl-tRNA strengthens the case for its involvement in peptide chain termination.

In considering possible mechanisms for the peptidyl-tRNA hydrolysis of peptide chain termination some observations by Caskey *et al.*, (1971) may be pertinent. The antibiotics lincocin with *E. coli* ribosomes and anisomycin with reticulocyte ribosomes completely inhibit peptide bond formation and ester formation while stimulating the peptidyl-tRNA hydrolysis in acetone. There is an 11-fold stimulation by anisomycin of peptidyl-tRNA hydrolysis in acetone using reticulocyte substrate and a 40% stimulation by lincocin using *E. coli* substrate. Recent reports by Innanen and Nicholls (1974) have confirmed these antibiotic effects on peptidyl transferase. Such modification of peptidyl transferase specificity could be operative in chain termination where RF, rather than lincocin or anisomycin, is the modifying component.

Antibody Inhibitors

Using antibodies directed against specific *E. coli* ribosomal proteins, Tate *et al.*, (1975) have developed additional evidence for peptidyl transferase participation in peptide chain termination. Antibodies against the *E. coli* 50S ribosomal proteins L11 and L16 inhibit *in vitro* termination as a result of affecting RF-dependent peptidyl-tRNA hydrolysis. The antibodies do not affect the interaction of RF with the ribosome. Peptidyl transferase activity has been restored to 50S core particles (0.8c cores) by addition of L11 and it was suggested that L11 was the ribosomal protein possessing peptidyl transferase enzyme activity (Nierhaus and Montejo, 1973). It seems likely, therefore, from these studies that the peptidyl transferase enzyme, L11, does have some role in the cleavage of the ester bond of peptidyl-tRNA at termination. The studies do not preclude that two different sites on L11 might be involved in peptide bond formation and in hydrolysis of peptidyl-tRNA or that the hydrolysis event involves a synergism between RF and the peptidyl transferase enzyme. L16 has been identified as the chloramphenicol-binding protein and has been deduced to be part of the A site moiety of the peptidyl transferase center (Nierhaus and Nierhaus, 1973). The inhibition of the peptidyl-tRNA hydrolysis partial reaction of *in vitro* termination by antibody to L16 suggests that the RF molecule may occupy a region of the A site during this event since the sites occupied by the other known components of the reaction (peptidyl-tRNA, P site; peptidyl transferase L11) are not affected by chloramphenicol. If L16 and L11 were neighbours, as is suggested from their roles in the peptidyl transferase centre, then this could support the concept that RF and the peptidyl transferase may functionally interact at chain termination.

On the basis of evidence already discussed, it is likely that the peptidyl transferase participates in peptide chain termination. Given this involvement, there are still at least two possible mechanisms for RF participation in peptidyl-tRNA hydrolysis. First, a nucleophilic group of the protein release factor could participate in the attack and transiently accept the nascent peptide in covalent linkage. Alternately, the role of RF may be only to modify the peptidyl transferase or restrict the choice of nucleophilic agent to promote hydrolysis without the direct participation suggested above. This would be similar to the anisomycin and lincocin stimulatory effects described. There is little evidence at the moment to choose between these or other models, except that preliminary attempts to demonstrate an RF-nascent peptide intermediate have been unsuccessful.

RF-Ribosomal Interaction

There is evidence using a number of approaches (Capecchi and Klein, 1969; Tompkins *et al.*, 1970) that, not only are ribosomes required, but the peptidyl-tRNA substrate must be in a configuration that permits reactivity with puromycin (P site). This implies that translocation must occur subsequent to formation of the last peptide bond and prior to termination. The P site requirement for release may also be viewed as additional evidence for participation of peptidyl transferase in hydrolysis. RF-1 and RF-2 participate in peptide chain termination by their interaction at the ribosomal A site (Tompkins *et al.*, 1970). These conclusions are further supported by the ability of RF to bind to ribosomes with occupied P sites. Both 30S and 50S ribosomal subunits are required for stable RF-ribosomal binding. The antibiotics streptomycin and tetracycline are effective inhibitors of *E. coli* RF binding.

The guanine nucleotide GTP is clearly involved in the binding of reticulocyte RF to reticulocyte ribosomes (Tate *et al.*, 1973). Furthermore, GTP, in interaction with *E. coli* RF-3, facilitates the displacement of RF-1 and RF-2 from ribosomes. These results are similar to those now reported for binding and release of IF-2 and EF-Tu from *E. coli* ribosomes (Lockwood *et al.*, 1972; Yokosawa *et al.*, 1973). It appears that the soluble factors indicated above cannot occupy the ribosome simultaneously. Tate *et al.*, (1973) have shown that *E. coli* ribosomes containing EF-G bound as a stable fusidic acid EF-G·ribosome complex will not interact with either RF-1 or RF-2. A number of other studies have shown that stable EF-G binding to the *E. coli* ribosome can exclude the ribosomal binding of aminoacyl-tRNA with and without EF-Tu (Miller, 1972; Richman and Bodley, 1972; Cabrer *et al.*, 1972; Richter, 1972). Conversely ribosomes containing aminoacyl-tRNA bound enzymatically with EF-Tu or non-enzymatically will not associate with EF-G. Collectively, the studies indicate that these soluble factors cannot be accommodated simultaneously on a single ribosome and suggest that the cycle of events in protein biosynthesis proceed in a stepwise fashion requiring the displacement of one factor from the ribosome before the protein factor involved in the next intermediate event can interact.

The ribosomal proteins required for peptide chain termination have been investigated by the use of: (1) antibiotics, (2) antibodies directed against ribosomal proteins, (3) ribosomes depleted of proteins, and (4) cross-linkage of RF to ribosomal proteins.

A summary of antibiotics which inhibit partial reactions of peptide chain termination has been described earlier in this article and elsewhere (Caskey and Beaudet, 1971). At the present time, we are able to make only limited use of this information in assignment of specific ribosomal proteins to antibiotic binding sites. Vazquez (1974) has summarized the effects of many of these antibiotics and described functional maps of the eukaryotic and prokaryotic ribosome on the basis of a variety of studies. Correlations of altered ribosomal components have been made for antibiotic resistance. For example, bacterial resistance to streptomycin and erythromycin has been attributed to mutations affecting the ribosomal proteins S4 (Traub and Nomura, 1968) and L4 (Otaka *et al.*, 1971), respectively. We may therefore assume that S4 and L4 are involved in peptide chain termination since these antibiotics are inhibitors of codon recognition and peptidyl-tRNA hydrolysis, respectively.

Antibodies have been developed to specific prokaryotic ribosomal proteins (Stoffler and Wittman, 1971) and used extensively to identify functional activities for these proteins. These methods have identified proteins occurring at the ribosomal subunit interface (Morrison *et al.*, 1973), neighbouring ribosomal proteins (Lutter et al., 1972), and ribosomal proteins involved in the binding of EF-Tu (Highland *et al.*, 1974a) and EF-G (Highland *et al.*, 1974b). Tate *et al.*, (1975) have employed these specific ribosomal antibodies to determine those ribosomal proteins involved in peptide chain termination. The L7/L12, S2 and S3 proteins were found to be required for RF binding. As discussed earlier L11 and L16 antibodies inhibited peptidyl-tRNA hydrolysis but not RF-ribosomal binding. Using predissociated ribosomes, the antibodies against L2, L4, L6, L14, L15, L17, L18, L20, L23, L26, and L27 inhibited all termination partial reactions. Since these antibodies had no effect on 70S ribosomes, it was concluded that they interfered with association of ribosomal subunits, which is requisite for chain termination.

In different studies (Brot *et al.*, 1974) have demonstrated the requirement of L7/L12 in peptide chain termination. The removal of L7/L12 eliminated the ability of RF to bind to the ribosome. This activity was fully restored by addition of purified L7/L12 to the depleted ribosomes. Thus, by two methods L7/L12 were found critical for RF-1 and RF-2 ribosomal binding. These results are similar to those found for fMet-tRNA and aa-tRNA ribosomal binding. Recently ribosomes depleted of L11 have been investigated for their ability to participate in the peptidyl-tRNA hydrolysis mediated by RF (Armstrong and Tate, 1978). These studies also indicated a requirement for L11, again suggesting its role in the peptidyl transferase reaction.

The interaction of RF-2 with ribosomal proteins has been determined by the formation of protein couples using diimido bifunctional coupling agents (Stoffler and Caskey, 1978). In all cases the couples were formed in response to codon-directed binding of RF-2 to the ribosome and were detected by radioimmunologic techniques. High levels of coupling were observed between RF-2 and the ribosomal proteins L2, L7/L12, L11, S17, and S18. A number of studies have independently predicted that S18 was located in the mRNA binding domain. Both S17 and S18 are known to bind to the 5′ and central regions of the 16S rRNA. Pongs and Rossner (1975) have prepared an affinity probe which is a derivative of the termination codon UGA. This codon was found to directly link to S18 and to a lesser extent S4. Thus, RF interacts at ribosomal mRNA translation sites similarly to aa-tRNA molecules. The role of L11 in peptidyl transferase activity and the requirement of L7/L12 for RF-ribosomal binding have been previously discussed.

Conclusion

A model for the intermediate events of peptide chain termination is presented in Figure 1. The model is based primarily upon data obtained from *in vitro* studies with both mammalian and bacterial systems and assumes a common mechanism for the two. Not all intermediate events and requirements have been demonstrated for each cell type, and the sequence of intermediate events should be regarded as tentative.

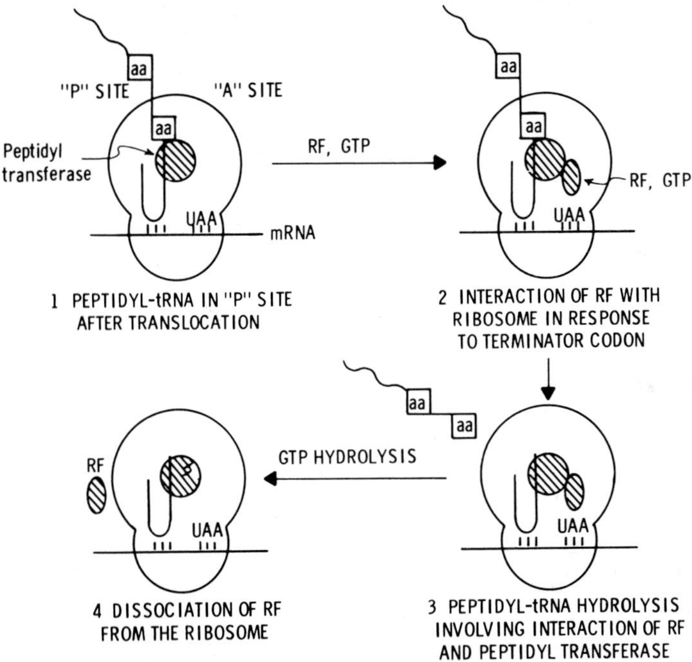

Fig. 1 A model for the intermediate steps of peptide chain termination.

Both mammalian and bacterial cells utilize the same terminator codons (UAA, UAG, and UGA). In bacterial cells these codons are recognized by protein release factors which are codon specific (RF-1, UAA or UAG; RF-2, UAA or UGA). In mammalian cells the RF apparently recognizes all three terminator codons. A separate protein factor RF-3, identified in bacterial extracts, has the capacity to facilitate the binding of RF to ribosomes and interacts with GDP and GTP. Since mammalian RF alone is stimulated by GTP, a functional RF-3 equivalent may not be involved. The data from both cell types favour the involvement of GTP and its γ-phosphate hydrolysis in RF binding and dissociation rather than peptidyl-tRNA hydrolysis. The release factors from both bacterial and mammalian sources cannot interact with ribosomes containing the translocation factors which are required for the event immediately preceding termination.

In *E. coli* the ribosomal proteins L7/L12, S2, S3 and the thiostrepton-binding protein are involved in the interaction of RF with the ribosome. Peptidyl-tRNA hydrolysis is inhibited by antibodies to L11 and L16 and requires both an active ribosomal peptidyl transferase and the RF. The exact mechanism of this complementation is uncertain. Following peptidyl-tRNA hydrolysis, RF dissociates from the ribosome, possibly dependent on the hydrolysis of GTP, and is free to recycle.

Acknowledgements

C.T. Caskey was supported by the Howard Hughes Medical Institute. J.M. Campbell was supported by the Robert A. Welch Foundation, Grant Q-533.

References

Armstrong, I.L. and Tate, W.P. (1978). *J. Mol. Biol.* (in press).
Beaudet, A.L. and Caskey, C.T. (1970). *Nature (London)* **227**, 38.
Beaudet, A.L. and Caskey, C.T. (1971). *Proc. Natl. Acad. Sci.* **68**, 619.
Brot, N., Tate, W.P., Caskey, C.T. and Weissbach, H. (1974). *Proc. Natl. Acad. Sci.* **71**, 89.
Cabrer, B., Vazquez, D. and Modolell, J. (1972). *Proc. Natl. Acad. Sci.* **69**, 733.
Campbell, J.M. (1978). Ph.D. Dissertation, Baylor College of Medicine.
Capecchi, M.R. (1967). *Proc. Natl. Acad. Sci.* **58**, 1144.
Capecchi, M.R. and Klein, H.A. (1969). *Cold Spring Harbor Symp. Quant. Biol.* **34**, 469.
Capecchi, M.R. and Klein, H.A. (1970). *Nature (London)* **226**, 1029.
Caskey, C.T., Tompkins, R., Scolnick, E., Caryk, T. and Nirenberg, M. (1968). *Science* **162**, 135.
Caskey, C.T., Beaudet, A.L., Scolnick, E.M. and Rosman, M. (1971). *Proc. Natl. Acad. Sci.* **68**, 3163.
Caskey, C.T. and Beaudet, A.L. (1971). *In:* Molecular Mechanisms of Antibiotic Action on Protein Biosynthesis and Membranes (eds. Muñoz, E., Garcia-Ferrandiz, F. and Vazquez, D.), p. 326. Proc. Symp., Granada, Spain, Elsevier, Amsterdam.
Caskey, C.T., Bosch, L. and Konecki, D.S. (1977). *J. Biol. Chem.* **252**, 4435.
Dalgarno, L. and Shine, J. (1973). *Nature New Biol.* **245**, 261.
Erdmann, V.A., Fahnestock, S., Higo, K. and Nomura, M. (1971). *Proc. Natl. Acad. Sci.* **68**, 2932.
Fahnestock, S., Neumann, H., Shashoua, V. and Rich, A. (1970). *Biochemistry* **9**, 2477.
Ganoza, M.C. (1966). *Cold Spring Harbor Symp. Quant. Biol.* **31**, 273.
Ganoza, M.C. and Tompkins, J.K.N. (1970). *Biochem. Biophys, Res. Commun.* **40**, 1455.
Goldstein, J.L. and Caskey, C.T. (1970). *Proc. Natl. Acad. Sci.* **67**, 537.

Goldstein, J., Milman, G., Scolnick, E. and Caskey, T. (1970a). *Proc. Natl. Acad. Sci.* **65**, 430.
Goldstein, J.L., Beaudet, A.L. and Caskey, C.T. (1970b). *Proc. Natl. Acad. Sci.* **67**, 99.
Highland, J.H., Ochsner, E., Gordan, J., Hasenbank, R. and Stoffler, G. (1974a). *J. Mol. Biol.* **86**, 175.
Highland, J.H., Ochsner, E., Gordon, J., Bodley, J.W., Hasenbank, R. and Stoffler, G. (1974b). *Proc. Natl. Acad. Sci.* **71**, 627.
Ilan, J. (1973). *J. Mol. Biol.* **77**, 437.
Innanen, V.T. and Nicholls, D.M. (1973). *Biochim. Biophys, Acta* **324**, 533.
Innanen, V.T. and Nicholls, D.M. (1974). *Biochim. Biophys. Acta* **361**, 221.
Klein, H.A. and Capecchi, M.R. (1971). *J. Biol. Chem.* **246**, 1055.
Konecki, D.S., Aune, K.C., Tate, W. and Caskey, C.T. (1977). *J. Biol. Chem.* **252**, 4514.
Lockwood, A.H., Sarkar, P. and Maitra, U. (1972). *Proc. Natl. Acad. Sci.* **69**, 3602.
Lutter, L.C., Zeichhardt, H. and Kurland, C.G. (1972). *Mol. Gen. Genet.* **119**, 357.
Menninger, J.R. (1971). *Biochim. Biophys, Acta* **240**, 237.
Miller, D.L. (1972). *Proc. Natl. Acad. Sci.* **69**, 752.
Milman, G., Goldstein, J., Scolnick, E. and Caskey, T. (1969). *Proc. Natl. Acad. Sci.* **63**, 183.
Monro, R.E., Staehelin, T., Celma, M.L. and Vazquez, D. (1969). *Cold Spring Harbor Symp. Quant. Biol.* **34**, 357.
Morrison, C.A., Garrett, R.A., Zeichhardt, H. and Stoffler, G. (1973). *Mol Gen. Genet.* **127**, 359.
Nierhaus, K.H. and Montejo, V. (1973). *Proc. Natl. Acad. Sci.* **70**, 1931.
Nierhaus, D. and Nierhaus, K.H. (1973). *Proc. Natl. Acad. Sci.* **70**, 2224.
Otaka, E., Itoh, T., Osawa, S., Tanaka, K. and Tamaki, M. (1971). *Mol. Gen. Genet.* **114**, 14.
Pongs, O. and Rossner, E. (1975). *Hoppe-Seyler's Z. Physiol. Chem.* **356**, 1297.
Prakash, V., Campbell, J.M., Caskey, C.T. and Aune, K.C. (1978). (In preparation).
Ratliff, J.C. and Caskey, C.T. (1977). *Arch Biochem. Biophys.* **181**, 671.
Richman, N. and Bodley, J.W. (1972). *Proc. Natl. Acad. Sci.* **69**, 686.
Richter, D. (1972). *Biochem. Biophys. Res. Commun.* **46**, 1850.
Scolnick, E., Tompkins, R., Caskey, T. and Nirenberg, M. (1968). *Proc. Natl. Acad. Sci.* **61**, 768.
Scolnick, E.M. and Caskey, C.T. (1969). *Proc. Natl. Acad. Sci.* **64**, 1235.
Scolnick, E., Milman, G., Rosman, M. and Caskey, T. (1970). *Nature (London)* **225**, 152.
Shine, J. and Dalgarno, L. (1975). *Eur. J. Biochem.* **57**, 221.
Smrt, J., Kemper, W., Caskey, T. and Nirenberg, M. (1970). *J. Biol. Chem.* **245**, 2753.
Stoffler, G. and Wittman, H.G. (1971). *J. Mol. Biol.* **62**, 407.
Stoffler, G. and Caskey, C.T. (1978). (Unpublished data).
Tate, W.P., Beaudet, A.L. and Caskey, C.T. (1973). *Proc. Natl. Acad. Sci.* **70**,

2350.
Tate, W.P., Caskey, C.T. and Stoffler, G. (1975). *J. Mol. Biol.* **93**, 375.
Tompkins, R.K., Scolnick, E.M. and Caskey, C.T. (1970). *Proc. Natl. Acad. Sci.* **65**, 702.
Traub, P. and Nomura, M. (1968). *Science* **160**, 198.
Vazquez, D. (1974). *FEBS Lett.* **40**, S63.
Vogel, Z., Zamir, A. and Elson, D. (1969). *Biochemistry* **8**, 5161.
Yokosawa, H., Inoue-Yokosawa, N., Arai, K., Kawakita, M. and Kaziro, Y. (1973). *J. Biol. Chem.* **248**, 375.

READING FRAME ERRORS ON RIBOSOMES

C.G. KURLAND

Department of Molecular Biology, Wallenberg Laboratory, University of Uppsala, Sweden.

Introduction

Ribosomes are of interest primarily because they synthesize proteins, and because they do so with such extreme accuracy. However, it would be difficult to glean this point from the current ribosome literature, which is so preoccupied with static structural details that the functional aspects of the problem are often obscured. A welcome contrast is provided by the literature concerning the suppression of genetic lesions at the translational level. Here, the analysis of serendipitous errors in translation focuses attention on precisely those structural and kinetic parameters which control the accuracy of the process. It is not surprising, therefore, that significant insights into the details of aa-tRNA selection by the ribosome have been gained through studies of the activities of suppressor tRNAs (Smith, chapter 6, this volume).

On the other hand, it is by no means obvious that the errors in protein synthesis are dominated by mistakes in the selection of aa-tRNA. Nevertheless, this seems to have been the tacit assumption of the field (see, for example, the introduction to the important paper by Atkins *et al.*, 1972). One consequence of this bias is that experimental clues to the mechanism of mRNA movement on the ribosome have not received adequate attention. It is the intent of the present chapter to try to redress this temporary oversight.

We can begin by noting that both aa-tRNA selection and orderly mRNA movement on the ribosome are equivalent processes from a formal point of view: both processes must be viewed as transport phenomena and the same sorts of kinetic parameters will determine their error frequencies (Kurland, 1978). This means that without additional information to the contrary, we can not expect the error frequency of mRNA movement to be significantly less than that for aa-tRNA selection. Now, the apparent error frequency for protein synthesis in *E. coli* is close to 10^{-4} mistakes per amino acid (Edelman and Gallant, 1978). This figure defines an approximate upper level for the error frequencies of both mRNA movement and aa-tRNA selection as well as for the charging of the aa-tRNAs. Unfortunately, the partitioning of the overall error frequency between these individual steps in protein synthesis is not yet known.

If we assume for the moment that the probability of an error in mRNA movement is independent of that for aa-tRNA selection, the consequences of these different sorts of mistakes will be quite different. Thus, the matching of noncognate aa-tRNA with a codon would result in a mistake limited to one particular amino acid position in the nascent protein. In contrast, the advance of the mRNA on the ribosome by two or four nucleotides instead of the usual three would result in an error that is propagated down the remaining length of the mRNA. Accordingly, a frameshift error will lead to the production of a radically altered translation product.

Since we would expect the frameshifted ribosome to come upon an out-of-phase termination codon after an average of twenty one amino acids has been translated, a likely outcome of a frameshift error would be the production of an abortively terminated (i.e. short) polypeptide. This suggests that translation error estimates obtained by the purification of 'nearly-normal' translation products and the subsequent measurement of amino acid replacements in the purified proteins would underestimate systematically the frequency of frameshift errors. As a consequence, the suppression of frameshift mutations has provided a virtually unique source of information concerning the accuracy of mRNA movements.

I begin this presentation with a discussion of two kinds of tRNA suppressors. Both of these provide important clues to the mechanism through which reading frame errors are generated on the ribosome. The elements of a model to account for the occurence of spontaneous reading frame errors is then developed and the data supporting this model are discussed.

The Hirsh UGA Suppressor

Nonsense codons are translatable by mutated tRNA species that can effectively compete with the release factors and thereby insert an amino acid at a position which otherwise would correspond to the codon following the terminus of an aborted polypeptide. The demonstration that such suppressor tRNAs are mutated in their anticodons so that a predictable triplet interaction with the nonsense codon is possible provided clear support for the simplest version of the adapter hypothesis (Smith, this volume). Here, the translation of a triplet nucleotide sequence into a cognate amino acid could be formally viewed as a consequence of simple base pair interactions between codon and anticodon. Nevertheless, the formal simplicity of this mechanism is deceptive. Thus, it has long been known that neither the stability nor the specificity of the aa-tRNA interaction with the codon programmed ribosome can be attributed simply to triplet-triplet nucleotide interactions (Kurland, 1970).

One anomalous nonsense suppressor has been particularly useful for the resolution of this problem. This is the UGA suppressor described by Hirsh (1971). This suppressor tRNA turned out to be a mutant form of $tRNA^{Trp}$ and it contains the normal anticodon (CCA) corresponding to the codon UGG. Surprisingly, the structural alteration of this tRNA was found in its D stem, where G is replaced by an A. It was subsequently shown that the wild type $tRNA^{Trp}$ and suppressor $tRNA^{Trp}_{UGA}$ manifest the same apparent affinities for both the UGG and UGA codons in the absence of the ribosome (Högenauer 1974; Buckingham, 1976).

These observations were difficult to reconcile with the prevailing view that all the specificity of the translation process resides in the base pair specificity of the codon-anticodon interaction. Thus, the data are consistent with two heretical conclusions: one is that sites far away from the anticodon can influence the codon specificity of a tRNA molecule. The other is that the apparent stability of the codon-anticodon interaction in the absence of the ribosome is not a measure of the codon specificity of a tRNA molecule on the ribosome.

The solution to this puzzle must satisfy one additional boundary condition. This is the well established fact that all of the sequence specificity for the elaboration of a protein resides in the nucleotide sequence of the corresponding mRNA. The dilemma confronting us then is to figure out how codons can be the sole determinants for the selection of the aa-tRNA while at the same time providing a role in the selection process for sites on the tRNA distant from the anticodon. To this can be added the complementary notion that the ribosome might contribute importantly to the proper matching of codon and cognate tRNA.

One way to reconcile these seemingly disparate components of the problem is evident in a simple physical model for the selection of aa-tRNA by the codon-programmed ribosome (Kurland *et al.*, 1975). Here, several postulates are made: First, it is assumed that tRNA can exist in several distinguishable conformational states.

Second, it is assumed that there is a particular conformational state for the tRNA which permits its anticodon to base pair with the codon while other sites on the tRNA are interacting with conjugate structures on the ribosome. Finally, and most critical, it is assumed that the occupation of the conformational state optimal for simultaneous interaction with codon and ribosome is dependent on the character of the codon-anticodon interaction: When the two triplets are properly matched the required conformational state is favoured but when codon and anticodon are mismatched the required state is disfavoured.

According to this conformational selection model the experiments performed to test the differences in codon affinity for the wild type and tRNA $^{Trp}_{UGA}$ species in the absence of the ribosome would not detect differences in conformational states of the two tRNAs. However, the ribosome could select one species over the other depending upon which tRNA molecule could more easily adapt the conformation favouring an interaction at the A site. In addition, this model explicity accounts for the different codon specificities of the two tRNA^{Trp} species in terms of the effect of the D stem alteration in the suppressor on the conformational coupling between the anticodon loop and the rest of the molecule. Thus, it was suggested that the tRNA$^{Trp}_{UGA}$ was so altered that it would assume the conformation favourable to ribosome binding when merely two out of three of its anticodon bases were properly matched with the codon (Kurland *et al.*, 1975).

The latter point provides an explicit way to test the conformational selection model with the aid of the Hirsh suppressor. This tRNA will translate the codons UGG and UGA. If the above suggestion is correct, it should also translate the cysteine codons UGU and UGC. Indeed, it could subsequently be shown with a synthetic polymer that by ommitting cysteine from the incubation mixtures the tRNA$^{Trp}_{UGA}$ could efficiently insert tryptophan at the UGU codon, while wild type tRNA^{Trp} could not do so as effectively (Buckingham and Kurland, 1977).

More recently Vacher and Buckingham (1978) have provided additional support for the conformational selection model. They reasoned as follows: the abnormal properties of the tRNA$^{Trp}_{UGA}$ can be thought of as resulting from a relaxation of the conformational coupling between the anticodon and other parts of the tRNA. Therefore, a further alteration of the tRNA$^{Trp}_{UGA}$ which tightens its structure, might make it behave more like the wild type species. Therefore, they studied the functional properties of tRNA$^{Trp}_{UGA}$ after U.V. induced crosslinking of the position 8 thio-U to the position 14 C of this tRNA, as well as the wild type species. Their results show that such internal crosslinking converts the $tRNA^{Trp}_{UGA}$ to a species that has the restricted coding properties of the wild type tRNA.

The generality of these effects remains to be tested with other tRNA species. Indeed, a systematic search for appropriate mutant tRNAs with properties similar to those of the Hirsh suppressor is under way (Murgola *et al.*, 1978). In the meantime, the foregoing results give us some confidence that one form (Kurland *et al.*, 1975) or another (Kurland, 1978) of the conformational selection mechanism for the tRNA is operative on the ribosome.

One of the objectives of this paper is to show that the application of similar ideas to the problems concerning the mechanism of orderly mRNA movement on the ribosome also has promise. Here, instead of the selection of competing tRNAs we will consider the problem of selecting competing binding states for the mRNA. It will be argued that distortions of the conformational coupling between the tRNA and mRNA, which is also dependent on the codon-anticodon interaction, leads to errors in the movement of the mRNA. Such reading frame errors are the basis of frameshift suppression and, therefore, we will now turn our attention to this class of suppressors.

The *Suf* D Suppressor

The demonstration that certain frameshift mutations can be suppressed by nonallelic mutations (Yourno, *et al.*, 1969) led to the identification of several different tRNA frameshift suppressors (Riddle and Roth, 1970; 1972a; 1972b). One of these, the product of the *suf* D locus, has been particularly well studied. It was shown to be a mutant form of $tRNA^{Gly}_{GGG}$ with an altered anticodon loop (Riddle and Carbon, 1973). In particular it was found that the suppressor $tRNA^{Gly}_{suf\,D}$ carries a repeat sequence of four C residues in place of the normal three Cs in its anticodon.

In parallel studies, one particular suf D suppressible frameshift mutant (his D3068) in histidinol dehydrogenase was shown to result from the insertion of an extra base in a region of its mRNA containing a glycine codon (Yourno 1972). These data taken together with the observed structural alteration of the $tRNA^{Gly}_{suf\,D}$ provided the basis for an attractive model to account for the mechanism of frameshift suppression (Riddle and Roth, 1973).

It had previously been suggested by Woese (1970) that the codon-anticodon interaction might provide a physical mechanism to maintain the reading frame of the mRNA during translation. In one version of his model Woese suggests that a conformational rearrangement of the anticodon loop from one stacked configuration to another could be accompanied by a coupled rearrangement and movement of the corresponding codon. In this way the reading frame of the mRNA would be advanced by precisely one codon.

Such a model would also account for the suppressor activity of $tRNA^{Gly}_{\mathit{suf}\,D}$ if it is assumed that the fourth C in its anticodon can base pair with a fourth nucleotide, presumably a G, in the mRNA. The creation of an aberrant codon-anticodon interaction between four base pairs might then lead to the advance of the mRNA reading frame by four bases and thereby compensate, i.e. suppress, the frameshift caused by an extra base in the mRNA (Riddle and Carbon, 1973).

Thus, the Woese model and the data concerning *suf* D suppression of frameshift mutations mutually support a rather simple mechanism to account for the movement of the mRNA reading frame relative to the ribosome. Here, the movement itself is viewed as a consequence of conformational rearrangements of tRNA and mRNA, while the reading frame of the mRNA is maintained by the base pairing specificity of the codon-anticodon interaction.

The seductive implication of this model, in particular when it is applied to the *suf* D suppressor, is that the length of the mRNA movement is determined by the maximum number of base pairs which can form between the mRNA and the anticodon loop. Nevertheless, the data obtained with the Hirsh suppressor contradict this simplistic interpretation. Thus, our account of the functions of this unusual tRNA molecule is that it is capable of making base pair interactions with only the first two bases of the UGA codon. On the other hand, it is a strong suppressor of UGA nonsense codons *in vivo* (Hirsh, 1971). This implies that following the acceptance of the $tRNA^{Trp}_{UGA}$ at the A site, the subsequent movements of the tRNA and mRNA more often than not are normal. If this is so, it follows that even with this tRNA the movement of mRNA corresponds most often to three nucleotides and not to two nucleotides as the simplest interpretation of the Woese model would suggest.

Finally, we note that the efficiency of the suf D frameshift suppressor is limited so that only 10% of the wild type activity is recovered through its function (Riddle and Roth, 1970; Yourno, 1972). This could have a number of different interpretations, one of which is that it may not be so easy to form a four basepair interaction with mRNA at the A site. Therefore, we have further cause to wonder how the extra base in the *suf* D suppressor anticodon effects the reading frame of the mRNA.

An Alternative Model

The data from the Hirsh suppressor suggests that there is a preferred step length of three nucleotides for the advance of the mRNA on the ribosome. We can reinforce the significance of this conclusion by relating it to other aspects of the translation process.

First, it had earlier been suggested by Eisinger (1971) and confirmed in depth by Grosjean *et al.*, (1976) that there is a significant advantage with respect to the stability as well as the specificity of triplet-triplet nucleotide interaction when both triplet sequences are fixed in a defined conformation. Hence, we infer that what has previously been called the codon presentation site on the ribosome (Kurland, 1977) will fix that codon being translated at the ribosomal A site in a well defined configuration to optimize its interaction with a cognate anticodon.

Second, it seems likely that the codon presentation site will be so arranged that there is minimum ambiguity in the identification of the triplet corresponding to the codon which is to be translated. In other words, we expect there to be steric constraints on the mRNA such that the selection of an aa-tRNA to match the codon in the presentation site will be minimally influenced by nucleotides in adjoining codons, and in particular, that there will be little opportunity for an interaction of the incoming tRNA with nucleotides in adjoining codons. Therefore, it seems unlikely *a priori* that mutated tRNA, such as $tRNA^{Gly}_{suf\,D}$, would be able to interact with a four nucleotide sequence in the mRNA during the codon-anticodon matching process at the A site. Accordingly, we consider an alternative way that mutational change in the anticodon loop of a tRNA might raise the frequency of reading frame errors while maintaining the usual triplet-triplet interaction with the mRNA at the A site.

The extra nucleotide in the anticodon loop of $tRNA^{Gly}_{suf\,D}$ can have two consequences. First, the anticodon loop of the mutant is distorted by the extra mass and may therefore not be able to adopt a completely standard configuration during movement. Second, the mutant will have two alternative ways of making a triplet interaction with the appropriate glycine codon: one would involve the first three Cs in the sequence, and the other would involve the second, third and fourth Cs. It is possible that the interaction of one of these triplets leads to a particularly distorted configuration for the codon-anticodon interaction that will subsequently lead to a reading frame error. Thus, we are led to the hypothesis that a significant departure from the normal geometry of the codon-anticodon interaction will enhance the error frequency during the movement of the mRNA.

This is the control thesis of the present study.

We can support this thesis by recalling that the translocation of peptidyl-tRNA from A to P site and the advance of the mRNA by one codon are concomitant or at least tightly coupled processes (Lucas-Lenard and Lipmann, 1971). To this may be added the fact that the distance the mRNA moves, circa 10 Å, is relatively short compared to the dimensions of the tRNA. Therefore, when one of the C-triplets in the anticodon of $tRNA^{Gly}_{suf\,D}$ is interacting with a glycine codon, a relatively small departure from the standard orientation of the tRNA at its binding site may create a significant strain on the codon-anticodon interaction. As a consequence when the mRNA is released from the codon presentation site, the ensuing movement may be accompanied by slippage, blockage or the formation of non-standard codon-anticodon interactions which lead to errors in the advance of the mRNA.

Given the current dearth of information about the details of the RNA movements on the ribosome, we are obliged to be somewhat vague about the nature of the molecular perturbations that could give rise to errors in the advance of the mRNA. Nevertheless, the present model in its most general form has the virtue that it can account for data that has until now been somewhat puzzling.

Our explanation of the frameshift suppressor activity of $tRNA^{Gly}_{suf\,D}$ takes into account the extra nucleotide in this tRNA's anticodon not by introducing an anomolous fourth base pair with the mRNA at the A site, but by considering the steric consequences of a triplet interaction in a non-standard configuration. If such a steric perturbation can influence the accuracy of the mRNA movement, it seems reasonable to expect that other ones could do so as well. Thus, the missense error of selecting an aa-tRNA which is mis-matched with the codon at the A site might also lead to reading frame errors because here too the codon-anticodon interaction is distorted. Similarly, a correctly matched anticodon in a tRNA which is not 'normal' for that codon might do the same. In other words, we suggest that whenever an aa-tRNA is accepted at the ribosomal A site with either a mis-matched anticodon or a perturbed structure elsewhere in the tRNA, the error frequency for the subsequent movements of the mRNA will be enhanced. It remains only to document that precisely such effects have been observed.

Ribosome-dependent Reading Frame Errors

A group of 31 frameshift mutants isolated and characterized by Newton (1970) were shown to be structural mutants of β-galactosidase.

These were classified as 'nonleaky' according to the criterion that the mutants could not grow on plates containing a lactose minimal medium. However, a more exacting analysis of sixteen of these mutants revealed low levels of β-galactosidase production corresponding to a mean value of 0.03% of the wild type enzyme level (Atkins *et al.*, 1972). Since these 16 mutants were randomly distributed along the cistron and met other relevant criteria, it could be concluded that the β-galactosidase produced by these strains reflected the spontaneous occurrence of reading frame errors that suppressed their respective frameshift mutations.

We assume for the sake of argument that most of the reading frame errors which successfully suppress a frameshift error occur within ten codons to either side of the insertion or deletion, and that there are equal probabilities for a (+1) and (-1) reading frame error. The mean value of 0.03% suppression would then correspond to a reading frame error frequency of approximately 3×10^{-5} per codon. This figure can then be compared with the 10^{-4} per codon missense error frequency (Edelman and Gallant, 1977) which must be partitioned at the very least between the tRNA charging and aa-tRNA selection errors. If this partitioning were equal, a 5×10^{-5} missense error rate and a 3×10^{-5} reading frame error rate indicate missense and reading frame errors are very approximately equally probable. Such agreement in their error frequencies is consistent with the suggestion that aa-tRNA selection and mRNA movement are similar processes (Kurland, 1978).

The spontaneous suppression frequencies of these sixteen frameshift mutants varied considerably from site to site (Atkins *et al.*, 1972). Here, the range was 0.05-0.0003% suppression, which corresponds to fluctuations in excess of two orders of magnitude. Such fluctuations suggest that the nature of the nucleotide sequences in the mRNA close to each of the frameshift errors (i.e. the context) determine the probability of a corrective reading frame error. One way to account for such effects is to assume that the efficiency of frameshift suppression depends on the distance of the mutation from repetitive sequences of nucleotides longer than three, which are both infrequent and particularly amenable to slippage during the RNA movements (Gorini, 1971; Atkins *et al.*, 1972). More subtle effects of context on the properties of the mRNA, which influence its movement directly or via its interaction with tRNA and ribosome, are of course possible. At any rate, it seems not at all surprising that the probability of a reading frame error should be dependent on the local nucleotide sequence of the mRNA.

That the suppression of these β-galactosidase frameshift mutants really occurs at the translational level was demonstrated by studying the effects of different ribosome mutants on the frequency of the suppression (Atkins *et al.*, 1972). Here, three sorts of effects are observed. First, mutants of the *str*A (*rps*L) type, which restrict missense and nonsense errors, also are found to restrict the frequency of frameshift suppression. Second, *ram*1 (*rps*D) mutants which enhance missense and nonsense errors, also enhance the frameshift suppression frequency. Finally, the addition of streptomycin to the restricted *rps*L mutants, relieves the restriction on the frameshift suppression by these mutants.

Atkins *et al.*, (1972), while noting the striking correlation between the ribosome phenotype's effect on the frequency of missense as well as nonsense suppression on the one hand and on the other that on frameshift suppression, remark that 'there is no evidence that the same factors (*Str*A, ***ram***, and Sm) controlling translation efficiency should also control a general slippage of the ribosome'. In contrast, the present model requires precisely such a correlation.

Additional support for the present interpretation comes from still another source. Mutations in the so-called stringent factor, which lead to the relaxation of the amino acid dependent control of RNA synthesis, also lead to increases in the missense error frequency of protein synthesis when a required amino acid is limiting growth (Lavalle and De Hauwer, 1968; Hall and Gallant, 1972; Edelman and Gallant, 1977). In addition, there is under these conditions an accumulation of small polypeptides, which seem to represent abortively terminated translation products (Hall and Gallant, 1972; O'Farrel and O'Farrel, 1978). As pointed out above, the occurrence of a reading frame error should lead to the production of such short polypeptides due to the presence of out-of-phase terminating codons. Therefore, these data suggest that mutations in stringent factor simultaneously effect the missense and reading frame error frequencies. Again, this correlation is required by the present model.

Discussion

The translation of a fully degenerate, punctuated code requires that each succeeding codon be identified solely by its position in the mRNA. There are no internal rules to eliminate the possibility that overlapping triplets from adjoining codons will be translated. Likewise, slippage of the mRNA at any codon will not be recognizable as a mistake by the translation apparatus. Therefore, it seems likely *a priori* that the 30S ribosomal subunit positions each succeeding codon in a configuration that unambigously identifies that triplet as the one to be translated at the A site. Nevertheless, the movement of tRNA and mRNA out of the A site will require the release of mRNA from its binding site. It is during this movement that the system would seem to be most vulnerable to reading frame errors.

Following Woese (1970), we assume that part of the mechanism for maintaining the reading frame during this movement is provided by the codon-anticodon interaction. To this we add two other considerations: one is that we assume that tRNA undergoes a conformational rearrangements very like those postulated for tRNA selection site. The other is that the conformation of the tRNA as a whole is coupled to the codon-anticodon interaction, as indicated by the data obtained with the $tRNA^{Trp}_{UGA}$. Accordingly, when either the codon-anticodon interaction or the conformation of the tRNA does not correspond to some standard configuration, we expect the movement of the mRNA to be perturbed.

In effect, we are treating the tRNA-mRNA complex as a single molecular species, the movement of which is governed by conformational reaarangements very like those postulated for tRNA selection at the A site. Unfortunately, we cannot be more specific about the details because so much fundamental information is missing. For example, we still do not know what role is played, if any, by the P site tRNA in determining the reading frame of the mRNA.

Nevertheless, the principal corollary of this view of RNA movement on the ribosome is verified by the general rule that whenever the error frequency of tRNA selection is increased, the reading frame error rate also increases. Indeed, the range of this correlation is surprisingly wide.

There is, however, one relevant circumstance that we have not yet discussed. It was recognized explicitly by Gorini (1971) that a nonsense suppressor tRNA is a 'nonstandard' tRNA, and is treated as such by the ribosome. Therefore, we would be encouraged to pursue the present line of analysis in more detail if it could be shown that nonsense suppression is accompanied by an enhanced reading frame error rate. Indeed, it may very well be that some of the context effects on nonsense suppression (Fluck *et al.*, 1977) reflect the site-dependent probabilities of reading frame errors occurring after the nonsense codon is translated by a suppressor tRNA.

Acknowledgements

I thank M. Laughrea, P. Jelenc, and I. Winkler for their criticism and advice.

References

Atkins, J.F., Elseviers, D. and Gorini, L. (1972). *Proc. Nat. Acad. Sci.* **69**, 1192.
Buckingham, R.H. (1976). *Nucleic Acids Res.* **3**, 965.
Buckingham, R.H. and Kurland, C.G. (1977). *Proc. Nat. Acad. Sci.* **74**, 5496.
Edelman, P. and Gallant, J. (1977). *Cell.* **10**, 131.
Eisinger, J. (1971). *Biochem. Biophys. Res. Commun.* **43**, 854.
Fluck, M.M., Salser, W. and Epstein, R.H. (1977). *Molec. Gen. Genet.* **151**, 137.
Gorini, L. (1971). *Nature New Biol.* **234**, 261.
Grosjean, H., Söll, D.G. and Crothers, D.M. (1976). *J. Mol. Biol.* **103**, 499.
Hall, B. and Gallant, J. (1972). *Nature New Biol.* **237**, 131.
Hirsh, D. (1971). *J. Mol. Biol.* **58**, 439.
Högenauer, G. (1974). *FEBS Lett.* **39**, 310.
Kurland, C.G. (1970). *Science.* **169**, 1171
Kurland, C.G. (1977). *In:* Molecular Mechanisms of Protein Biosynthesis (eds. H. Weissbach and S. Pestka), 81. Academic Press, London and New York.
Kurland, C.G. (1978). *Biophys. J.* **22** (in press).
Kurland, C.G., Rigler, R., Ehrenberg, M. and Blomberg, C. (1975). *Proc. Nat. Acad. Sci.* **72**, 4248.
Lavalle, R. and De Hauwer, A. (1968). *J. Mol. Biol.* **37**, 269.
Newton, A. (1970). *J. Mol. Biol.* **49**, 589.
O'Farrel, P.Z. and O'Farrel, S.F. (1978). *Cell* (in press).
Riddle, D.L. and Carbon, J. (1973). *Nature New Biol.* **242**, 230.
Riddle, D.L. and Roth, J.R. (1970). *J. Mol. Biol.* **54**, 131.
Riddle, D.L. and Roth, J.R. (1972a). *J. Mol. Biol.* **66**, 483.
Riddle, D.L. and Roth, J.R. (1972b). *J. Mol. Biol.* **66**, 495.
Vacher, J. and Buckingham, R.H. (1978). (submitted for publication).
Woese, C. (1970). *Nature (London).* **226**, 817.
Yourno, J. (1972). *Nature New Biol.* **239**, 219.
Yourno, J., Barr, D. and Tanemura, S. (1969). *J. Bact.* **100**, 453.

SUPPRESSOR tRNAs IN PROKARYOTES

J.D. SMITH

MRC Laboratory of Molecular Biology,
University Medical School, Cambridge, U.K.

Introduction

Suppression of missense and chain terminating codons or frameshift mutants during translation results from a mutation altering one of the components of the translational system so that the mutant codon is misread. Unlike mutations altering the decoding site on the ribosome, which allow a comparatively non-specific codon misreading (for review see Steege and Söll, 1978), mutations giving suppressor tRNAs alter the tRNA decoding properties in a very specific way. This article discusses how bacterial suppressor tRNAs have helped in understanding tRNA function. It is not possible here to cite all the extensive literature on suppressors so that in several instances I have referred to recent reviews. Frameshift suppressors are not reviewed here because they are discussed in C.G. Kurland's article (Chapter 5, this volume).

Most of the suppressor tRNAs result from a mutation in a tRNA structural gene which alters the anticodon allowing the misreading. Two interesting exceptions, discussed later are: 1) the *E. coli* $tRNA^{Trp}$ UGA suppressor where a base substitution in the D stem of the tRNA results in an ambiguous codon reading (Hirsh, 1971); and 2) a recessive suppressor where altered codon reading appears to result from an incompletely modified anticodon base. Base substitutions in the tRNA anticodon give specific switches in codon recognition so that the mutant suppressor tRNA can no longer translate its original codon. In *E. coli* the majority of codons are translated by more than one tRNA species or by tRNAs coded by multiple identical genes so that a suppressor mutation in one of these does not impair normal reading. An anticodon change in a tRNA uniquely responsible for a coding function would be lethal to the cell and such suppressors can only be isolated where the tRNA gene has been duplicated, as in a merodiploid, or where the gene has been inserted into a bacteriophage. Examples are the $tRNA^{Trp}$ UAG (su_7^+) suppressor (Söll and Berg, 1969; Yaniv *et al.*, 1974) and one of the $tRNA_2^{Gly}$ AGA missense suppressors (Hill *et al.*, 1969).

As would be expected most suppressor tRNAs are derived by mutations giving single base substitutions so that the set of nonsense suppressors is limited and can be predicted from the genetic code. UAG suppressors could possibly be derived from leucine, serine, tyrosine, tryptophan, glutamine, lysine and glutamic acid tRNAs. Suppressors reading UAG as leucine, serine, tyrosine and glutamine have been isolated in *E. coli* (and bacteriophage) and the anticodon changes of some of these are listed in Table I. The amino acid inserted in response to the UAG codon does not necessarily specify the tRNA from which the suppressor was derived, as in su_7^+ the anticodon mutation also changes the aminoacylation specificity of $tRNA^{Trp}$ so that it can be charged with glutamine. Only some of the theoretically possible UGA and UAA suppressors have been identified.

Certainly the compatibility of an altered anticodon with tRNA function must determine which suppressors can be isolated. This is of interest since it is now feasible to construct suppressor tRNAs artificially; either by cutting and splicing the tRNA using RNA ligase or by directed mutagenesis on tRNA genes isolated in phages or plasmids. One limitation may be the importance of the anticodon loops in aminoacyl tRNA synthetase recognition by some tRNAs, another results from the importance of modified bases in codon-anticodon recognition.

Anticodon-codon Interactions: the Role of Modified Bases

The first two bases in the codon pair with anticodon bases 2 and 3 by Watson-Crick base pairing (with the exception of the GUG reading by the initiator tRNA). To explain how a single tRNA could read codons for one amino acid differing only in the third base Crick (1966) proposed that the first base of the tRNA anticodon, in addition to pairing by DNA type hydrogen bonding could also pair with base 3 of the codon by defined non-standard 'wobble' hydrogen-bonded pairing chosen by stereochemical considerations. These were G-U, U-G and I-A/C or U. At that time the only modified base identified in the wobble position of a tRNA anticodon was I (yeast $tRNA_1^{Ala}$). This type of three codon recognition is frequent in eukaryotic tRNAs, but in *E. coli* inosine has only been found in $tRNA_1^{Arg}$

TABLE I

Some Nonsense Suppressors in E. coli *and T4*

	Suppressor locus	Map position (min., 0-100)	Amino acid inserted	tRNA	Anticodon change	Remarks
E. coli	SU_1 amber	43	serine			
	SU_2 amber	15	glutamine			
	SU_3 amber	27	tyrosine	Tyr 1	QUA → CUA	Q is a modified G residue
	SU_3 ochre	mutant of ($\phi 80psu_3A2P$)	tyrosine	Tyr 1	QUA → $\overset{*}{U}$UA	$\overset{*}{U}$ is an unidentified uridine derivate
	SU_4 ochre	27	tyrosine	probably Tyr 1		
	SU_5 ochre	16	lysine			
	SU_6 amber		leucine			
	SU_7 amber	83	glutamine and tryptophan	Trp	CCA → CUA	
	SU_8 ochre	84				
	SU_M ochre	88	tyrosine	Tyr 2		
	UGA	83	tryptophan	Trp	none (G24 → A)	
T4	Arg UGA		arginine	T4 Arg	$\overset{*}{U}$CU → $\overset{*}{U}$CA	$\overset{*}{U}$ is an unidentified uridine derivative
	Gln ochre		glutamine	T4 Gln	$\overset{*}{U}$UG → $\overset{*}{U}$UA	" "
	Gln amber		glutamine	T4 Gln	$\overset{*}{U}$UG → CUA	" "
	Ser amber		serine	T4 Ser	$\overset{*}{U}$GA → CUA	" "

(For references see Steege and Söll, 1978).

which translates CGU, CGC and CGA (Murao *et al.*, 1972).

Recently, other modified bases in the anticodon wobble base position have been identified (for reviews and literature citations see Nishimura, 1972; McCloskey and Nishimura, 1977). In *E. coli* a different three base wobble is found with the tRNAs valine 1, serine 1 and an alanine tRNA where the first anticodon base is uridine-5-oxyacetic acid (or its methyl ester) and which recognizes codons ending in A, G and U. In *Bs. subtilis* 5-methoxyuridine in these tRNAs permits the same codon recognition. These two modifications of uridine allow U-U recognition in the tRNA wobble position presumably by altering the conformation of the anticodon on the ribosome. U-U pairing involves a displacement of the bases relative to standard base pairing which is appreciably greater than U-G or I-A pairing (Crick, 1966).

The 5-substituted 2-thiouridine derivatives, 5-methylaminomethyl-2-thiouridine and 5-(methoxycarbonylmethyl)-2-thiouridine in the first anticodon base, pair preferentially with A in the third codon position. These are only found in tRNAs such as *E. coli* $tRNA_1^{Gln}$ where the codons ending in A and G (CAA and CAG) code for one amino acid (glutamine) while the codons ending in U and C (CAU and CAC) code for another (histidine). CAG is read by a different tRNA having the anticodon CUG. Nishimura (1972) has suggested that these base modifications are necessary to avoid U-U pairing which with these codons would give mistranslation. Apart from suppressor tRNAs, unmodified U is only found in the first anticodon base in a yeast $tRNA^{Leu}$ (anticodon UAG) where no ambiguous reading is possible. One could explain the restrictions on U-G and U-U recognition by failure to base pair because of the 2-thio substitution. However, the role of the modification at the 5 position may be equally important since yeast $tRNA_3^{Arg}$ which preferentially reads AGA has as first anticodon base 5-(methoxycarbonylmethyl)uridine. The modified guanine Q (or its derivatives) replaces guanine as the first anticodon base of tRNAs reading codons with A in the second position in *E. coli* and in some eukaryotes (but not in yeast). It has the wobble characteristic of G although G-U pairing is favoured over G-C pairing. Its significance is unknown.

So far the association between codon recognition and the identity of the anticodon wobble base follows a logical pattern although the mode of recognition of the modified uridines on the ribosome is unclear. Surprisingly, there are examples where, at least *in vitro,* codon recognition is quite independent of the first anticodon base, so that only the first two codon bases appear to be specifically recognised. The four valine codons GUU, GUC, GUA and GUG can all be recognised on *E. coli* ribosomes in protein synthesis by the valine tRNAs with anticodons VAC, GAC and IAC (Mitra *et al.*, 1977; V is uridine-5-oxyacetic acid). Codon directed ribosome binding shows that rabbit liver $tRNA_1^{Val}$, with the anticodon IAC is also recognised on *E. coli* ribosomes by all four valine codons (Jank *et al.*, 1977). In fact much earilier experiments by Bergquist *et al.*, (1968) suggested that recognition of the four glycine codons on *E. coli* ribosomes was independent of the first anticodon base in the glycine tRNA. Whether this recognition, called 'two out of three' by Lagerkvist (1978), occurs during *in vivo* translation is not clear. Obviously this type of codon recognition would result in missense where codons differing in the third base specify different amino acids. The $tRNA^{Trp}$ UGA suppressor and one missense $tRNA^{Gly}$ suppressor offer examples of this kind of misreading and are discussed later and also in the article by C.G. Kurland, (chapter 5, this volume).

The nucleoside residue adjacent to the third base of the anticodon is always a purine which in most tRNAs is modified. Comparison of tRNA sequences shows interesting correlations between the modified base and the codons recognised (for reviews and literature citations see Nishimura, 1972; McCloskey and Nishimura, 1977). In tRNAs translating codons beginning with U this base is almost always an adenosine with a bulky hydropholic group, for example N^6 isopentenyl adenosine; in tRNAs translating codons beginning with A it is an adenosine hypermodified with a hydrophilic group, for example the N^6 substituted carbamylthreonine adenosine, t^6 A. Exceptions are rat liver $tRNA^{Tyr}$ which has t^6 A in this position and *Bs. subtilis* $tRNA_m^{Met}$ in which the base is m^2 A.

This correlation could be considered as a consequence of the specificity of recognition of the anticodon region by the modifying enzymes. However, at least in some tRNAs it is clear that the base modification is essential for tRNA function. *E. coli* $tRNA_1^{Tyr}$ su_3^+ in which the 2-thiomethyl-6-isopentenyladenosine is replaced by an unmodified adenosine residue is inactive in ribosome binding although it can be aminoacylated (Gefter and Russell, 1969). Similar results have been reported with *E. coli* $tRNA^{Ile}$ lacking the t^6 A modifications (Miller *et al.*, 1976) and the functional importance of these hypermodified bases has been confirmed in studies on chemically modified tRNAs (for review see Goddard, 1977).

In tRNAs recognising codons beginning with C or G the purine adjacent to the anticodon is either unmodified or methylated suggesting a special function of the two kinds of hypermodified A residues in the recognition on the ribosome of U or A as the first anticodon base.

Nonsense and Missense Suppressors

Amber and Ochre Suppressors

Table I presents the properties of some *E. coli* and T_4 nonsense suppressors. All amber suppressors exclusively translate UAG and those sequenced have the anticodon CUA. Su_7^+ is derived by mutation of the single gene for $tRNA^{Trp}$ and is recessive lethal (Söll and Berg, 1969). The anticodon change in su_7^+ tRNA alters the aminoacyl tRNA synthetase recognition so that this suppressor tRNA can be charged with both glutamine and tryptophan (Yaniv *et al.*, 1974; Celis *et al.*, 1976), glutamine being the preferred amino acid. T_4 codes for 8 tRNAs (arg, ile, thr, ser, pro, gly, leu and gln) whose genes map in a segment near the lysozyme *e* gene (McClain, 1977; Velten *et al.*, 1976). In most strains of *E. coli* these tRNAs are not essential for phage growth; three have been converted to nonsense suppressors (Table I).

E. coli and T_4 ochre suppressors translate both UAA and UAG codons, unlike the yeast ochre suppressors which are UAA specific. The ochre suppressors derived by mutation from T_4 $tRNA^{Gln}$ (Seidman *et al.*, 1974) and *E. coli* $tRNA_1^{Tyr}$ (Altman *et al.*, 1971; Altman, 1976) both have the anticodon NUA where N is an unidentified uridine derivative, not necessarily the same in the two tRNAs. The modification of this uridine residue appears to be essential for UAA reading by the T_4 glutamine suppressor tRNA (Colby *et al.*, 1976); Siedman *et al.*, 1974). In neither of these suppressors has the N base been identified so the basis of the dual recognition of UAA and UAG codons is unclear. Under the conditions they have been studied the modifications of U to N is incomplete in both the tRNAs, so that UAG could be recognised by the unmodified tRNA. (Seidman *et al.*, 1974; Altman, 1976).

UGA Suppressors

UGA suppressors might be expected to arise by single anticodon base changes from leucine, serine, cysteine, tryptophan, arginine or glycine tRNAs. The only identified suppressor of this kind is T_4 $tRNA^{Arg}$ UGA in which the anticodon is changed from $\overset{*}{U}$CU to $\overset{*}{U}$CA, $\overset{*}{U}$ being an unidentified modified uridine residue (Kao and McClain, 1977).

The *E. coli* $tRNA^{Trp}$ UGA suppressor is derived in an unexpected way by a mutation in the single $tRNA^{Trp}$ gene which changes G residue 24 to A so that in the D stem of the tRNA a G-U base pair is changed to an A-U pair (Hirsh, 1971). *E. coli* tRNA ^{Trp} has the anticodon CCA but in addition to translating the tryptophan codon UGG also translates UGA with very low efficiency so that UGA mutants are leaky (Hirsh and Gold, 1971). A low frequency readthrough of UGA terminator codons has recently been demonstrated in the *in vivo* translation of four of the φX 174 bacteriophage genes which use UGA as the termination signal (Pollock *et al.,* 1978). The A24 mutation results in a greatly increased efficiency of misreading the UGA codon (50% transmission) while still allowing translation of the tryptophan codon so that the suppressor mutation is not lethal. Buckingham and Kurland (1977) have found that in the absence of competing tRNA, the cysteine codon UGU is also misread by the suppressor tRNA which has a 'two out of three' reading similar to that reported for the valine tRNA's. In the absence of ribosomes, the tryptophan suppressor tRNA does not show an increased affinity for the UGA triplet (Hogenauer, 1974; Buckingham, 1976) so any tRNA conformational change necessary for misreading must involve interactions at the ribosome binding site. This question is discussed in the article by C.G. Kurland, (chapter 5, this volume).

Reeves and Roth (1971) isolated a recessive UGA suppressor (*sup* K) in *Salmonella.* Nonsense suppressors specifying mutants in a tRNA structural gene are dominant in merodiploids because the wild type tRNA does not compete in nonsense translation. Recessive suppressors would be expected to arise in a different way. *Sup* K strains are deficient in a specific tRNA methylase activity (Reeves and Roth, 1975; Pope *et al.,* 1978) identified as that converting the wobble anticodon base uridine-5-oxyacetic acid to its methyl ester in alanine and serine tRNAs. Unfortunately it is not yet known which tRNA is involved in the suppression. tRNA^{Ser} , would be a possible candidate if the lack of complete modification allowed a two base recognition of UGA, using the first and third positions of the anticodon VGA. The *sup* K suppressor has also been reported to suppress certain frameshift mutants (Ryasaty and Atkins, 1968).

Missense Suppressors

Missense suppressors could arise by mutations altering tRNA decoding, or mutations in a tRNA gene changing the specificity of aminoacylation, analagous to the su_3^+ mischarging mutants. Another possibility would be a change in the specificity of an aminoacyl tRNA synthetase. However all the characterised missense suppressors are mutations in tRNA structural genes altering the anticodon (for review see Hill, 1975). They are different mutants of the three tRNA^{Gly} species in *E. coli* and originate from Yanofsky's work on missense mutants derived from glycine codons in the tryptophan synthetase A protein gene (Yanofsky *et al.*, 1966).

The anticodon changes and properties of these suppressors are listed in Table II. The $tRNA_3^{Gly}$ *ins* mutation is unusual in that it corrects a translational defect caused by a missense suppressor. *E. coli* has a single copy of the gene for $tRNA_2^{Gly}$ which normally translates the glycine codon GGA. The $tRNA_2^{Gly}$ suppressor which reads AGA (see Table II) is probably the only glycine tRNA efficiently reading GGA so that missense mutants of this gene, for example su_{36}, are semilethal. The reason why they are not completely lethal may be because another $tRNA^{Gly}$ can mis-read GGA inefficiently. The *ins* mutation converting one of the $tRNA_3^{Gly}$ genes to a GGA reader overcomes this defect (Table II).

An important finding with the glycine missense suppressors is that changes in the second two bases of the anticodon result in a greatly reduced rate of aminoacylation by glycine tRNA synthetase indicating that this part of the tRNA structure is directly involved in the specific recognition by the synthetase.

Arrangement of tRNA Genes in *E. coli*

E. coli has about 55 tRNA genes. Over 40 of these have been located with varying degrees of precision on the chromosome. The suppressor tRNA genes listed in Tables I and II were mapped by conventional methods. Ikemura and Ozeki (1977) have identified the approximate map position of over 20 tRNA genes by two ingenious indirect methods depending, either on gene dosage effects from F′ carried tRNA genes, or amplified synthesis on induction of λ prophages inserted near the tRNA gene. In addition the sequence analysis of multimeric tRNA precursors has enabled small clusters of tRNA genes to be identified. Figure 1 shows the collected results of tRNA gene mapping. The overall conclusion is that in *E. coli* many of the genes are arranged in small clusters, some of which are tandem and co-transcribed. Although several of these clusters comprise two or more identical tRNA genes (e.g. 2 tyr_1, 3 gly_3, 6 leu_1) which might have been expected to have evolved by gene duplication, other clusters include totally unrelated tRNA genes (e.g. tyr_2, gly_2 and thr_3). This type of gene arrangement has consequences in the transcription and processing of tRNAs which are discussed in the article by S. Altman, (chapter 10, this volume).

Mutants of Suppressors and tRNA Function

The suppressors can be studied genetically by selection of mutants

TABLE II

Anticodon changes in E. coli *glycine tRNA missense suppressors.*

Gene	tRNA	Suppressed allele	Codon read as glycine	Anticodon	Reduced rate of aminoacylation	Reference
gly U	gly 1	none (wild type)	GGG (gly)	CCC		
"	"	trp A461	GAG (glu)	CUC	slightly	Hill *et al.*, 1974
"	"	trp A36	AGA (arg)		yes	Carbon *et al.*, 1969
"	"	trp A46	GAA/G (glu)		yes	Hill *et al.*, 1974
gly T	gly 2	none (wild type)	GGA/G (gly)	$\overset{*}{U}$CC		
"	"	trp A36	AGA (arg)	$\overset{*}{U}$CC[(a)]	yes	Roberts and Carbon, 1974, 1975
gly V	gly 3	none (wild type)	GGU/C (gly)	GCC		
"	"	ins (see text)	GGA/G (gly)	UCC	no	Squires and Carbon, 1971
"	"	trp A78	UGU (cys)			Fleck and Carbon, 1975
"	"	trp A58	GAU/C (asp)			Guest and Yanofsky, 1965
"	"	trp A46	GAA/G (glu)			Guest and Yanofsky, 1965
"	"	trp A128	AGU/C (ser)			Murgola and Yanofsky, 1974
gly W	gly 3	none (wild type)	GGU/C (gly)	GCC		
"	"	trp A58	GAU/C (asp)			Fleck and Carbon, 1975
"	"	trp A78	UGU (cys)	GCA[(b)]	yes	Carbon and Fleck, 1974

(a) $\overset{*}{U}$ is a modified uridine residue. The base on the 3′ side of the anticodon is changed from A to t^6A.
(b) The base on the 3′ side of the anticodon is changed from A to ms^2i^6A.

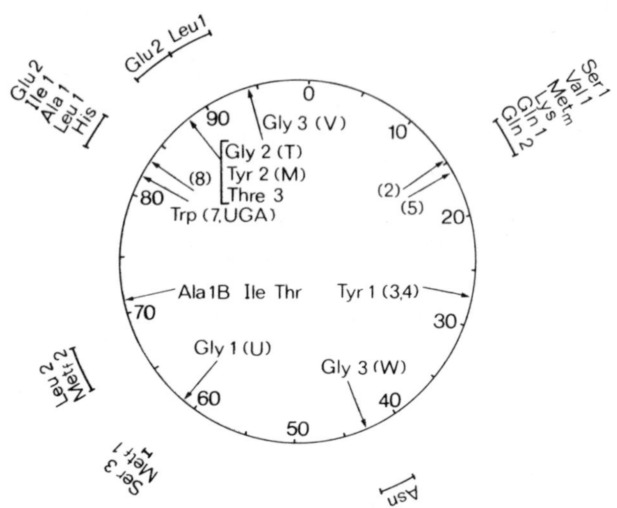

Fig. 1 Positions of tRNA genes on the *E. coli* genetic map. Those placed outside the map circle are from the results of Ikemura and Ozeki (1977). The numbers and letters in brackets refer to the suppressors listed in Tables I and II.

which quantitatively or qualitatively altered suppressor activity. Together with sequence and functional analysis of the mutant tRNAs this provides an approach to the study of tRNA structure and function where the effects of single base changes in a tRNA can be analysed. The most extensive study of this kind is with *E. coli* $tRNA_1^{Tyr}$ su_3^+ amber suppressor tRNA where mutants giving 24 different base substitutions in the tRNA molecule have been isolated (for reviews see Smith, 1972, 1976). Many suppressor defective mutants have also been characterised in the T_4 nonsense tRNA suppressors (for review see McClain, 1977). Leupold and his colleagues (reviewed by Hawthorne and Leupold, 1974) have isolated suppressor defective mutants in the haploid yeast *Schizosaccharomyces pombei* and established fine structure genetic maps of several tRNA genes.

The su_3^+ amber suppressor is derived by changing the anticodon from QUA to CUA (where Q is a modified guanosine residue) as a result of a mutation in one of two identical $tRNA_1^{Tyr}$ genes almost adjacent on the *E. coli* chromosome (Goodman *et al.*, 1968). Many of the studies of this gene were made possible by the elimination of one of the gene copies by unequal recombination, and by the isolation of a non-defective bacteriophage carrying a single suppressor copy of the $tRNA_1^{Tyr}$ gene (Russell *et al.*, 1974). Using cells infected with this phage the suppressor tRNA synthesis can be specifically amplified by a gene dosage effect. Mutants with altered suppressor properties are screened by plating the phage on suitable *E. coli* lac^-_{amber} strains using indicator plates for β-galactosidase activity. Alternatively suppressor defective mutants may be selected in *E. coli* su^+_3 gal K^-_{amber} gal E^- non-suppressible strains where suppression of the galactokinase amber results in cell lysis during growth on a galactose medium (Abelson *et al.*, 1970).

The aim of isolating single base substitution mutants in which one tRNA function is specifically altered is made complicated by the fact that the majority of the nucleoside residues are involved in the tRNA structure so that in many cases a sequence alteration will not lead to a localised structural change. A surprising consequence was the finding that mutant tRNAs with base changes expected to alter the tRNA conformation were synthesised in reduced yield, or in some mutants not at all, because of failure to process the tRNA precursor efficiently. Indeed this led to the discovery that in *E. coli* the primary tRNA transcripts contain additional terminal sequences specifically removed by processing enzymes. One of these enzymes, RNase P, which cleaves the precursor adjacent to the 5′ terminal nucleotide of the mature tRNA sequence recognises the tRNA conformation in the precursor so that mutant tRNAs with altered conformation are processed inefficiently or incorrectly. This is discussed in detail in the article by S. Altman (Chapter 10, this volume).

In consequence many mutant tRNAs cannot be isolated in sufficient yield for biochemical studies. The sequence changes in the tRNAs are given in Figure 2. The mutants are defined by the substituting base numbered from the 5′ -end. The A15 mutant tRNA is of interest because G15 would be expected to pair with C57 (B.F.C. Clark Chapter 1, this volume). Both chemical modification and physiochemical studies have shown that this tRNA has an altered conformation (Cashmore, 1971; Altman *et al.*, 1974). This results in a failure of this tRNA to participate efficiently in protein synthesis due to a block in a step after the tRNA is bound to the A-site on the ribosome (Abelson *et al.*, 1974).

Two mutants alter base modification by changing the base which is modified so that the changes C→D at positions in the D loop presumably result from C→U changes in the precursor and the uridine residue can be modified to dihydrouridine.

The T_4 glutamine ochre tRNA suppressor defective mutants which have base changes in the D stem and in the anticodon stem of the tRNA result in failure to modify the uridine residue in the wobble anticodon position (Table I), and the modification of the purine adjacent to the anticodon (normally m^2 A). It is difficult to decide why these mutant tRNAs are defective in UAA reading but it is most

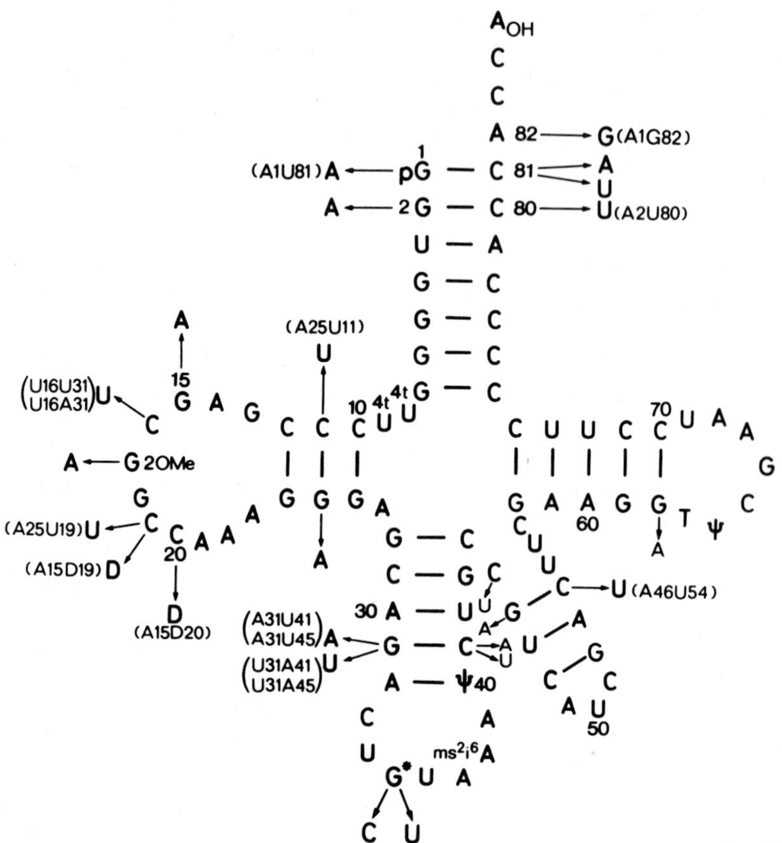

Fig. 2 The sequence of su_3^+ tyrosine tRNA showing base substitutions of mutants with defective or altered suppressor activity. Double mutants are indicated in brackets.

likely that one of these two base modifications is essential for suppression.

Several temperature sensitive su_3 mutants have been isolated (Smith *et al.*, 1971; Galluci *et al.*, 1971; Ozeki *et al.*, 1969). Two of these, A2 and A81 result from base substitutions at the end of the acceptor stem. Nagata and Horiuchi (1973) and Oeschger and Woods (1976) have isolated mutants of su_1^+ which are temperature sensitive in their suppressor activity. These offer interesting possibilities for studies in the timing of gene expression and also for the direct selection of amber mutants.

The double mutants in Figure 2 were selected as revertants of suppres- sor defective mutants having increased suppressor activity, and in severa the second mutation converts a mismatched base pair in one of the tRN stems to a base pair different from that in su_3^+ tRNA. These double mutants are synthesised in normal amounts confirming that precursor processing depends on recognition of a correct tRNA structure by RNase P. Several base pairs in the acceptor, D and anticodon stems may be changed in this way without any apparent effects on tRNA function. A different type of double mutant is where the two base changes are not related to each other in the tRNA structure. Base substitutions in the anticodon stem may be compensated by specific single base changes in the D loop. These compensating changes may have the effect of disfavouring an alternative precursor structure which cannot be properly processed (Smith, 1976).

A useful approach to the question of how tRNAs are recognised by aminoacyl tRNA synthetases comes from mutant suppressor tRNAs where the mutation alters the specificity of synthetase recognition. Mutants of su_3^+ amber suppressor tRNA which are charged with the wrong amino acid can be isolated by a simple selection method (Hooper *et al.*, 1972; Shimura *et al.*, 1972; Smith and Celis, 1973; Ghysen and Celis, 1974). Some amber mutants are not suppressed by su_3^+ because translation of the amber codon as tyrosine gives an inactive protein. Mutants of su_3^+ which suppress these amber mutants do so by reading the codon as an amino acid other than tyrosine by mischarging of the mutant tRNA. The su_3^+ mutants selected by this technique all specify tRNAs which mischarge with glutamine (Figure 2 and Table III). The A1 mutant tRNA is charged with glutamine tRNA synthetase at a much reduced rate compared with *E. coli* $tRNA^{Gln}$. The findings with these mutants may be summarised as follows: 1) Mischarging with glutamine results from substitutions of each of 4 bases at the acceptor stem terminus (Figure 2) locating a specific region of this tRNA involved in recognition by glutamine and tyrosine tRNA synthetases; 2) with the exception of the G82 mutant, mischarging with glutamine does not depend on changes giving sequence homology with $tRNA^{Gln}$. Apparently a change in conformation of the tRNA acceptor stem is important as suggested by the fact that A2 tRNA mischarges with glutamine while the double mutant A2U80 in which tRNA bases 2 and 80 are paired normally does not; 3) the mutants also give some information about the importance of this region of the tRNA structure in aminoacylation by tyrosine tRNA synthetase. Mutant tRNAs A1 and G82 and the double mutant A1G82 (Inokuchi *et al.*, 1974) only translate UAG as glutamine *in vivo*, while the other mutants give an ambiguous translation as glutamine and tyrosine. G82 cannot be reverted to give a tyrosine inserting suppressor so that the identity of the base adjacent to the CCA terminus is critical in recognition of tRNA by tyro-

TABLE III

Properties of the Su_3 Mischarging Mutants (and related mutants)

Suppressor	Temperature Sensitivity	Relative amount of tRNA synthesized	*In vivo* efficiency of suppressors (%)	UAG translated *in vivo* as:	Ratio Tyr/Gln
Su_3^+	tr	100	50-55	tyr	
Alp [(a)]	ts	40	42	gln	
A2P [(a)]	ts	25	10-15	tyr, gln	>90/10
U80	tr	80		tyr (possibly another amino acid)	
A81	ts	25	10-15	tyr, gln	>95/5
U81	tr	>90	15-20	tyr, gln	80/20
G82	tr	>90	15-20	gln	
A1 U81 (double mutant)	tr	>90		tyr, gln	70/30
A2 U80 (double mutant)	tr	>90	50	tyr	
A1 G82 (double mutant)	tr	40	49	gln	

(a) A1P and A2P have a second mutation (P) in the 5′ precursor segment which increases the rate of processing of these tRNAs. Mutants are designated by the substituting base numbered from the 5′ end (see Figure 2).

sine tRNA synthetase (Celis *et al.*, 1977).

A number of experimental results show that the anticodon loop of some, but not all, tRNAs is involved in specific synthetase interactions (for review see Goddard, 1977). Most suppressor tRNAs originate by mutations changing the tRNA anticodon so that, in some, synthetase recognition might be expected to be altered. An interesting example is su_7^+ where the $tRNA^{Trp}$ anticodon change from CCA to CUA, which converts the tRNA to an amber suppressor, also changes the tRNA synthetase recognition so that the tRNA is charged with glutamine (Yaniv *et al.*, 1974) as well as tryptophan (Celis *et al.*, 1976). Combined with the results from the su_3^+ mischarging mutants this shows that recognition by glutamine tRNA synthetase involves two defined regions of the tRNA; the anticodon and the acceptor stem.

Earlier results on the glycine missense suppressors (for review see Hill, 1975) showed that base substitutions in the second and third anticodon bases of all three $tRNA^{Gly}$ species greatly change the kinetics of aminoaceylation with glycine tRNA synthetase by both reducing the affinity of binding and decreasing the V_{max} of aminoacylation but without apparently changing the specificity of aminoacylation (Table II).

The synthetases are an unusually diverse set of enzymes in their structure and there appears to be no common way by which each recognizes its cognate tRNA (C. Bruton, Chapter 2, this volume). The isolation and characterisation of mischarging suppressor tRNAs provides a way of delineating those regions of different tRNA molecules important in these interactions.

References

Abelson, J.N., Gefter, M.L., Barnet, L., Landy, A., Russell, R.L. and Smith, J.D. (1970). *J. Mol. Biol.* **47**, 15.

Altman, S. (1976). *Nucleic Acids Res.* **3**, 441.

Altman, S., Brenner, S. and Smith, J.D. (1971). *J. Mol. Biol.* **56**, 195.

Altman, S., Bothwell, A.L.M. and Stark, B.C. (1974). *Brookhaven Symp. Biol.* **26**, 12.

Berquist, P.L., Burns, D.J.W. and Plinston, C.A. (1968). *Biochemistry* **7**, 1751.

Buckingham, R.H. (1976). *Nucleic Acids Res.* **3**, 965.

Buckingham, R.H. and Kurland, C.G. (1977). *Proc. Nat. Acad. Sci.* **74**, 5496.

Carbon, J., Squires, C. and Hill, C.W. (1969). *Cold Spring Harbor Symp. Quant. Biol.* **34**, 505.

Carbon, J. and Fleck, E.W. (1974). *J. Mol. Biol.* **85**, 371.

Cashmore, A.R. (1971). *Nature (London)* **230**, 236.

Celis, J.E., Coulondre, C. and Miller, J.H. (1976). *J. Mol. Biol.* **104**, 729.

Celis, J.E., Squire, M., Kaltoft, K. and Riisom, E. (1977). *Nucleic Acids Res.* **4**,

2799.
Colby, D.S., Schedl, P. and Guthrie, C. (1976). *Cell* **9**, 449.
Crick, F.H.C. (1966). *J. Mol. Biol.* **19**, 548.
Fleck, E.W. and Carbon, J. (1975). *J. Bacteriol.* **122**, 492.
Gallucci, E., Pacchetti, G. and Zangrossi, S. (1971). *Mol. Gen. Genet.* **106**, 362.
Gefter, M.L. and Russell, R.L. (1969). *J. Mol. Biol.* **39**, 145.
Ghysen, A. and Celis, J.E. (1974). *J. Mol. Biol.* **83**, 333.
Goddard, J.P. (1977). *Prog. Biophys. molec. Biol.* **32**, 233.
Goodman, H.M., Abelson, J., Landy, A., Brenner, S. and Smith, J.D. (1968). *Nature (London)* **217**, 1019.
Guest, J.R. and Yanofsky, C. (1965). *J. Biol. Chem.* **240**, 679.
Hawthorne, D.C. and Leupold, U. (1974). *Curr. Top. Microbiol. Immunol.* **64**, 1.
Hill, C.W. (1975). *Cell* **6**, 419.
Hill, C.W., Foulds, J., Soll, L. and Berg, P. (1969). *J. Mol. Biol.* **39**, 563.
Hill, C.W., Combriato, G. and Dolph, W. (1974). *J. Bacteriol.* **117**, 351.
Hirsh, D. (1971). *J. Mol. Biol.* **58**, 439.
Hirsh, D. and Gold, L. (1971). *J. Mol. Biol.* **58**, 459.
Högenauer, G. (1974). *FEBS Lett.* **39**, 310.
Hooper, M.L., Russell, R.L. and Smith, J.D. (1972). *FEBS Lett.* **22**, 149.
Ikemura, T. and Ozeki, H. (1977). *J. Mol. Biol.* **117**, 419.
Inokuchi, H., Celis, J.E. and Smith, J.D. (1974). *J. Mol. Biol.* **85**, 187.
Jank, P., Shindo-Okada, N., Nishimura, S. and Gross, H.J. (1977). *Nucleic Acids Res.* **4**, 1999.
Kao, S. and McClain, W.H. (1977). *J. Biol. Chem.* **252**, 8254.
Lagerkvist, U. (1978). *Proc. Nat. Acad. Sci.* **75**, 1759.
McClain, W.H. (1977). *Acc. Chem Res.* **10**, 418.
McClain, W.H. and Seidman, J.G. (1975). *Nature (London)* **257**, 106.
McCloskey, J.A. and Nishimura, S. (1977). *Acc. Chem. Res.* **10**, 403.
Miller, J.P., Hussoin, Z. and Schweizer, M.P. (1976). *Nucleic Acids Res.* **3**, 1185.
Mitra, S.K., Lustig, F., Åkesson, B., Lagerkvist, U. and Strid, L. (1977). *J. Biol. Chem.* **252**, 471.
Murao, K., Tanabe, T., Ishii, F., Namiki, M. and Nishimura, S. (1972). *Biochem. Biophys. Res. Comm.* **47**, 1332.
Murgola, E.J. and Yanofsky, C. (1974). *J. Bacteriol.* **117**, 444.
Nagata, T. and Horiuchi, T. (1973). *Mol. Gen. Genet.* **123**, 77.
Nishimura, S. (1972). *Prog. Nucleic Acid Res. Mol. Biol.* **12**, 49.
Oeschger, M.P. and Woods, S.L. (1976). *Cell* **7**, 205.
Ozeki, H., Inokuchi, H., Aono, H. (1969). *Japan J. Genet.* **44**, 406.
Pollock, T.J., Tessman, I. and Tessman, E.S. (1978). *Nature (London)* **274**, 34.
Pope, W.T., Brown, A. and Reeves, R.H. (1978). *Nucleic Acids Res.* **5**, 1041.
Reeves, R.H. and Roth, J.R. (1975). *J. Bacteriol.* **124**, 332.
Reeves, R.H. and Roth, J.R. (1971). *J. Mol. Biol.* **56**, 523.
Roberts, J.W. and Carbon, J. (1974). *Nature (London)* **250**, 412.
Roberts, J.W. and Carbon, J. (1975). *J. Biol. Chem.* **250**, 5530.
Russell, R.L., Abelson, J.N., Landy, A., Gefter, M.L., Brenner, S. and Smith, J.D. (1970). *J. Mol. Biol.* **47**, 1.
Ryasaty, S. and Atkins, J. (1968). *J. Mol. Biol.* **34**, 541.

Seidman, J.G., Comer, M.M. and McClain, W.H. (1974). *J. Mol. Biol.* **90**, 677.

Shimura, Y., Aono, H., Ozeki, H., Sarabhai, A., Lamfrom, H. and Abelson, J. (1972). *FEBS Lett.* **22**, 144.

Smith, J.D. (1972). *Ann. Rev. Genet.* **6**, 235.

Smith, J.D. (1976). *Prog. Nucleic Acid Res. Mol. Biol.* **16**, 25.

Smith, J.D., Barnett, L., Brenner, S. and Russell, R.L. (1971). *J. Mol. Biol.* **54**, 1.

Smith, J.D. and Celis, J.E. (1973). *Nature New Biol.* **243**, 66.

Soll, L. and Berg, P. (1969). *Nature (London)* **223**, 1340.

Steege, D.A. and Söll, D.G. (1978). Biological Regulation and Development Vol. 1 (ed. R. Goldberger)., Plenum Press.

Squires, C. and Carbon, J. (1971). *Nature New Biol.* **233**, 274.

Velten, J., Fukuda, K. and Abelson, J. (1976). *Gene* **1**, 93.

Yaniv, M., Folk, W.R., Berg, P. and Soli, L. (1974). *J. Mol. Biol.* **86**, 245.

Yanofsky, C., Ito, J. and Horn, V. (1968). *Cold Spring Harbor Symp. Quant. Biol.* **31**, 151.

THE USE OF SUPPRESSED NONSENSE MUTATIONS TO GENERATE ALTERED REPRESSORS

J.H. MILLER, C. COULONDRE, U. SCHMEISSNER, A. SCHMITZ, M. HOFER AND D. GALAS

Department of Molecular Biology, University of Geneva, Switzerland.

Introduction

We have previously described a set of correlated nonsense mutations in the *lac I* gene of *E. coli* (the structural gene for the *lac* repressor) in which the wild-type codon is known for 85 out of 90 mutational sites (Miller *et al.*, 1977; 1978a; Coulondre and Miller, 1977a). This offers a unique opportunity to generate a family of altered *lac* repressor molecules by the use of nonsense suppressors. At each position in the protein specified by an amber codon, five different residues can be inserted by characterized suppressors; serine, glutamine, tyrosine, leucine, and lysine (see Gorini, 1970). In addition, lysine, glutamine and tyrosine can be inserted in response to the ochre codon, and tryptophan at UGA sites. Since ochre codons can be converted to amber codons, this permits the exchange of 3-5 residues at each of the 90 positions corresponding to a nonsense site. (The *lac* repressor consists of 360 amino acids; Beyreuther *et al.*, 1973, 1975 and Farabaugh, 1978). Because the wild-type amino acid is known in virtually every case, this produces a set of over 300 altered repressors with known sequence changes. Figure 1 shows the residues in the repressor which can be replaced by this technique.

This type of study offers several advantages over the analysis of missense mutations. First of all, specific amino acid replacements can be identified without having to sequence every altered protein. Moreover, at each site under study a hierarchy of amino acid exchanges can be compared. Very rare substitutions are subject to investigation. For instance, some replacements produced by this method would require two or even three base changes if achieved by direct mutagenesis. Also, selection for different phenotypes is not required, since each amber mutation is detected by screening only for the i^- character

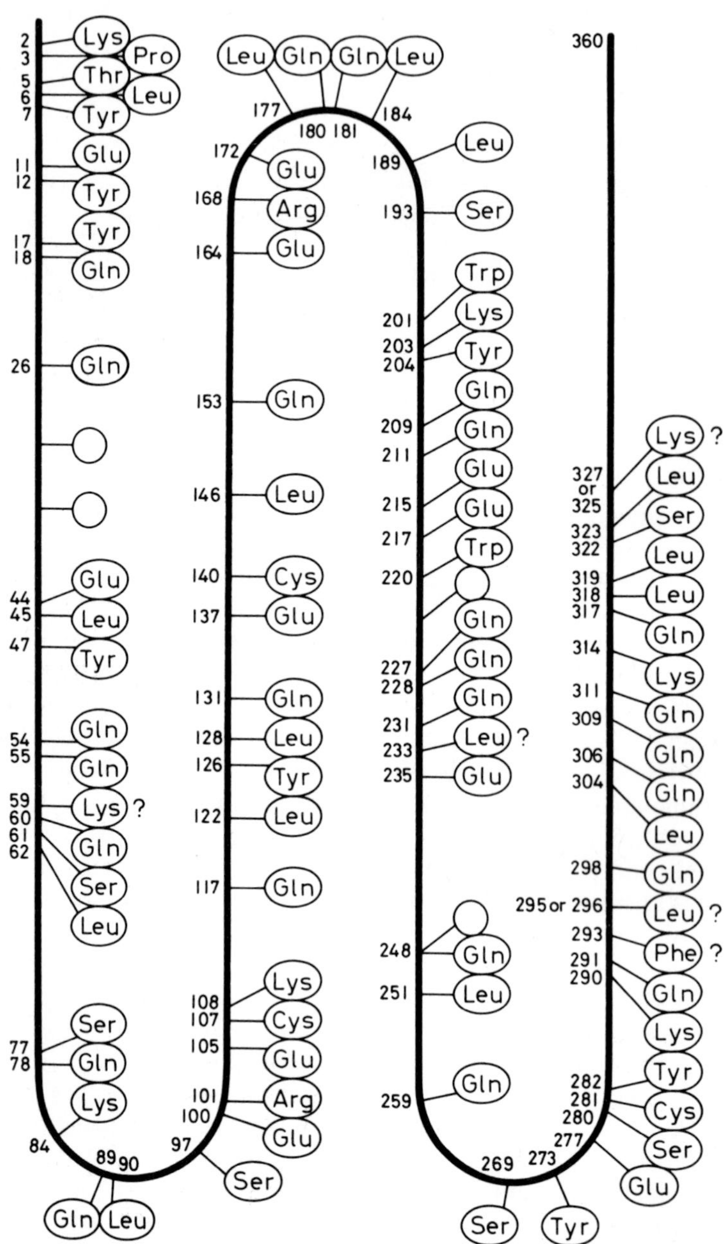

Fig. 1 Substitution sites in the *lac* repressor. The 90 substitution sites in the *lac* repressor which result from nonsense mutations in the *lacI* gene are shown. In each case the wild-type amino acid is given.

of the resulting polypeptide fragment in an su⁻ strain. Therefore, replacements leading to no measurable change in repressor activities are also amenable to analysis.

Effect of Amino Acid Replacements in the *Lac* Repressor

Figure 2 summarizes in schematic form the results of quantitative assays for several activities performed on many of the suppressed derivatives (see Miller *et al.*, 1978a for a more detailed analysis of these results). In the amino-terminal portion of the protein, this work complements the peptide analysis of repressors resulting from missense substitutions (see Reviews by Beyreuther, 1978, and Weber and Geisler, 1978), showing for instance that whereas tyrosines 17 and 47 and, to a lesser extent tyrosine 7, are essential for operator binding, tyrosine 12 can be replaced by a variety of amino acids without noticeably damaging repression *in vivo.* Substitution of tyrosine for proline at position 3 results in a tight-binding repressor which would be extremely difficult to obtain otherwise, since 3 base changes would be needed to convert the CCA codon to UAU or UAC (Steege, 1977; Schmitz *et al.*, 1978). However, glutamine or lysine at this position does not increase operator binding relative to wild-type.

Beyreuther (1978) and coworkers have shown that exchanges at positions 74 and 75 lead to i^s repressors, which can no longer be induced *in vivo.* Suppressed nonsense mutations allow us to now identify certain replacements at positions 78, 84, 97, 193, 220, 248, 273, and probably 293 and 296, among others, as also resulting in i^s repressors of various strengths. Temperature sensitive proteins, temperature sensitive mutants with reverse induction curves (i^r) *in vivo,* and mutants failing to synthesize tetramers also appear readily in this analysis.

Conclusion

In general, the effects of the substitutions are in good agreement with the findings of Perutz and coworkers (Perutz and Lehmann, 1968) for the haemoglobin system concerning the nature of the residue being replaced. Polar residues (43-44 of the 53 tested in the *lac* repressor) can usually be replaced by both polar and non-polar amino aicds, whereas non-polar residues are generally sensitive to substitution by polar residues (only 13 of 32 non-polar residues can be replaced by at least one of the three polar residues tested). Overall, many amino acid substitutions in the repressor are effectively neutral. Of the 302 replacements scored, 58% do not result in a detectable change in repression *in vivo* or in the response to inducers.

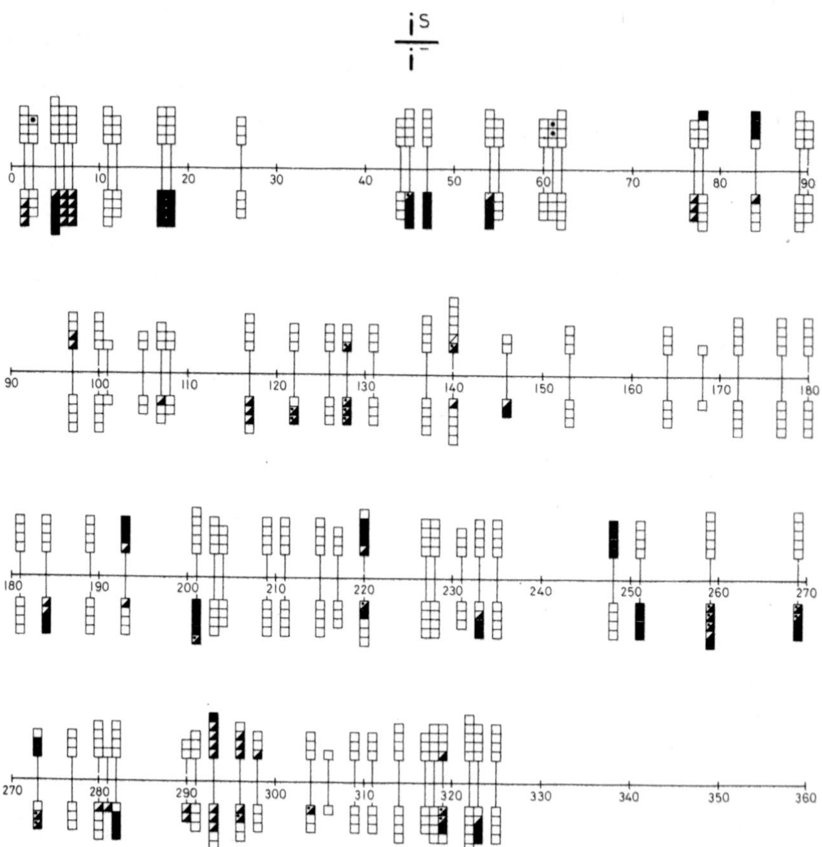

Fig. 2 Effects of amino acid replacements in the *lac* repressor. At each position corresponding to a known residue between 1 and 6 substitutions have been scored. These are indicated by a set of boxes. The i^s phenotype (failure to be induced *in vivo*) is considered above the horizontal line, which corresponds to the length of the protein. Most i^s repressors have reduced IPTG binding affinities. Replacements which do not result in an i^s repressor are represented by open boxes above the line. Amino acid exchanges producing i^s repressors are depicted by filled in boxes. Boxes which are half-filled in represent weak i^s effects, and those with a dot in the unfilled portion indicate temperature sensitive i^s proteins. For example, 5 different substitutions at residue 220 result in 3 i^s proteins, one weak i^s protein, and one protein which is not i^s. The asterisks (*) at positions 3 and 61 indicate partial i^s character due to repressors which bind operator more tightly than wild-type. The i^- phenotype is considered in a similar manner below the line. (These are repressors which can no longer bind to operator, either due to a direct destruction of the operator binding site, or else to an indirect effect on the general structure of the protein). Filled in boxes represent molecules which cannot repress *in vivo.* Partially filled in boxes indicate weak i^- effects, and a dot depicts temperature-sensitivity. Open boxes indicate replacements which do not affect the ability to bind operator.

By directing the substitution of specific residues in the repressor, nonsense suppression can greatly facilitate spectroscopic studies. Changes in fluorescence after tryptophan substitution (Sommer *et al.*, 1976) have been monitored, as has the [^{19}F] NMR spectrum of 3-fluorotyrosine containing repressor after specific tyrosine substitution (Lu *et al.*, 1976; P. Lu, unpublished results).

The set of correlated nonsense sites has allowed studies of the specificity of mutagens with a degree of precision never before achieved (Coulondre and Miller, 1977b; Coulondre *et al.*, 1978). Also, the mutations at known sites have been used to mark deletion intervals (Schmeissner *et al.*, 1977a,b) on the physical map, thus facilitating the placement of other types of mutations as a prelude to their sequence analysis (Farabaugh *et al.*, 1978; Calos *et al.*, 1978a,b).

Acknowledgements

This work was supported by a grant from the Swiss National Fund (F.N. 3.179.77) to J. H. M.

References

Beyreuther, K. (1978). The Operon. Cold Spring Harbor Laboratory (in press).

Beyreuther, K., Adler, K., Geisler, N. and Klemm, A. (1973). *Proc. Nat. Acad. Sci.* **70**, 3576.

Beyreuther, K., Adler, K., Fanning, E., Murray, C., Klemm, A. and Geisler, N. (1975). *Eur. J. Biochem.* **59**, 491.

Calos, M., Galas, D. and Miller, J.H. (1978a). *J. Mol. Biol. (submitted).*

Calos, M., Johnsrud, L. and Miller, J.H. (1978b). *Cell* **13**, 411.

Coulondre, C. and Miller, J.H. (1977a). *J. Mol. Biol.* **117**, 525.

Coulondre, C. and Miller, J.H. (1977b). *J. Mol. Biol.* **117**, 577.

Coulondre, C., Miller, J.H., Farabaugh, P.J. and Gilbert, W. (1978). *Nature, (London)* (in press).

Farabaugh, P.J. (1977). *Nature, (London)* (in press).

Farabaugh, P.J., Schmeissner, U. Hofer, M. and Miller, J.H. (1978). *J. Mol. Biol.* (submitted).

Lu, P., Jarema, M., Mosser, K. and Daniel, W.E., Jr. (1976). *Proc. Nat. Acad. Sci.* **73**, 3471

Miller, J.H., Ganem, D., Lu, P. and Schmitz, A. (1977). *J. Mol. Biol.* **109**, 275.

Miller, J.H., Coulondre, C., Hofer, M., Schmeissner, U., Sommer, H., Schmitz, A. and Lu, P. (1978). *J. Mol. Biol.* (submitted).

Perutz, M.F. and Lehmann, H. (1968). *Nature, (London)* **219**, 202

Schmeissner, U., Ganem, D. and Miller, J.H. (1977a). *J. Mol. Biol.* **109**, 303.

Schmeissner, U., Ganem, D. and Miller, J.H. (1977b). *J. Mol. Biol.* **117**, 572.

Schmitz, A., Coulondre, C. and Miller, J.H. (1978). *J. Mol. Biol.* (in press).
Sommer, H., Lu, P. and Miller, J.H. (1976). *J. Biol. Chem.* **251**, 3774.
Steege, D.A. (1977). *Proc. Nat. Acad. Sci.* **74**, 4163
Weber, K. and Geisler, N. (1978). The Operon. Cold Spring Harbor Laboratory (in press).

USE OF THE ISO-1-CYTOCHROME *C* SYSTEM FOR INVESTIGATING NONSENSE MUTANTS AND SUPPRESSORS IN YEAST

F. SHERMAN, B. ONO† AND J.W. STEWART

Department of Radiation Biology and Biophysics,
University of Rochester School of Medicine and Dentistry,
New York, U.S.A.

Introduction

The UAA, UAG and UGA chain terminating codons in the yeast *Saccharomyces cerevisiae* can be translated in mutants containing alterations of components of the protein synthesizing system. The major type of mutants that is best understood and that is the main topic of this report is comprised of the nonsense suppressors that determine altered tRNAs. Also described herein is the ψ^+ non-Mendelian genetic determinant that modifies the expression of certain of the tRNA suppressors, the UAA suppressors, and that by itself causes a low degree of suppression of at least some UAA mutations. Although not discussed in this chapter, mention should be made of the so-called 'omnipotent suppressors' which have been suggested to have altered ribosomes that result in misreading (Hawthorne and Leupold, 1974; Gerlach, 1976; Smirnov *et al.,* 1976).

So far the only method in yeast for unequivocally determining which nonsense codon is recognized by a suppressor and which amino acid is inserted relies on the use of *cycl* mutants that contain defined alteration in the gene determining the primary structure of iso-1-cytochrome *c.* In the first part of this article we have summarized the results that were used to establish the nucleotide sequence of *cycl* mutants containing nonsense codons. In conjunction with the defined *cycl* mutants, we are attempting to uncover and characterize all nonsense suppressors in yeast. The iso-1-cytochrome *c* system also provides a convenient way for determining the efficiency of action of the suppressors. In the second part of this chapter we have summarized

† Present address: Department of Pharmaceutical Technology, Faculty of Pharmaceutical Sciences, Okayama University, Okayama, 700 Japan.

all of the UAA and UAG suppressors so far characterized with the iso-1-cytochrome *c* system.

Identification of cycl Mutants Having UAA and UAG Nonsense Codon

Mutationally altered forms of iso-1-cytochrome *c* have been used to deduce the nucleotide sequence of numerous *cycl* mutants. Over 200 independently derived *cycl* mutants have been isolated with a spectroscopic scanning procedure (Sherman, 1964), a benzidine staining procedure (Sherman *et al.*, 1968) and an enrichment procedure using chlorolactate medium (Sherman *et al.*, 1974). The majority of the mutants, *cycl-17* through *cycl-211*, were obtained with use of chlorolactate medium that is the most effective means for obtaining *cycl* mutants which either lack iso-1-cytochrome *c* or contain nonfunctional iso-1-cytochrome *c*. The chlorolactate enrichment procedure relies on the fact that mutants containing approximately 5% of the normal level or normal activity of total cytochrome *c* are defective in the utilization of lactate, although they can still utilize other non-fermentable substrates such as glycerol or ethanol for growth. These partially deficient mutants, presumably because of the defect in lactate utilization, are resistant to the toxic action of the chlorolactate analogue. Thus chlorolactate medium, which contains a non-fermentable carbon source such as glycerol or ethanol as well as chlorolactate, is particularly useful for obtaining *cycl* mutants, since they lack iso-1-cytochrome *c* but still retain the normal low-level amount of iso-2-cytochrome *c*. The *cycl* mutants are the major class of cytochrome *c* deficient mutants arising on chlorolactate mutant, although one also observes substantial numbers of *cyc2* and *cyc3* mutants, which are partially deficient in both iso-1-cytochrome *c* and iso-2-cytochrome *c*.

Revertants of the *cycl* mutant can be selected by growth on lactate medium. The two major types of *cycl* revertants include intragenic revertants, which almost always have the normal level of iso-1-cytochrome *c*, and revertants that have increased levels of iso-2-cytochrome *c* but that are still deficient in iso-1-cytochrome *c*. For unknown reasons, revertants containing nonsense suppressors that act on UAA and UAG *cycl* mutants rarely arise on lactate medium and those that do, appear to have low efficiencies of action (Sherman *et al.*, 1973). Intragenic revertants are conveniently distinguished from extragenic revertants by a battery of tests that include examinations of cytochrome *c* content, growth on lactate medium and genetic analysis (Sherman *et al.*, 1974).

It has been possible to deduce the nucleotide changes in many of the *cycl* mutants from altered iso-1-cytochromes *c* in intragenic revertants and from the assumption that mutations occur primarily by single base-pair changes. The types of *cyc1* mutants that have been characterized from mutationally altered iso-1-cytochrome *c* include nonsense mutants (see below), frameshift mutants (Sherman and Stewart 1973; Stewart and Sherman 1974), missense mutants (Putterman *et al.,* 1974; Schweingruber *et al.,* 1978a, b) and initiator mutants (Stewart *et al.,* 1971).

Cyc1 mutants were inferred to have either UAA or UAG mutations if all or almost all of the intragenic revertants contained iso-1-cytochromes *c* with single replacements of several of the amino acids whose codons differed from either UAA or UAG by single bases. Replacements by any of the six amino acids, glutamine, leucine, tyrosine, lysine, glutamic acid and serine can arise from UAA mutants by single base-pair substitutions; all of these replacements plus tryptophan can arise from UAG mutants. Examples of the distribution of amino acid replacements are shown in Figure 1 for the UAA mutant *cyc1-9* and for the UAG mutant *cyc1-179.* As shown in Figure 1, the revertants chosen for study arose spontaneously or were induced by a variety of different mutagens that would be expected to produce different types of lesions. Excluding low frequencies of revertants that arose by multiple base-pair changes, all of the *cyc1-9* and *cyc1-179* revertants contained single amino acid replacements that occured by single base pair substitutions of, respectively, UAA and UAG. Furthermore, the sequencing of the end of the gene by amino acid changes in frameshift mutants (Stewart and Sherman, 1974) and by sequencing of the region of the mRNA encompassing residue position 2 (Szostak *et al.,* 1977), indicates that the *cyc1-9* and *cyc1-179* mutants arose from the wild-type by single base-pair changes. Thus the formation by a single-base pair change and finding all six replacements that have codons differing from UAA by one base-pair change clearly establishes that the *cyc1-9* contains an UAA codon corresponding to position 2. Similarly, the formation and reversion of the *cyc1-179* mutants established that it contains a UAG codon corresponding to position 9; only lysine residues were lacking among the sample of 52 *cyc1-179* revertants but this residue is found at the UAG site in normal iso-1-chtochrome *c.*

Listed in Table I are the other *cyc1* mutants at other sites that gave rise to revertants having patterns of amino acid replacements consistent with UAA or UAG mutants. The lack of tryptophan replacements in UAA revertants and the tryptophan replacements in UAG revertants were not the sole criteria for differentiating the two types of nonsense mutants. It is possible that tryptophan at certain sites results in nonfunctional iso-1-cytochrome *c,* or that the required mutation at the specific A·T base-pair occurs infrequently. Indeed, the rarity of certain of the acceptable replacements is revealed by the lack of serine residues at the UAA site in 55 revertants from the *cyc1-72* mutant (Sherman and Stewart, 1974) even though serine is compatible with function (Liebman *et al.*, 1975a), and by the lack of lysine residue at the UAG site in revertants from the *cyc1-179* mutant (Stewart and Sherman 1972; Sherman and Stewart, 1974) even though lysine is found at this position in normal iso-1-cytochrome *c.* Supporting evidence for the UAA and UAG assignment comes from the consistency of the pattern of suppression and to a lesser degree from the induced rates of reversion by 4-nitroquinoline-1-oxide (NQO). As will be discussed below, certain suppressors appear to act exclusively on UAA mutants while others appear to act exclusively on UAG mutants. In addition, the two types of nonsense mutants could be distinguished by the higher rate of NQO-induced reversion of UAG mutants, since NQO selectively mutates G·C base pairs (Prakash and Sherman, 1974). The frequencies of intragenic reversion induced by NQO and ethyl methanesulfonate, presented in Table II, clearly indicated that NQO reverted all of the UAG mutants at a higher frequency than the UAA mutants whereas such a distinction is not revealed from the frequencies of reversion induced with ethyl methanesulfonate. Thus, we believe the complete agreement by the three independent criteria summarized in Table I, i.e., the lack of tryptophan replacements in UAA revertants, the specificity of certain nonsense suppressors, and the higher NQO-induced reversion of UAG mutants, establish the assignments of the UAG and UAA *cyc1* mutants listed in the Table.

Nonsense Suppressors

Mutant genes that suppress two or more nutritional markers have been the subject of many studies since Hawthorne and Mortimer

Fig. 1 (See opposite). The amino acid sequences and the corresponding mRNA sequences relating to the formation and reversion of the *cyc1-9* and *cyc1-179* nonsense mutants. The *cyc1-9* and *cyc1-179* revertants contained replacements of all amino acids having codons that differ by one base from respectively, UAA and UAG codons, except for the lack of *cyc1-179* revertants containing lysine, which is found at the corresponding site in normal iso-1-cytochrome *c.* Not shown are the two *cyc1-9* revertants and the four *cyc1-179* revertants that arose by multiple base changes. The revertants were induced by the following mutagens: UV, ultra-violet light; X-rays; α-part; polonium-210 α-particles; EMS, ethyl methane-sulphonate, DES, diethyl sulphate; MMS, methyl methanesulphonate; NIL, 1-nitrosoimidazolidone-2; NTG, N-methyl-N′-nitro-N-nitrosoguanidine; and NA, nitrous acid (from Stewart *et al.*, 1972; Stewart and Sherman, 1972; Sherman and Stewart, 1973; 1974; Stewart and Sherman, in preparation).

```
                    2                                     9
(Met)Thr-Glu-Phe-Lys-Ala-Gly-Ser-Ala-Lys-Lys-
 AUG ACU GAA UUC AAG GCC GGU UCU GCU AAG AAA
          |                                  |
          | cyc1-9                           | cyc1-179
          ↓                                  ↓
         UAA                                UAG
        ochre                              amber
```

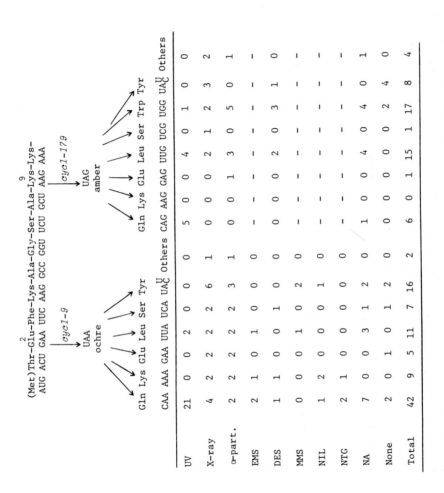

	Gln	Lys	Glu	Leu	Ser	Tyr		Gln	Lys	Glu	Leu	Ser	Trp	Tyr	
	CAA	AAA	GAA	UUA	UCA	UA^U_C	Others	CAG	AAG	GAG	UUG	UCG	UGG	UA^U_C	Others
UV	21	0	0	2	0	0	0	5	0	0	4	0	1	0	0
X-ray	4	2	2	2	2	6	1	0	0	0	2	1	2	3	2
α-part.	2	2	2	2	2	3	1	0	0	1	3	0	5	0	1
EMS	2	1	0	1	0	0	0	-	-	-	-	-	-	-	-
DES	1	1	0	0	1	0	0	0	0	0	2	0	3	1	0
MMS	0	0	0	1	0	2	0	-	-	-	-	-	-	-	-
NIL	1	2	0	0	0	1	0	-	-	-	-	-	-	-	-
NTG	2	1	0	0	0	0	0	-	-	-	-	-	-	-	-
NA	7	0	0	3	1	2	0	1	0	0	4	0	4	0	1
None	2	0	1	0	1	2	0	0	0	0	0	0	2	4	0
Total	42	9	5	11	7	16	2	6	0	1	15	1	17	8	4

TABLE I

Identification of ochre (UAA) and amber (UAG) mutants by amino acid replacements and the corroboration of these assignments by suppressibility and NQO reversion

		Residue	Amino acid replacements							Suppressed by:			
										SUP2-o	*SUP7*-a	NQO	References for amino
Mutant	Lesion	position	Trp	Gln	Leu	Tyr	Lys	Glu	Ser	etc.	etc.	revert.	acid replacements
cyc1-2	UAA	21	0	+	+	+	+	0	+	+	0	0	Stewart *et al.*, unpublished; see Sherman and Stewart, 1974
cyc1-9	UAA	2	0	+	+	+	+	+	+	±	0	0	Stewart *et al.*, 1972
cyc1-72	UAA	66	0	+	+	+	+	+	0	+	0	0	Stewart *et al.*, unpublished; see Sherman and Stewart 1974
cyc1-140	UAA	93	0	0	+	+	+	+	0	+	0	0	Stewart *et al.*, unpublished
cyc1-45	UAA	(102)	..	..	..	..	..	..	..	+	0	0	
cyc1-94	UAA	(79)	..	..	..	..	..	..	..	+	0	0	
cyc1-156	UAA	(26)	..	..	..	..	..	..	..	+	0	0	
cyc1-76	UAG	71	+	+	+	+	+	0	+	0	+	+	Stewart and Sherman 1973
cyc1-179	UAG	9	+	+	+	+	0†	+	+	0	+	+	Stewart and Sherman 1972; Sherman and Stewart 1973
cyc1-84	UAG	64	+	0	+	+	0	0	0	0	+	+	Schweingruber *et al.*, 1978
cyc1-177	UAG	(32)	..	..	..	..	..	..	..	0	+	+	

† A residue of lysine is at the amber site in the normal iso-1-cytochrome *c*.

The residue positions indicated in parenthesis were estimated by deletion mapping (Sherman *et al.*, 1975). The results of suppressibility is from Sherman *et al.*, (1974) and the results of NQO reversion is from Prakash and Sherman (1974).

TABLE II

Reversion frequencies (revertants per 10^7 survivors) of cyc1 *strains after treatment with 4-nitroquinoline-1-oxide (NQO) and ethyl methanesulphonate (EMS)*

Mutant	Nonsense codon	Reversion freq. NQO	Reversion freq. EMS
cyc1-2	UAA	1	16
cyc1-9	UAA	0	8
cyc1-45	UAA	0	13
cyc1-72	UAA	0	17
cyc1-94	UAA	0	19
cyc1-140	UAA	5	9
cyc1-156	UAA	6	14
cyc1-76	UAG	178	22
cyc1-84	UAG	35	29
cyc1-177	UAG	79	15
cyc1-179	UAG	92	8

From Prakash and Sherman (1974)

(1963) first reported their occurrence in yeast. The suppressors are conveniently classified by the pattern of suppression of numerous nutritional markers. Whether or not a particular marker is effectively suppressed by a particular suppressor depends upon the nonsense codon in the mutant, the functioning of the gene product with the inserted amino acid and the required level of the gene product that is determined by the efficiency of suppression. In addition, the acquisition and characterization of suppressors are dependent on genetic factors that modify their expression. The iso-1-cytochrome *c* system has been used to systematically examine a wide range of suppressors that were isolated by reversion of various nutritional markers in various genetic backgrounds.

Numerous chromosomal genes lower the expression of suppressors (the anti-suppressors) (Hawthorne and Leupold, 1974; Liebman *et al.*, 1976; McCready and Cox, 1973; Ono *et al.*, 1979); others increase the level of expression (the allo- or ana-suppressors) (Hawthorne and Leupold, 1974; Cox, 1977; Ono *et al.*, 1979). Of considerable importance is the non-Mendelian genetic determinant, ψ, that was first shown by Cox (1965) to modify the expression of certain nonsense suppressors. Normal laboratory strains of *S. cerevisiae* have been observed to be either ψ^+ or ψ^-. The efficiency of suppression of certain UAA suppressors is higher in ψ^+ strains compared to ψ^- strains as indicated by the patterns of suppression of nutritional markers, by the inviability and retardation of growth of certain suppressed ψ^+ strains (Cox, 1971), and by the levels of iso-1-cytochrome *c* in suppressed *cyc1* mutants (Liebman *et al.*, 1975a, b). The ψ^+ determinant in the absence of any known suppressors also weakly suppresses some UAA markers (Young and Cox, 1975; Liebman *et al.*, 1975b). This determinant also increases the efficiency of suppression of certain frameshift suppressors (Culbertson *et al.*, 1977), although it has not yet been found to modify the efficiency of UAG suppressors. It has been suggested that the ψ determinant does not reside on a nuclear chromosome because the meiotic progeny from $\psi^+ \times \psi^-$ crosses are predominantly ψ^+, indicating that the ψ^+ determinant is distributed to all meiotic cells (Cox, 1965). Similarly, the haploid progeny of heterokaryon $\psi^+ \times \psi^-$ crosses can have interchanges of nuclear markers and of the ψ^+ or ψ^- state, suggesting that the ψ^+ determinant is located outside of the nucleus (Fink and Conde, 1977). Analyses with other types of cytoplasmic mutants indicate that the ψ determinant is not related to mitochondrial DNA (Young and Cox 1972) nor to the double-stranded RNA which determines the 'killer' phenotype (Cox, unpublished, cited in McCready *et al.*, 1977). Unlike chromosomal genes, high rates of $\psi^+ \rightarrow \psi^-$ mutations can be induced by incubating cells in hypertonic media (Singh, Helms and Sherman, in preparation). At present, the molecular identity of the ψ determinant is unknown, although it is believed to be a self-replicating cytoplasmic element that directly or indirectly affects a component of the translational machinery.

Numerous suppressors have been isolated in ψ^+ and ψ^- strains by reversion of two or more nutritional markers and many of these have been characterized with the iso-1-cytochrome *c* system. In some studies only relatively few suppressors having certain patterns of suppression were examined with the iso-1-cytochrome *c* system; in other comprehensive studies, extensive numbers were isolated, systematically assigned to their respective loci and the amino acid replacements determined. The comprehensive studies that have been used with the iso-1-cytochrome *c* system include the high-efficiency UAA suppressors which were isolated in a ψ^- strain and which were found to insert tyrosine (Gilmore *et al.*, 1971), the high-efficiency UAG suppressors which were found to insert tyrosine (Liebman *et al.*, 1976), the moderately efficient UAA suppressors, which were isolated in a ψ^+ strain and were found to insert serine (Ono *et al.*, 1979), and the low efficiency UAA suppressors which were isolated in a ψ^+ and which were found to insert leucine (Ono *et al.*, in preparation).

In addition, the iso-1-cytochrome *c* was used to characterize single representative from two other groups; one of the lower efficiency UAG suppressors, *SUP52,* was shown to insert leucine (Liebman *et. al.,* 1977) and one of the recessive-lethal UAG suppressors, *SUP-RL1,* was shown to insert serine (Brandriss *et al.,* 1976). These groups of nonsense suppressors characterized with the *cyc1* system, summarized in Table III, are described in detail below.

Tyrosine-inserting UAA Suppressors

Gilmore (1967) reported one of the first systematic investigations attempting to determine the number of loci that gave rise to suppressors with certain patterns of suppression of nutritional markers. Mutations at seven distinct loci, *SUP2* through *SUP8,* were uncovered after considering revertants that efficiently suppressed the set of five markers *trp5-48, arg4-17, his-5-2, lys1-1* and *ade2-1* in ψ^- strains. The numbers of each of these independently-derived suppressors, referred to as class I, set 1 suppressors, are listed in Table IV. Further studies by Hawthorne and Mortimer (1968) revealed a related eighth locus, *SUP11;* the *SUP11* suppressors denoted as class II suppressor by Hawthorne and Mortimer (1968), were similar to the class I, set 1, suppressors described by Gilmore (1967) except that the *SUP11* suppressors were slightly less effective in suppressing the *ade2-1* and *ade5, 7-63* markers. Subsequently, Ono *et al.,* (1979) could assign all of the 49 highly efficient UAA suppressors isolated in a ψ^+ strain to these eight loci (Table IV). Furthermore, Strömnaes and Mortimer (personal communication) have analyzed hundreds of suppressors which act simultaneously on the five UAA markers defining the class I, set 1 suppressors and each of these mapped at one of the eight loci. Thus, there appears to be no more than eight genes that can give rise to these suppressors classified as class I, set 1 by Gilmore (1967) or as class I and class II by Hawthorne and Mortimer (1968). Consistent with the genetic evidence for eight genes, is the finding of eight *EcoRI* fragments of different size that hybridize to tyrosine tRNA (Olson *et al.,* 1977).

Each of the eight suppressors acted efficiently on the defined *cyc1-2* mutant, which contains a UAA codon corresponding to amino acid residue position 21 in iso-1-cytochrome *c.* The *cyc1-2* strains containing any of the eight suppressors produced 5% to 10% of the normal amount of iso-1-cytochrome *c* and all eight iso-1-cytochromes *c* contained a residue of tyrosine at the position corresponding to the site of the UAA codon (Gilmore *et al.,* 1971). Several of these eight suppressors were unable to produce detectable suppression of the

TABLE III

The major groups UAA and UAG suppressors examined with the cyc1 system

Types	Loci	No. of Loci	Insertion	Reference
UAA suppressors				
Recessive lethals	None	—	—	
High efficiency	*SUP2* etc.	Eight	Tyrosine	Gilmore *et al.*, (1971)
Low efficiency	*SUP16*, *SUP17*	Two	Serine	Ono *et al.*, (1979)
Very low efficiency	*SUP26* etc.	Six	Leucine	Ono *et al.* (in preparation)
	Others	Unknown	Unknown	
UAG suppressors				
Recessive lethal	*SUP-RL1*	Unknown	Serine	Brandriss *et al.*, (1976)
	Others	Unknown	Unknown	
High efficiency	*SUP2* etc.	Eight	Tyrosine	Liebman *et al.*, (1976)
Low efficiency	*SUP52*	Unknown	Leucine	Liebman *et al.*, (1977)
	Others	Unknown	Unknown	

TABLE IV

Numbers of independent mutations of each of the loci producing tyrosine-inserting suppressors, serine-inserting UAA suppressors, and leucine-inserting UAA suppressors

Tyrosine-inserters				Serine-insertors		Leucine-insertors	
	UAA						
Locus	ψ^-	ψ^+	UAG	Locus	No.	Locus	No.
SUP2	6	0	9	*SUP16*	51	*SUP26*	88
SUP3	5	2	7	*SUP17*	44	*SUP27*	16
SUP4	3	0	5			*SUP28*	7
SUP5	2	2	6			*SUP29*	24
SUP6	2	1	1			*SUP32*	10
SUP7	3	0	2			*SUP33*	10
SUP8	1	9	4				
SUP11	0	35	7				

References: Tyrosine-inserting UAA suppressors isolated in a ψ^- strain, Gilmore (1967); Tyrosine-inserting UAA suppressors isolated in a ψ^+ strain, Ono *et al.*, (1979); Tyrosine-inserting UAG suppressors, Liebman *et al.*, (1976); Serine-inserting UAA suppressors, Ono *et al.*, (1979); Leucine-inserting UAA suppressors, Ono *et al.*, (in preparation).

UAG mutant *cyc1-179,* although tyrosine was an acceptable replacement in intragenic revertants from this mutant. These results suggested that the gene product of the eight suppressors is tyrosine tRNA and that each of the genes could be mutated to produce altered tRNAs which suppressed UAA but not UAG codons.

Tyrosine-inserting UAG Suppressors

A comprehensive examination of UAG suppresors was undertaken by Liebman *et al.*, (1976), involving the isolation of 1088 suppressors that acted on two or more of the UAG nutritional markers *met8-1, aro7-1, trp1-1, ade3-26* and *ilv1-1.* All of these suppressors were examined for their ability to suppress the *cyc1-179* mutant which contains a UAG codon corresponding to position 9 in iso-1-cytochrome *c,* and only 43 of them caused production of more than 50% of the normal amount of iso-1-cytochrome *c* in the *cyc1-179* strains. Forty-one of these 43 UAG suppressors were analyzed genetically and shown to be allelic to one or another of the eight suppressors which cause the insertion of tyrosine at UAA codons (Table IV). Suppressors from six of the eight loci were tested and found to cause insertion of tyrosine at the site of the UAG codon in the *cyc1-179* mutant. Thus highly efficient UAA and UAG suppressors causing tyrosine insertion can arise by mutation of any of the same eight genes. Because many amino acid replacements in *cyc1-179* revertants can lead to functional iso-1-cytochrome *c*, and because serine-inserting and leucine-inserting, as well as tyrosine-inserting suppressors act on the markers used in the selection procedure, it is noteworthy that of the 1088 suppressors isolated from wild-type genes, only tyrosine inserters acted efficiently on the *cyc1-179* mutant. Apparently the tyrosine-inserting UAG suppressors, just like the tyrosine-inserting UAA suppressors, are the most efficient suppressors compatible with survival in haploid strains.

Direct evidence that the gene products of the tyrosine-inserting UAG and UAA suppressors are tyrosine tRNAs with altered anticodons was first provided by the observation that the tyrosine-inserting UAG suppressor, *SUP5*-a, contained an altered anticodon CψA similar to the bacterial amber suppressor, and that this same anticodon was observed whether the *SUP5*-a suppressor was derived from the wild-type gene *sup5+* or from the UAA suppressor *SUP5*-o (Piper *et al.*, 1976). Furthermore, recent sequence analysis of the cloned *SUP4*-o gene established that the triplet coding for the anticodon of this tyrosine-inserting UAA suppressor was TTA instead of the normal GTA, thus indicating the unmodified anticodon in the tRNA is UAA (Goodman *et al.*, 1977).

Serine-inserting UAA Suppressors

The initial searches for UAA suppressors that led to the discovery of the eight tyrosine-inserting suppressors described above were carried out with ψ^- strains. In ψ^+ strains, the tyrosine-inserting UAA suppressors cause lethality or retarded growth due to over-suppression, while less efficient suppressors that are responsive to the ψ^+ determinant are manifested. Ono *et al.*, (1979) undertook a comprehensive search for UAA suppressors in a ψ^+ strain that contained the UAA markers *trp5-48, can1-100, ura4-1, his5-2, lys1-1* and *leu2-1* as well as the *cyc1-72* marker which contains a UAA mutant codon corresponding to position 66 in iso-1-cytochrome *c*. Over 200 revertants that suppressed three or more UAA markers were isolated and grouped into classes on the basis of their patterns of suppression of the UAA markers. Genetic analysis of representative numbers indicated that the classes of revertants resulted from the interaction of suppressors, which had characteristic efficiencies and patterns of suppression, and of genetic modifiers, which affected these suppressors. All but two of the 218 suppressors could be assigned to two major classes, the suppressors with high efficiencies and the suppressors with low efficiencies. In spite of the fact that a ψ^+ strain was used, 40% of the suppressors, constituting the highly efficient suppressors, were allelic to one or another of the known tyrosine-inserting suppressors. However, 35 of the 49 highly efficient suppressors that were tested belonged to the *SUP11* locus (Table IV), suggesting that the *SUP11* suppressors may be less inhibitory to ψ^+ strains than the other tyrosine-inserting suppressors. The suppressors with lower efficiencies, constituting 59% of the total number of revertants, were easily distinguished from the tyrosine-inserting suppressors by their suppression of the UAA marker *his4-1176* and by their lack of suppression of the UAA marker *ura4-1* (Table V).

TABLE V

Differentiation of the various classes of UAA suppressors by the suppressibility of discriminating markers

	Tyrosine High efficiency *SUP11* etc.	Serine Moderate efficiency *SUP16, SUP17*	Leucine Low efficiency *SUP26* etc.
leu2-1	+	+	+
his5-2	+	+	−
can1-100	+	−	−
ura4-1	+	−	+
his4-1176	−	+	−
cyc1-72†	60	30	10

† Approximate percent of the normal amount of iso-1-cytochrome *c* in ψ^+ strains.

Genetic analysis of 95 of the lower efficiency suppressors led to the placement of 51 and 44 suppressors to, respectively, the *SUP16* and *SUP17* loci (Table IV). The *SUP16* locus, which is on the left arm of chromosome XVI, was shown to be identical to the *SUQ5* locus described by Liebman *et al.*, (1975a) and probably to the *SUP15* locus described by Hawthorne and Mortimer (1968).

Sequence analysis of iso-1-cytochrome *c* from *cyc1-72 SUP16* and *cyc1-72 SUP17* strains demonstrated that both of these suppressors cause the insertion of serine at the UAA site. The insertion of serine by *SUP16* and by *SUQ5* (Liebman *et al.*, 1975a) corroborated the genetic evidence that they are indentical. The demonstration that serine is inserted at the UAA site by *SUQ5* in the ψ^- as well as the ψ^+ strain establishes that the suppressor and not ψ^+ is the origin of the serine insertion.

Leucine-inserting UAA Suppressors

Among the UAA suppressors isolated from a ψ^+ strain, Ono *et al.*, (1979) uncovered 2 out of 218 suppressors that could be distinguished from the tyrosine-inserting suppressors and the serine-inserting suppressors. These exceptional suppressors had a very low efficiency of action on the *cyc1-72* allele and they suppressed the *ura4-1* marker, which was not suppressed by the serine-inserting suppressors, and they did not suppress the *lys1-1* marker, which was suppressed by the tyrosine- and serine-inserting suppressors. Some of the discriminating markers used to differentiate the three major classes of UAA suppressors are listed in Table V. Both of the exceptional suppressors mapped at the same locus which was denoted *SUP26.*

In order to isolate additional suppressors having characteristics of the *SUP26* suppressors, Ono, Stewart and Sherman (in preparation) selected suppressors from the previously described strain (ψ^+ *cyc1-72 trp5-48 can1-100 ura4-1 his5-2 lys1-1 leu2-1*) on a medium which lacked uracil and leucine but which contained canavanine. The absence of uracil prevented the occurrence of serine-inserting suppressors because they do not suppress the *ura4-1* marker, and the presence of canavanine prevented the occurrence of the tyrosine-inserting suppressors because they suppressor the *can1-100* marker. A total of 155 revertants that were analyzed genetically could be assigned to the following six unlinked loci with a distribution shown in Table IV: *SUP26, SUP27, SUP28, SUP29, SUP32* and *SUP33.* The position of the *SUP29* suppressor on chromosome X and the pattern of suppression suggests that this suppressor may be allelic to the *SUP30* suppressor that was previously described by Hawthorne and Mortimer (1968). All six of these suppressors had a similar pattern of suppression of nutritional markers and all acted relatively inefficiently on the *cyc1-72* allele. Five of the suppressors, *SUP26, SUP27, SUP28, SUP29* and *SUP32* were shown to cause insertion of leucine in iso-1-cytochrome *c* at the position corresponding to the site of the UAA codon in the *cyc1-72* mutant. While iso-1-cytochrome *c* from the remaining suppressor, *SUP33*, has not yet been examined, the similarity in the pattern of suppression suggests that it also may cause insertion of leucine at UAA sites.

Leucine-inserting UAG Suppressor

In the comprehensive search and characterization of 1088 UAG suppressors, Liebman *et al.*, (1976) uncovered 76 suppressors that were less efficient in suppressing the *cyc1-179* marker although they still suppressed the same nutritional markers, *met8-1, aro7-1, trp1-1, ilv1-1* and *ade3-26,* as the highly efficient tyrosine-inserting suppressors. Some of these suppressors with lower efficiencies were found to produce higher levels of iso-1-cytochrome *c* when coupled with the UAG allele *cyc1-76.* One of these suppressors, *SUP52,* caused the production of 15 to 20% of the normal level of iso-1-cytochrome *c* when in combination with the *cyc1-76* mutant. The suppressed iso-1-cytochrome *c* contained a residue of leucine at the position corresponding to the site of the UAG codon (Liebman *et al.*, 1977). The map position of *SUP52,* a few map units from the centromere on the left arm of chromosome X, suggests that it may be identical to the *SUP51* suppressor reported by Hawthorne and Mortimer (1968), although there may be a slight difference in their pattern of suppression (Liebman *et al.*, 1977). Genetic analysis indicated that at least one other locus could give rise to suppressors phenotypically similar to the *SUP52* suppressor. The number of different loci and the kinds of amino acids inserted by the low-efficiency UAG suppressors are currently under investigation.

Serine-inserting UAG Suppressor

A search for different classes of UAG suppressors that may insert different amino acids was undertaken with diploid strains of yeast (Brandriss *et al.*, 1975). Most of the suppressors isolated in a diploid strain homozygous for the UAG markers *trp-1* and *aro7-1* could not be maintained in haploid strains but were reasonably stable in the heterozygous condition. Genetic analysis established that these UAG suppressors were lethal in the hemizygous condition, suggesting that the suppressors are determining essential tRNAs coded for by only a single gene in the haploid genome. One of the recessive lethal suppressors, *SUP-RL1,* which mapped on the right arm of chromosome III, was shown to cause the insertion of serine in iso-1-cytochrome *c* at the position that correspond to the site of the UAG codon in the mutant *cycl-179* (Brandriss *et al.*, 1976). This suppressor was shown to produce approximately 50% of the normal amount of iso-1-cytochrome *c* in disomic strains heterozygous for the suppressor. This level of suppression is nearly equal to the approximately 75% level observed with the tyrosine-inserting UAG suppressors. There is evidence from a cell-free synthesizing system that the *SUP-RL1* suppressor determines a mutated tRNA which is capable of translating a Qβ synthetase mRNA containing an UAG mutation

(Gesteland *et al.*, 1976).

Discussion

In this chapter we have described the UAA and UAG suppressors that have been characterized with the *cyc1* mutants having defined UAA and UAG mutations in the iso-1-cytochrome *c* structural gene. The major classes of suppressors are summarized in Table III and each of the suppressors is listed in Table VI.

TABLE VI

The UAA and UAG suppressors examined with the cyc1 *system*

	Amino acid replacement			
	UAA	UAG	Efficiency	Map position
SUP2	Tyrosine	Tyrosine	High	Chromo. IV R
SUP3	Tyrosine	(-)	High	Chromo. XV L
SUP4	Tyrosine	Tyrosine	High	Chromo. X R
SUP5	Tyrosine	Tyrosine	High	Frag. 8
SUP6	Tyrosine	Tyrosine	High	Chromo. VI R
SUP7	Tyrosine	Tyrosine	High	Chromo. X L
SUP8	Tyrosine	(-)	High	Chromo. XIII R
SUP11	Tyrosine	Tyrosine	High	Chromo. VI R
SUP16	Serine		Low	Chromo. XVI R
SUP17	Serine		Low	Chromo. IX L
SUP26	Leucine		Very low	Chromo. XII R
SUP27	Leucine		Very low	Chromo. IV R
SUP28	Leucine		Very low	Unknown
SUP29	Leucine		Very low	Chromo. X L
SUP32	Leucine		Very low	Unknown
SUP33	(-)		Very low	Unknown
SUP52		Leucine	Low	Chromo. X L
SUP-RL1		Serine	High	Chromo. III R

The map position of the tyrosine-inserting suppressors are from Hawthorne and Mortimer (1968) and Mortimer and Hawthorne (1973). The remaining map positions are from the references cited in Table III.

The suppressors were isolated by the co-reversion of two or more nutritional markers. One of the most striking conclusions is that the groups of suppressors having the same pattern of suppression of nutritional markers and having similar efficiencies of action all cause the insertion of the same amino acid. While the *SUP52* suppressor, a leucine-inserting UAG suppressor, could not be distinguished from the tyrosine-inserting UAG suppressors on the basis of suppression of nutritional markers, the *SUP52* suppressor had a distinctly lower efficiency of action on the *cyc1-179* allele (Liebman *et al.*, 1977). However there are slight variations in the efficiencies of suppression of certain of the suppressors within a single group; among the tyrosine-inserting UAA suppressors, *SUP11* appears to act with a slightly lower efficiency, and among the serine-inserting UAA suppressors, *SUP17* appears slightly less efficient than the *SUP16* suppressor (Ono *et al.*, 1979).

The maximum number of loci giving rise to suppressors having a particular pattern of suppression should be revealed after analyzing extensive numbers of revertants. There appear to be only eight loci that can yield tyrosine-inserting suppressors and each of these loci can yield both UAA suppressors and UAG suppressors (Table IV). Similarly, there appear to be only two serine-inserting UAA suppressors and only six leucine-inserting UAA suppressors. These suppressors presumably arise by single base-pair changes and the suppressors within a group appear to arise from redundant genes that determine the same tRNA at approximately the same level of expression.

The suppressibility of markers and the assignment of the suppressors to the various groups are determined by a multitude of genetic factors that include the nature of the mutant codon, the acceptability of the amino acid that is inserted and the efficiency of suppression which in turn is determined by the position of the nonsense codon in the marker and by intrinsic properties of the suppressor that can be modified by other genetic factors. A suppressor would not be manifested it its efficiency of expression was too high, resulting in inviability or retarded growth, or if the tRNA is a unique coding species or if it carries out another vital function possibly unrelated to translation. In addition, suppressors with very low efficiencies of expression would not be expected to be detected. The range of suppressors can be extended by the use of diploid strains for uncovering recessive-lethal suppressors such as *SUP-RL1* and by the use of genetic factors that increase or decrease the expression of suppressors. Thus the suppressors that have been uncovered probably are those that arose by single base-pair changes and those that are expressed within the range compatible with detection and cell viability. So far only UAA and UAG suppressors causing replacements of tyrosine, serine and leucine have been uncovered. It is unclear why suppressors causing insertions of glutamine, lysine, glutamic acid, and for UAG suppressors, tryptophan, have not appeared; suppressors arising by single base-pair changes and inserting these amino acids should be expected to decode UAA and UAG codons.

Suppressors acting solely on UAG codons are observed in both yeast and bacteria and these arise by single base-pair changes consistent with the 'wobble' hypothesis. A yeast tyrosine-inserting UAG suppressor was shown to contain a CψA anticodon similar to the anticodon of a tyrosine tRNA amber suppressor from *Escherichia coli* (Piper *et al.*, 1976). Also the leucine-inserting UAG suppressor, *SUP52* and serine-inserting UAG suppressor, *SUP–RL1*, could have arisen from tRNAs that decode, respectively, UUG leucine codons and UCG serine codons. However, in contrast to UAG suppressors, the ochre suppressors acting on both UAG and UAA codons are observed only in bacteria and the suppressors acting solely on UAA codons are observed only in yeast. While the nature of the changes in tRNAs of yeast UAA suppressors are unknown, the alterations are believed to reside at the anticodon at least in the tyrosine-inserting suppressors. This was first demonstrated by showing that a tyrosine-inserting UAG suppressor, *SUP5*-a contained the same altered anticodon CψA whether it was derived by mutation from the wild-gene or by mutation from the UAA suppressor *SUP5*-o. More direct demonstration involved the determination of the DNA sequence of the cloned *SUP4*-o gene, which was found to have a TTA sequence at the position corresponding to the GTA sequence at the anticodon of the normal tRNA, indicating that the unmodified anticodon in the suppressor tRNA is UUA (Goodman *et al.*, 1977). The UAA-specific suppressors could be attributed to anticodons having the sequence IUA, as first suggested by Bock (1967), or to SUA, as first suggested by Gilmore *et al.*, (1971), where S could be a 2-thio-5-carboxymethyluridine derivative (Yoshida *et al.*, 1971) or possibly a 5-carboxymethyluridine derivative (Kuntzel *et al.*, 1975). Assuming the properties of known tRNAs apply to suppressors, then suppressors with IUA anticodons would decode the UAA nonsense codon and the two UAU and UAC tyrosine codons while suppressors with SUA anticodons would decode only the UAA nonsense codon. Thus suppressors with the SUA anticodon, referred to as 'sepia' suppressors (Gilmore *et al.*, 1971) best explains the specificity of the tyrosine-inserting UAA suppressors. However, it is not yet clear whether the serine-inserting and leucine-inserting suppressors are sepia suppressors or even if they have alterations at the anticodon. If suppressors with the anticodon IUA, referred to as 'topaz' suppressors, arose from serine tRNAs with IGA anticodons, the serine-inserting UAA suppressors would be expected to insert serine at tyrosine codons as well as at UAA codons. Since none of the serine-inserting UAA suppressors significantly retard growth as would be expected with general replacements of serine for tyrosine, it would have to be assumed that insertion of serine at tyrosine sites occurs at a low efficiency. While yeast contain serine tRNAs having the IGA anticodon (Zachau *et al.*, 1966), there are no known serine tRNAs (Nishimura, 1974) that can give rise to SUA anticodons by single base-pair changes. A more direct analysis is required to deduce the alteration responsible for the serine-inserting UAA suppressors and to determine if they are indeed serine tRNAs that normally decode UCA triplets.

The leucine-inserting UAA suppressors can be derived by single base changes at the anticodon of only tRNAs that decode solely UUA triplets and that would be predicted to have SAA anticodons, where S is a 2-thio-5-carboxymethyluridine or 5-carboxymethyluridine methyl ester. While such a leucine tRNA has not been clearly identified in yeast, Piper and Wasserstein (1977) have reported an uncharacterized leucine tRNA that may decode either or both of the leucine codons UUA and UUG. It still should not be excluded that the leucine-inserting UAA suppressors or other yeast suppressors arose by mutation external to the anticodon. Of the 18 suppressor loci listed in Table VI all but two are at distinct and separate sites. *SUP29,* a leucine-inserting UAA suppressor and *SUP52,* a leucine-inserting UAG suppressor, showed no recombination in 46 tetrads from a *SUP29* × *SUP52* cross (Ono, Stewart and Sherman, in preparation). This close linkage can be interpreted in either of two ways: there are two adjacent genes that determine two different leucine tRNAs, and one of the genes can give rise to UAA suppressors and the other to UAG suppressors; alternatively the *SUP29* and *SUP52* suppressors were derived from the same gene but not in a way that can be accounted for by changes at the anticodon.

Acknowledgements

This investigation was supported in part by Public Health Service Research Grant GM12702 from the National Institutes of Health and in part by the U.S. Department of Energy at the University of Rochester Department of Radiation Biology and Biophysics. This paper has been designated Report No. UR-3490-1417.

References

Brandriss, M.C., Söll, L. and Botstein, D. (1975). *Genetics* **79**, 551.

Brandriss, M.C., Stewart, J.W., Sherman, F. and Botstein, D. (1976). *J. Mol. Biol.* **102**, 467.

Bock, R.M. (1967). *J. Theoret. Biol.* **16**, 438.

Capecchi, M.R., Hughes, S.H. and Wahl, G.M. (1975). *Cell* **6**, 269.

Cox, B.S. (1965). *Heredity* **20**, 505.

Cox, B.S. (1971). *Heredity* **26**, 211.

Cox, B.S. (1977). *Genet. Res.* **30**, 187.

Culbertson, M.R., Charnas, L., Johnson, T. and Fink, G.R. (1977). *Genetics*

86, 745.

Fink, G.R. and Conde, J. (1977) *In:* International Cell Biology, 1976-1977. (1st Intern. Cong. Cell Biol. Boston, Mass.) (eds. Brinkley, B.R. and Porter, K.R 414, Rockefeller Univ. Press, New York.

Gerlach, W.L. (1976). *Molec. gen. Genet.* **144**, 213.

Gesteland, R.F., Wolfner, M., Grisafi, P., Fink, G., Botstein, D. and Roth, J.R. (1976). *Cell* **7**, 381.

Gilmore, R.A. (1967). *Genetics* **56**, 641.

Gilmore, R.A., Stewart, J.W. and Sherman, F. (1971). *J. Mol. Biol.* **61**, 157.

Goodman, H.M., Olson, M.V. and Hall, B.D. (1977). *Proc. Nat. Acad. Sci.* **74**, 5453.

Hawthorne, D.C., and Leupold, U. (1974). *Current Topics Microbiol. Immunol.* **64**, 1.

Hawthorne, D.C. and Mortimer, R.K. (1963). *Genetics* **48**, 617.

Hawthorne, D.C. and Mortimer, R.K. (1968). *Genetics* **60**, 735.

Kuntzel, B., Weissenbach, J., Wolff, R.E., Tumaitis-Kennedy, T.D., Lane, B.G. and Dirheimer, G. (1975). *Biochimie* **57**, 61.

Liebman, S.W., Sherman, F. and Stewart, J.W. (1976). *Genetics* **82**, 251.

Liebman, S.W., Stewart, J.W. and Sherman, F. (1975a). *J. Mol. Biol.* **94**, 595.

Liebman, S.W., Stewart, J.W., and Sherman, F. (1975b). *Genetics* **80**, 53.

Liebman, S.W., Stewart, J.W., Parker, J.H. and Sherman, F. (1977). *J. Mol. Biol.* **109**, 13.

McCready, S.J. and Cox, B.S. (1973). *Molec. gen. Genet.* **124**, 305.

McCready, S.J., Cox, B.S. and McLaughlin, C.S. (1977). *Mol. gen. Genet.* **150**, 265.

Mortimer, R.K. and Hawthorne, D.C. (1973). *Genetics* **74**, 33.

Nishimura, S. (1974). *In:* Biochemistry of Nucleic Acids (K. Burton, ed.), 289. Butterworth, London.

Olson, M.V., Montgomery, D.L., Hopper, A.K., Page, G.S., Horodyski, F. and Hall, B.D. (1977). *Nature* **267**, 639.

Ono, B., Stewart, J.W. and Sherman, F. (1979). *J. Mol. Biol.* (in press).

Piper, P.W. and Wasserstein, M. (1977). *Eur. J. Biochem.* **80**, 103.

Piper, P.W., Wasserstein, M., Engbaek, F., Kaltoft, K., Celis, J.E., Zeuthen, J., Liebman, S.W. and Sherman, F. (1976). *Nature (London)* **262**, 757.

Prakash, L. and Sherman, F. (1974). *Genetics* **77**, 245.

Putterman, G.J., Margoliash, E. and Sherman, F. (1974). *J. Biol. Chem.* **249**, 4006.

Schweingruber, M.E., Stewart, J.W. and Sherman, F. (1978). *J. Mol. Biol.* **118**, 481.

Schweingruber, M.E., Stewart, J.W. and Sherman, F. (1978). *J. Biol. Chem.* (in press).

Smirnov, V.N., Surguchov, A.P., Fominykch, E.S., Lizlova, L.V., Saprygina, T.V. and Inge-Vechtomov, S.G. (1976). *FEBS Letters* **66**, 12.

Sherman, F. (1964). *Genetics* **49**, 39.

Sherman, F. and Stewart, J.W. (1973). *In:* The Biochemistry of Gene Expression in Higher Organisms (eds. Pollak, J.K. and Lee, J.W.), 56. Australian and New Zealand Book Co. PTY. LTD., Sydney, Australia.

Sherman, F. and Stewart, J.W. (1974). *Genetics* **78**, 97.
Sherman, F., Stewart, J.W., Parker, J.H., Inhaber, E., Shipman, N.A., Putterman, G.J., Gardisky, R.L. and Margoliash, E. (1968). *J. Biol. Chem.* **243**, 5446.
Sherman, F., Liebman, S.W., Stewart, J.W. and Jackson, M. (1973). *J. Mol. Biol.* **78**, 157.
Sherman, F., Stewart, J.W., Jackson, M., Gilmore, R.A. and Parker, J.H. (1974). *Genetics* **77**, 255.
Sherman, F., Jackson, M., Liebman, S.W., Schweingruber, A.M. and Stewart, J.W. (1975). *Genetics* **81**, 75.
Stewart, J.W. and Sherman, F. (1972). *J. Mol. Biol.* **68**, 429.
Stewart, J.W. and Sherman, F. (1973). *J. Mol. Biol.* **78**, 169.
Stewart, J.W. and Sherman, F. (1974). *In:* Molecular and Environmental Aspects of Mutagenesis (eds. Prakash, L., Sherman, F., Miller, M.W., Lawrence, C.W. and Taber, H.W.), 102, C.C. Thomas Pub. Inc., Springfield, Ill.
Stewart, J.W., Sherman, F., Shipman, N.A. and Jackson, M. (1971). *J. Biol. Chem.* **246**, 7429.
Stewart, J.W., Sherman, F., Jackson, M., Thomas, F.L.X. and Shipman, N. (1972). *J. Mol. Biol.* **68**, 83.
Szostak, J.W., Stiles, J.I., Bahl, C.P. and Wu, R. (1977). *Nature (London),* **265**, 61.
Yoshida, M., Takeishi, K. and Ukita, T. (1971). *Biochim. Biophys. Acta,* **228**, 153.
Young, C.S.H. and Cox, B.S. (1971). *Heredity* **26**, 413.
Young, C.S.H. and Cox, B.S. (1972). *Heredity* **28**, 189.
Zachau, H.G., Dütting, D. and Feldman, H. (1966). *Hoppe Seyler's Z. Physiol. Chem.* **347**, 212.

CHARACTERISATION AND SEQUENCE ANALYSIS OF NONSENSE SUPPRESSOR tRNAs FROM *SACCHAROMYCES CEREVISIAE*

P.W. PIPER

Imperial Cancer Research Fund Laboratories,
London, U.K.

Introduction

Nonsense suppressor mutations of *Saccharomyces cerevisiae* have been characterised from their ability to suppress auxotrophic markers that are known to correspond to either UAA, UAG or UGA nonsense mutations. Each of these mutations, with only a few exceptions, causes suppression at only one of the nonsense codons. UAA (ochre) and UAG (amber) suppressors have also been characterised from analyses on the cytochrome-1-*c* of strains bearing ochre or amber mutations in the gene for this protein, and it has been found that these ochre and amber suppressors each insert one of just three amino acids during translational suppression of UAA or UAG codons (Gilmore *et al.*, 1971; Sherman *et al.*, 1973; Liebman *et al.*, 1975; Liebman *et al.*, 1976; Brandriss *et al.*, 1976; Sherman, this volume). These amino acids are tyrosine, serine and leucine. The amino acid insertion specificities of yeast UGA suppressors (Hawthorne, 1976) are as yet unknown.

Evidence that nonsense suppression in yeast is mediated, as in prokaryotes, by altered transfer RNA molecules has come from two different lines of experimentation. On the one hand it has been demonstrated that tRNA isolated from suppressor-carrying yeast strains can permit the insertion of amino acids in response to nonsense codons in *in vitro* protein synthesising systems (Capecchi *et al.*, 1975; Gesteland *et al.*, 1976). On the other it has been shown that suppressor mutations can determine anticodon base changes in tRNAs from nucleotide sequencing studies of suppressor tRNAs and genes that encode them (Piper *et al.*, 1976; Goodman *et al.*, 1977; Piper, 1978). Both approaches have been used in establishing the molecular identity of the tRNA products of individual suppressor genes as well as the tRNA structural changes determined by the mutation that produces the nonsense suppressor function. It is also important to know the molecular nature of a suppressor tRNA if this molecule is to be purified for microinjection into the cells of higher eukaryotes with the intention of using it to detect nonsense mutant clones from these cells (Kaltoft *et al.,* 1976; Celis, 1977; Capecchi *et al.*, 1977; Celis *et al.*, this volume).

Separation of Yeast tRNAs on Two-dimensional Polyacrylamide Gels

The ability of *S. cerevisiae* ochre and amber suppressors to insert tyrosine, serine or leucine in response to the UAA or the UAG codon is an indication that these suppressors are probably derived by mutation of tyrosine, serine and leucine-accepting tRNAs of this organism. This is assuming that the suppressor mutations do not produce misacylation *in vivo* of the tRNAs whose structure they alter, a situation that does not hold for certain *Escherichia coli* tRNA mutations such as the SU_7^+ $tRNA^{Trp}$ (Yaniv *et al.*, 1974; Soll, 1974; Celis *et al.*, 1976) and glutamine-mischarging mutants derived from SU_3^+ $tRNA^{Tyr}$ (Celis *et al.*, 1973). To investigate whether yeast ochre and amber suppressors might be derived from the tyrosine, serine and leucine tRNAs we developed a fairly straightforward method for purifying in small amounts from a culture of a given yeast strain, those tyrosine, serine and leucine tRNAs that recognise a codon differing in not more than one base from the ochre or the amber codon. Such a procedure has to be relatively rapid so as to enable ready comparisons between different strains.

All nonsense codons commence with uridine, so that nonsense suppressing tRNAs will have adenosine as the final anticodon nucleoside. Most yeast tRNAs with adenosine as the 3′ terminal nucleoside of their anticodons have a hydrophobic base like 6-isopentenyladenine to the 3′ side of this adenosine, a modification that caused the tRNA to be retained strongly by benzoylated-DEAE cellulose (BD cellulose). Because of this those yeast tRNAs having a cognate codon differing from UAA or UAG in the second or third base position are amongst the late-eluting species on BD cellulose columns. The protocol we eventually adopted for purifying these molecules involved taking the total tRNA of each small ^{32}P-labelled yeast culture and isolating from it those species that were not eluted from a BD-cellulose column by a buffer containing 0.85*M* NaCl, 0.01*M* $MgCl_2$, 0.01*M* sodium acetate pH 5.0 (Piper *et al.*, 1976; Piper and Wasserstein, 1977; Piper, 1978). This sub-fraction was then resolved into individual tRNAs by two-dimensional polyacrylamide gel electrophoresis, the relative migration of the various species giving rise to the characteristic gel pattern illustrated in Figure 1. Amongst the tRNAs on these gels were identified the only tyrosine tRNA of yeast, as well as serine and leucine species which together can decode all the serine and leucine codons differing in a single base from either UAA or UAG (Piper and Wasserstein, 1977).

By employing this procedure for purifying these tRNAs in conjunction with RNA fingerprinting we are in a position to investigate whether certain specific amber or ochre suppressor mutations determine structural changes in these tyrosine, serine or leucine tRNAs. It is also possible to purify small, non-radioactive amounts of these tRNAs on the two-dimensional gels in order to either test each species individually for an ability to function as a suppressor tRNA in an *in vitro* protein synthesising system or else to label these RNAs *in vitro* for use as specific hybridisation probes in selecting *E. coli* clones containing the DNA of yeast tRNA genes inserted into plasmids (Beckmann *et al.*, 1977). A suppressor tRNA derived by mutation of a single base in a fraction of one of these species might be expected to comigrate on these gels with the wild-type tRNA. At least in the instances of amber suppressors derived from the *SUP5* and *SUPRL 1* loci the gels do not resolve the wild-type tRNA and the suppressor derived from it.

The yeast suppressors whose nature has been analysed in this was fall into the classes of tyrosine-inserting and serine-inserting suppressors listed in Table 1 and are derived from genetic loci that can be mutated to produce both the ochre and the amber suppressing phenotypes. The loci that give rise to tyrosine-inserting suppressors can each give either an ochre or an amber suppressor by a single step mutation (Liebman *et al.*, 1976), whereas those loci giving serine-inserting suppressors that are not recessive-lethal yield only ochre suppressors on single mutation, the amber suppressors from these loci being the result of double mutation (Sherman, 1978). The *SUPRL 1* locus, however, gives a recessive lethal serine-inserting amber suppressor by single mutation and an ochre suppressor by double mutation.

Characterisation of Tyrosine-inserting Suppressor tRNAs

The first positive characterisation of a yeast suppressor tRNA (Piper *et al.*, 1976) was performed on the amber suppressor *SUP5-a*, one of a class of eight unlinked highly efficinet suppressors that cause the insertion of tyrosine at UAG sites (Liebman *et al.*, 1976). A parallel series of eight allelic suppressors cause tyrosine to be inserted at UAA sites (Gilmore *et al.*, 1971). Olsen *et al.*, (1977)

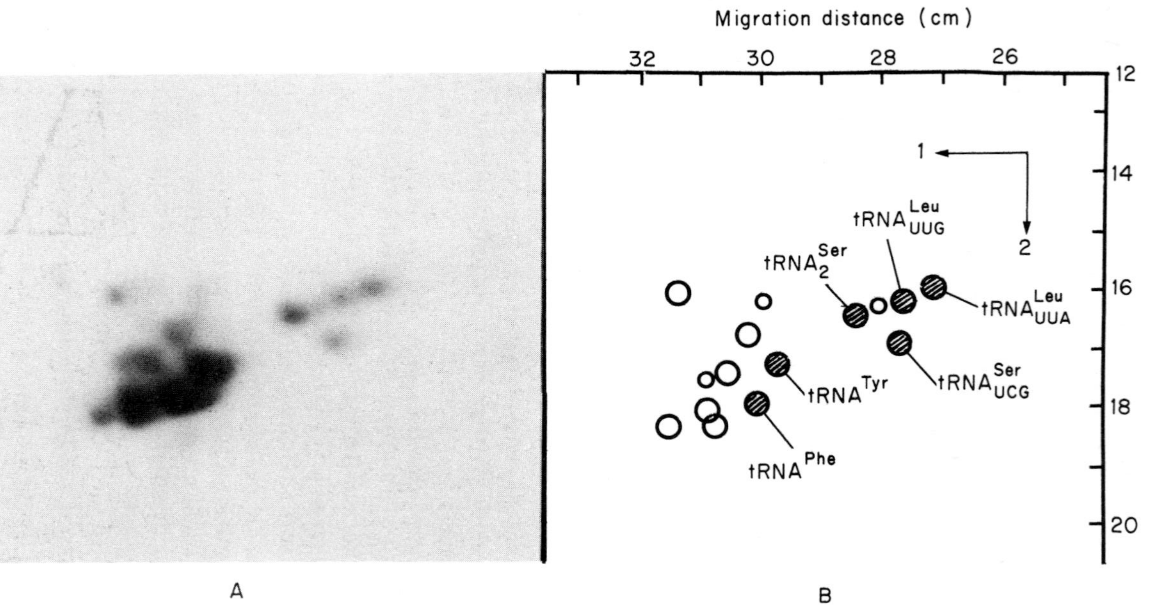

Fig. 1 Separation by two-dimensional polyacrylamide gel electrophoresis of the yeast tRNAs retained by BD-cellulose in 0.85*M* NaC1, 0.01*M* $MgC1_2$, 0.01*M* sodium acetate pH 5.0. (*A*) Autoradiogram of the gel; (*B*) a map of the individual major tRNA species. Open circles in (*B*) represent unidentified species of tRNA and shaded circles identified tRNAs. The scale in (*b*) indicates the distances of migration in the first dimension (1) 10% gel and in the second dimension (2) 20% gel. Each of the identified tRNAs except $tRNA^{Phe}$, has a cognate codon that is related to either UAA or UAG by a single base substitution.

TABLE I

Classes of amber and ochre suppressors of Saccharomyces cerevisiae *and the transfer RNAs from which they are derived*

Class and amino acid insertion specificity	Act on	Genetic loci that mutate[1] to these suppressors	tRNA species mutated[2] to give suppressor	References
tyrosine-inserting, highly efficient[3]	UAA or UAG	SUP2, SUP3, SUP4, SUP5, SUP6, SUP7, SUP8 and SUP11	$tRNA^{Tyr}$ (SUP4 and SUP5)	Piper *et al.*, 1976 Goodman *et al.*, 1977
serine-inserting moderately efficient	UAA or UAG	SUP16 (SUQ5), SUP17 as in Sherman, 1978	a $tRNA^{Ser}$ that co-purifies with minor $tRNA^{Ser}_{UCG}$	Waldron, Gesteland, and Piper, unpublished
serine-inserting, moderately efficient, and recessive-lethal	UAA or UAG	SUPRL1 (SUP6)	minor $tRNA^{Ser}_{UCG}$ (SUPRL1)	Piper, 1978
leucine-inserting, very low efficiency	UAA or UAG	SUP26 etc. as in Sherman, 1978	Unknown	------------------

[1] Alternative designations in brackets.
[2] The specific suppressor loci for which an identification of the tRNA giving rise to the suppressor has been performed are indicated in brackets.
[3] An efficient suppressor is defined as one that can appreciably suppress nonsense mutations in the structural gene for iso-1-cytochrome *c*.

recently have shown that yeast tRNA^{Tyr} hybridises to eight endonuclease EcoR_1 fragments of yeast DNA and have been able to correlate these fragments with individual tyrosine-inserting suppressor loci. The tyrosine-inserting suppressors of *S. cerevisiae* constitute a minor fraction of the total tRNA^{Tyr} of strains bearing these suppressors. The nature of the mutant gene product can be investigated by sequencing tRNA^{Tyr} associated with these strains.

When the tRNA^{Tyr} of *SUP5-a* and *sup*$^+$ strains was examined by fingerprint analysis (Piper *et al.*, 1976) a difference in the T_1 ribonuclease digests became apparent (Figure 2a, b, c.). A large oligonucleotide (market C in Figure 2b and c) was present in the digest

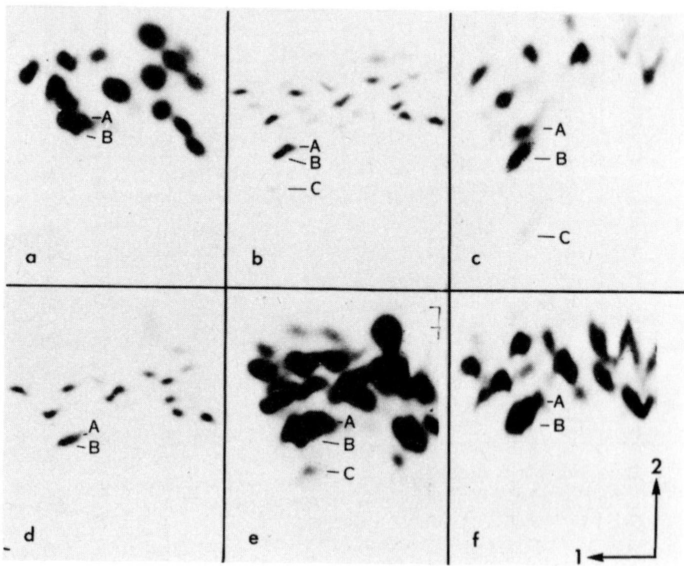

Fig. 2 T_1 ribonuclease digests of tyrosine tRNA from different suppressing and nonsuppressing strains (in brackets). a, *sup*$^+$ (SL210-3A); b and c, two different cultures of *SUP5-a* (L-133); d, *SUP5-o* (SL171-2C); e, *SUP5-o→a* (L-58) and f, *SUP5-a→a'* (L-133a). Oligonucleotide separation was by electrophoresis on cellulose acetate in the first dimension (1) and ascending chromatography on DEAE-cellulose thin layers in the second dimension (2) using a 3% dialysed homomixture (Piper *et al.*, 1976). A and B identify respectively the oligonucleotides pCUCUGp and ψAi^{6}AAψCUUGp C, ACUCψAi^{6}AAψCUUGp, is the anticodon fragment of the amber suppressor tRNA^{Tyr}, present only in fingerprints (b), (c) and (e). Other differences between the fingerprints are due to differences in the resolving power of homomixtures and fluctuations in the extent to which T_1 ribonuclease has cleaved the fragment Cm$^2_2$GCAAGp into Cm$^2_2$G>p and CAAGp.

of the *SUP5-a* tRNA^{Tyr} but was absent from the corresponding digest of the *sup*$^+$ tRNA^{Tyr}. Since this fragment is not one of the normal digestion products of yeast $tRNA^{Tyr}$, its nucleotide sequence was determined. It was degraded by pancreatic ribonuclease to $Ai^6 AA\psi p$, ACp, Gp, Cp, ψp and Up (Figure 3a). Also after reacting the U, G and ψ residues with a carbodimide blocking reagent and analysing the products of subsequent pancreatic ribonuclease digestion it was possible to show that it contained the sequences (C)UC and (C)UUG (not illustrated). The final structure was obtained from the sequences of fragments derived by partial U_2 ribonuclease digestion (Figure 3d)

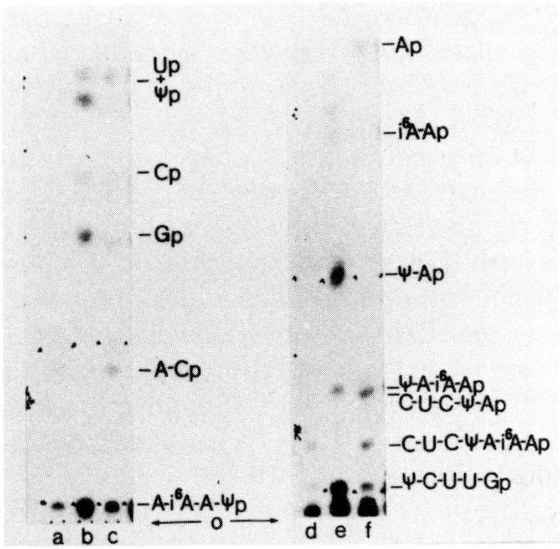

Fig 3 Pancreatic and U_2 ribonuclease digests of the anticodon oligonucleotides of the UAG suppressor $tRNA^{Tyr}$. a and c, Pancreatic ribonuclease digests of $ACUC\psi Ai^6 AA\psi CUUGp$, the anticodon loop fragments of the amber suppressor $tRNA^{Tyr}$, derived, respectively, from strains *SUP5-a* and *SUP5-0→*a. b, Marker digest of $\psi Ai^6 AA\psi CUUGp$. Separation was by electrophoresis, 1kV for 3 h, on DEAE paper at pH 3.5. d and f, U_2 ribonuclease digests of the anticodon loop fragment of amber suppressor $tRNA^{Tyr}$ derived from *SUP5-a* and *SUP5-o→a* respectively. e, Marker U_2 digest of $\psi Ai^6 AA\psi CUUGp$. Digestion was with 0.05 ml of U_2 ribonuclease (0.1 U ml^{-1}), 0.05 *M* sodium acetate, pH 4.5 for 4 h at 37 C, and separation was by electrophoresis, 1 kV for 6 h on DEAE paper in 7% formic acid.

which were consistent only with the nucleotide sequence: $ACUC\psi Ai^6AA\psi CUUGp$. From the structure of $tRNA^{Tyr}$, illustrated in Figure 5a, it is apparent that this oligonucleotide was originally derived from a $tRNA^{Tyr}$ with the anticodon CψA instead of the normal GψA. Such a molecule would be expected to function as a UAG suppressor, the anticodon CUA having been identified in amber suppressing tRNAs of *E. coli* (Goodman *et al.*, 1968; Yaniv *et al.*, 1974) phage T_4 (Comer *et al.*, 1975) and, as is described below, also in the recessive-lethal yeast *SUPRL1* suppressor.

The weak intensity of the mutant anticodon spot C in Figures 2b, c indicates that only 5-8% of the total $tRNA^{Tyr}$ of the *SUP5-a* strain existed in the form of the amber suppressor, a level consistent with this species being the product of one of the eight genes for tyrosine tRNA. As such it comprised no more than 0.15-0.35% of the total tRNA of the cell and was present at a level comparable to that of the suppressor tRNA in a diploid strain heterozygous for the *SUPRL1* amber suppressor (Piper, 1978).

The $tRNA^{Tyr}$ from other *SUP5* suppressor alleles has also been investigated. The allele *SUP5-o→a* was obtained (Liebman *et al.*, 1976) by mutation of the *SUP5-o* ochre suppressor in contrast to the *SUP5-a* allele which was obtained directly from the sup^+ wild type. However *SUP5-o→a* and *SUP5-a* are phenotypically indistinguishable and sequencing of their $tRNA^{Tyr}$ (Figures 2b, c, e and 3a, c, d, f.) indicated that they are identical. In the fingerprint of $tRNA^{Tyr}$ from a strain bearing a less efficient amber suppressor *SUP5-a→a′* which arose by intragenic mutation of *SUP5-a*, the anticodon fragment of the amber suppressor $tRNA^{Tyr}$ was undetectable (Figure 2f) and it is possible that the second site mutation at the *SUP5* locus caused reduced synthesis of suppressor tRNA in much the same way as certain mutations of the su_3^+ $tRNA^{Tyr}$ gene of *E. coli* impair tRNA maturation (Anderson and Smith, 1972).

Tyrosine-inserting ochre suppressors of yeast are also, at least in certain cases, forms of $tRNA^{Tyr}$ with an altered first anticodon base. Goodman *et al.*, (1977) have cloned the suppressor gene from a yeast strain bearing the *SUP4-o* ochre suppressor and analysed it by DNA sequencing. The sequence of this gene differs from that of the wild-type $tRNA^{Tyr}$ genes by virtue of a G·C → T·A transversion at the base pair that codes for the 'wobble' base of the $tRNA^{Tyr}$ anticodon. The nature of the 'wobble' base in yeast ochre suppressors has been the subject of some speculation since these suppressors act only on UAA and not on UAG codons, unlike bacterial ochre suppressors which recognise both UAA and UAG. There are not very many anticodon structures possible for the *S. cerevisiae* tRNAs that have acquired the ability to read UAA but not UAG by anticodon mutation but which do not produce misreading of the genetic code (Piper and Wasserstein, 1977). The available information would suggest that the 'wobble' nucleoside of a UAA-specific suppressor tRNA may be 5-carboxymethyluridine methyl ester (mem^5U), its 2-thiolated derivative (mcm^5S) or, although only in a mutant $tRNA^{Tyr}$, inosine. The ochre suppressor gene sequenced by Goodman *et al.*, (1977) must be transcribed into a $tRNA^{Tyr}$ with a uridine derivative (U*) at the 'wobble' position of the anticodon although it is possible that other tyrosine-inserting ochre suppressors possess inosine and arise from a G·C → A·T transition in the DNA. In a fingerprint analysis of the $tRNA^{Tyr}$ of a *SUP5-o* strain (Figure 2d) we were unable to detect the fragment of sequence: ACUU*ψ Ai^6 AAψCUUGp that would be expected to migrate on the fingerprints in Figure 2 close to the corresponding fragment (C) from the amber suppressor. Unfortunately T_1 ribonuclease cleaves after inosine and the limitations of our analysis systems have so far prevented a positive identification of inosine as the 'wobble' nucleoside of the *SUP5-o* suppressor that we investigated (Piper *et al.*, 1976).

Characterisation of Serine-inserting Suppressor tRNAs

Hawthorne and Leupold (1974) and Brandriss *et al.*, (1975) have described yeast amber suppressors derived in diploid-strains that could be maintained in the heterozygous state but not in haploid cells. One of these recessive-lethal suppressors, *SUPRL1-a* (alternatively designated *SUP61-a*), causes the insertion of serine at the UAG codon in a *cyc1* mutant (Brandriss *et al.*, 1976). This led to the suggestion that the *SUPRL1* suppressor might be derived from a UCG-decoding serine tRNA, since a single anticodon base change(CGA → CUA) should give such a molecule the ability to read UAG codons, and UCG is the only serine codon related to UAG by a single base change. Also purified total $tRNA^{Ser}_{UCG}$ extracted from a two-dimensional gel of *SUPRL1/sup*$^+$ strain tRNAs can cause *in vitro* translational readthrough of UAG codons (Gesteland *et al.*, 1976; Piper, 1978). It was further proposed by Brandriss *et al.*, (1976) that this mutant tRNA might be coded for by only a single gene in the haploid genome, and that its mutation to the suppressor causes lethality in haploid cells through a loss of the ability to translate the codon UCG. If this were the case, then *SUPRLI* would be analogous to certain recessive lethal suppressors isolated in merodiploids of *E. coli* and *Salmonella typhimurium* (Soll and Berg, 1969; Miller and Roth, 1971; Soll, 1974; Yaniv *et al.*, 1974). We have performed structural studies on the tRNA species present in a gel spot (Figure 1) that contains the UCG-reading tRNA of *S. cerevisiae* designed to test these proposals concerning the nature of the SUPRL1 suppressor (Piper, 1978).

The total tRNA present in gel spot *d* (Figure 1), loosely defined '$tRNA^{Ser}_{UCG}$', was isolated from haploid and diploid yeast strains lacking *SUPRL1-a* is heterogeneous in structure and exists as major and minor nucleotide sequences, the former being present in approximately three times the amount of the latter tRNA. These forms will be referred to as the *major* $tRNA^{Ser}_{UCG}$ and the *minor* $tRNA^{Ser}_{UCG}$ respectively. Although these sequences differ from each other at only a few nucleotide positions, they are not base modification variants of the same sequence and must therefore be encoded by more than one gene in the haploid genome. The complete nucleotide sequence of the *major* $tRNA^{Ser}_{UCG}$ was derived (Figure 4b) but it was only possible for us to deduce unequivocally the structure of the *minor* species in the vicinity of the anticodon since we were unable to separate this form from the *major* $tRNA^{Ser}_{UCG}$ and other sequence differences shown in Figure 5b are therefore tentative. Although the codon recognition of these tRNAs remains to be established biochemically, genetic evidence (see below) indicates that the *minor* species reads UCG, whereas it is as yet uncertain whether the species termed here *major* $tRNA^{Ser}_{UCG}$ recognises UCG, UCA, or both of these codons.

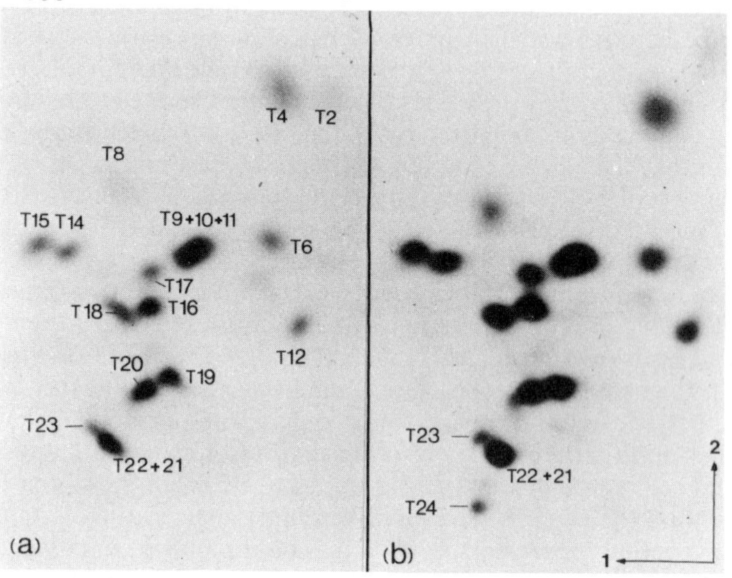

Fig. 4 T_1 ribonuclease digests of (a) wild-type $tRNA^{Ser}_{UCG}$ and (b) $tRNA^{Ser}_{UCG}$ from strain MBD16. Fractionation was by electrophoresis at pH 3.5 on cellulose acetate in the first dimension (1) and homochromatography using a 3% dialyzed homo-mixture in the second dimension (2). The sequences of the numbered oligonucleotides are listed in Piper (1978).

Figure 3a illustrates a fingerprint of a T_1 ribonuclease digest of wild type $tRNA^{Ser}_{UCG}$. The structure of the numbered oligonucleotides and their derivation is documented elsewhere (Piper, 1978). A small number of the oligonucleotides on this fingerprint are present in low yield since they are derived only from the *minor* $tRNA^{Ser}_{UCG}$

whereas those fragments originating only from the *major* sequence or derived from both species are present in much higher yield. The corresponding fingerprint of $tRNA^{Ser}_{UCG}$ from a strain heterozygous for the recessive-lethal *SUPRL1-a* suppressor, MDB16 (Brandriss *et al.*, (1976) is shown in Figure 3b. A distinct and reproducible difference between this fingerprint and that from the wild-type $tRNA^{Ser}_{UCG}$ was the presence, in fairly low yield, of an additional large oligonucleotide designated T24 in Figure 3b. The sequence of T24: $ANUCUAi^6AA\psi$-CUCUUGp (Piper, 1978), shows that it cannot originate from a mutant form of the *major* sequence that differs in only a single base from the wild-type species (Figure 4b). Instead this structure indicates a derivation from a tRNA having the same anticodon region structure as the *minor* $tRNA^{Ser}_{UCG}$ but for the substitution of U for G at position 35 and the lack of a ribose methylation at position 44. This, and the absence of other changes on the fingerprints, provide a strong indication that this altered tRNA is the *SUPRL1-a* suppressor, that it is derived from the *minor* $tRNA^{Ser}_{UCG}$, and that it has the ability to read the codon UAG through a G to U anticodon base change (Figure 4b).

Differences were also found amongst the minor pancreatic ribonuclease digestion products of the $tRNA^{Ser}_{UCG}$ of MBD16 that were consistent with this interpretation (Piper, 1978), although a direct demonstration of the G·C → T·A transversion in the DNA postulated to give rise to the *SUPRL1-a* mutation must await the cloning and sequencing of the suppressor gene. So far we have been unable to separate the *major* and *minor* components of $tRNA^{Ser}_{UCG}$ from the presumed suppressor species in our *SUPRL1/sup*$^+$ strain tRNA, and this has prevented a direct demonstration that it is specifically the latter tRNA that is reading the UAG codon when total *SUPRL1/sup*$^+$ $tRNA^{Ser}_{UCG}$ is added to *in vitro* protein synthesising systems.

In the T_1 ribonuclease digests of $tRNA^{Ser}_{UCG}$ from a *SUPRL1/sup*$^+$ strain (Figure 3b) the oligonucleotide T24 and the corresponding fragment from the wild-type *minor* $tRNA^{Ser}_{UCG}$, $Ai^6AA\psi CUCUU_m Gp$ (T23), were present in equimolar yields, indicating that the suppressor species was present in an amount equal to that of the *minor* $tRNA^{Ser}_{UCG}$. In conjunction with the sequence data this has led to propose that *SUPRL1/sup*$^+$ strains carry one gene for the wild-type *minor* $tRNA^{Ser}_{UCG}$ as well as one copy of this gene that has mutated so as to code for the *SUPRL1-a* suppressor. In turn, haploid cells will have only one gene for this tRNA. Other genes will in turn code for the *major* $tRNA^{Ser}_{UCG}$ and the abundance of this species relative to that of the *minor* form further suggests that it may be encoded as more than one gene in the haploid genome.

The suppressor mutation in *SUPRL1/sup*$^+$ cells affects the

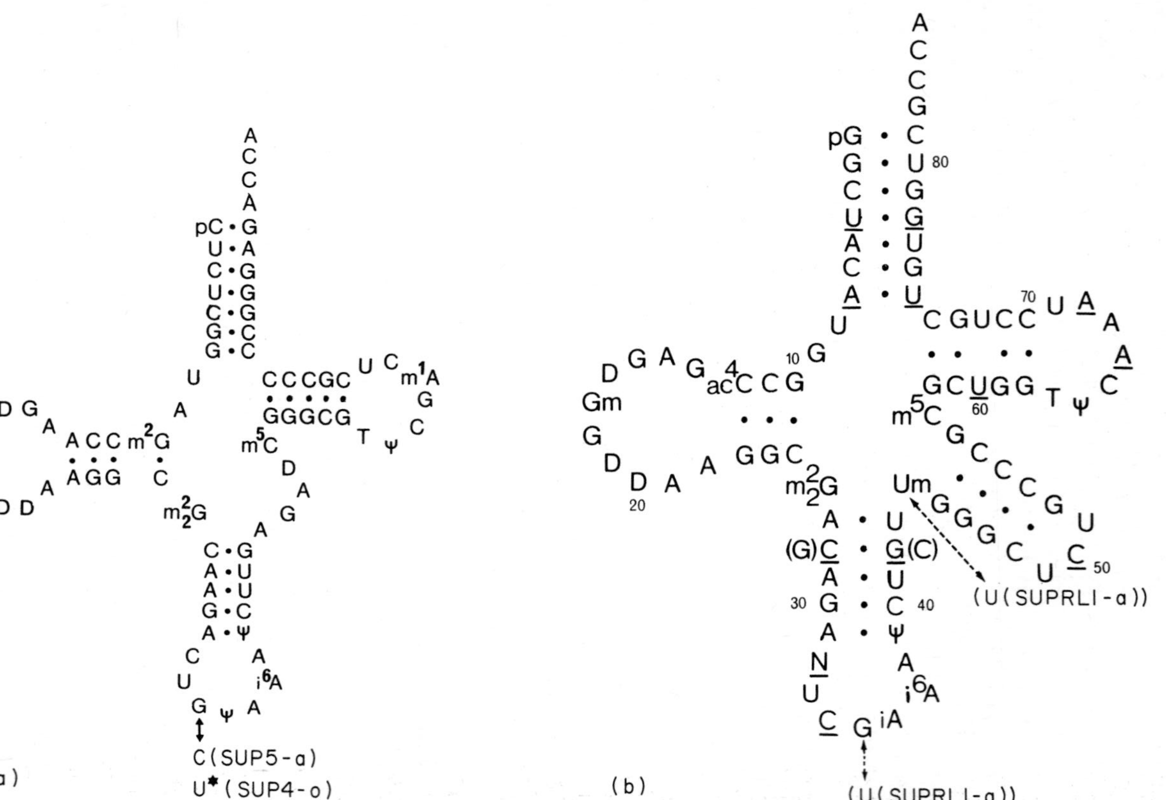
(a)
C(SUP5-a)
U*(SUP4-o)
(b)
(U(SUPRLI-a))
(U(SUPRLI-a))

structure of a fraction of only one of the *S. cerevisiae* $tRNA^{Ser}$ species, and whether the lethality it causes in haploid cells is due to the total alteration of a unique coding tRNA, presumably depends on whether there are other UCG reading tRNAs in the cell. The tRNA it co-purifies with, here called *major* $tRNA^{Ser}_{UCG}$, was previously thought to be another UCG reading tRNA (Piper, 1978), but more recent evidence, discussed below, indicates that it can mutate by single mutation to give ochre suppressors and might therefore correspond to a UCA-reading or UCA and UCG-reading tRNA. Since one half of the *minor* $tRNA^{Ser}_{UCG}$ is altered in MBD16, haploid and homozygous *SUPRL1* strains, if they were viable, might be expected to totally lack the wild-type form of this molecule. The absence of a wild-type copy of the *minor* $tRNA^{Ser}_{UCG}$ gene is presumably the condition that causes lethality in such cells. However it might be suggested that these cells are inviable due to the amount of suppressor or efficiency of suppression being too high and that the suppressor may have been reduced to a level compatible with viability in *SUPRL1/sup*$^{+}$ strains by a gene dosage effect. If the recessive lethality of *SUPRL1* is due to over-suppression, revertants in which the suppressor has either been lost or its efficiency substantially reduced would not be expected to exhibit the haplo-lethality. One such revertant has been isolated by Dr. Susan Liebman (personal communication) from the *SUPRL1/sup*$^{+}$ strain DBD 339 (Brandriss *et al.*, 1975). Although devoid of detectable suppressor activity it still exhibits the lethal effect of *SUPRL1*, never yielding more than two live spores upon tetrad analysis, thereby showing that the lethality caused by *SUPRL1* in haploid cells is not due to excessive suppression of amber codons. Instead this lethality is probably a direct consequence of the *minor* $tRNA^{Ser}_{UCG}$ losing its ability to decode UCG codons.

Only one recessive-lethal serine-inserting suppressor has been characterised in *S. cerevisiae* and it would appear from the searches of Hawthorne and Leupold (1974) and Brandriss *et al.*, (1975) that

Fig. 5 (see opposite) (a) The primary structure of yeast $tRNA^{Tyr}$ showing the base substitutions present in the *SUP5* amber suppressor (Piper *et al.*, 1976) and *SUP4* ochre suppressor (Goodman *et al.*, 1977). (b) The nucleotide sequence of the *major* $tRNA^{Ser}_{UCG}$ of *S. cerevisiae*. The differences in structure between the *major* and *minor* $tRNA^{Ser}_{UCG}$ species were determined by homology for the region of nucleotides 26 to 45 inclusive, and the differences for the *minor* form in this region are represented by nucleotides in brackets. Sequence analysis indicated that approximately half of the *minor* species from the *SUPRL1* amber suppressor-bearing strain MBD16 had the substitution of U for the G at position 35 and U for Um at position 44 (indicated by broken arrows). A line under the symbol for a nucleotide is used to indicate those residues that differ between the UCU, UCC and UCA coding species, $tRNA^{Ser}_{2}$ (Zachau *et al.*, 1966) and the *major* form of $tRNA^{Ser}_{UCG}$.

there are not many genetic loci in this organism that can mutate to suppressors causing recessive-lethality. Certain other serine-inserting ochre and amber suppressors do not produce such lethality. The tRNA product of one of these suppressor genes has been extensively purified using column chromatography by Dr. C. Waldron (personal communication) using the *in vitro* assay systems for ochre and amber suppressors developed by Dr. R. Gesteland and his collaborators (Gesteland *et al.*, 1976). He has found that purified *SUP16* (alternatively designated *SUQ5*) ochre or amber suppressing tRNAs co-purify with one of the serine tRNAs of the yeast cell, but that this species is not the one that reads the UCU, UCC and UCA codons($tRNA_2^{Ser}$). We have electrophoresed two preparations of Dr. Waldron's purified *SUP16* ochre suppressor on two-dimensional polyacrylamide gels and find that it is revealed by staining of the gels as a single spot almost exactly coinciding in position with purified ^{32}P-labelled $tRNA^{Ser}_{UCG}$ (unpublished results).

We have fingerprinted $tRNA_2^{Ser}$ and $tRNA^{Ser}_{UCG}$ from strains carrying the *SUP16* ochre and amber suppressors and were unable to detect an altered species of $tRNA^{Ser}$ (Piper, 1978). However the ability to detect mutational changes in a small fraction of gel-purified tRNA even when it is known from suppressor assays to contain the suppressor is limited by the resolution of the fingerprints and not too much significance should be placed on these negative results. Furthermore more recent work has indicated that the $tRNA^{Ser}_{UCG}$ analysed in our sequence study was devoid of an additional tRNA species that comigrates with $tRNA^{Ser}_{UCG}$ when total yeast tRNA is separated on the two-dimensional gels, our samples having been partially purified by column chromatography before gel electrophoresis during which stage this tRNA appears to have been removed. The sequence and codon specificity of this additional species is as yet unknown, yet its T_1 ribonuclease digest fingerprints differ between strains possessing the *SUP16* ochre, the *SUP16* amber and wild-type nonsuppressing phenotypes (unpublished results). It is probably not a UCG-reading species since it is difficult to envisage how such a molecule could mutate to a UAA-reading tRNA by what is apparently a single step mutation (Sherman, 1978) giving rise to the *SUP16-o* and *SUP17-o* suppressors, and it is possible that it is a $tRNA^{Ser}$ that selectively reads the UCA codon and that these suppressors are generated by anticodon mutation of this tRNA. Preliminary RNA sequence analysis indicates that in the nucleotide 26 to 45 region (Figure 4b), except possibly at the first anticodon base, the component lost in the previous study has the same structure as the *major* $tRNA^{Ser}_{UCG}$. Further work will be necessary to establish whether or not it is identical to this species and whether the column fractionation step of our previous study led to a selective loss of the

SUP16-o suppressor-bearing fraction of this tRNA.

The Effects of the ψ^+ Factor and of Antisuppressor Mutations on the Structure and Function of Yeast Suppressors

The ψ^+ factor (Cox, 1971) is a curious genetic element sometimes present in *S. cerevisiae* which is inherited in non-Mendelian fashion and which is known not to be associated with the mitochondrial DNA or the two-micron circular DNA although it is presumably a gene function of a cytoplasmic self-replicating nucleic acid. It enhances several-fold the efficiency of tRNA ochre and some frameshift suppressors but is without effect on UAG or UGA suppressors, and renders the tyrosine-inserting ochre suppressors so efficient as to make them generally lethal to the cell (Liebman *et al.*, 1975). As its mechanism of action was unknown we investigated whether it exerted its effect by altering the molecular structure of certain of the tRNA molecules of the cell, and not just ochre suppressors, and examined the modified nucleotide composition of tRNA fractions from isogenic ψ^+ and ψ^- strains. Purified *SUP16-o* suppressor tRNAs from ψ^+ and ψ^- strains are known to have comparable efficiencies as ochre suppressors in mammalian *in vitro* protein synthesis systems (C. Waldron, unpublished). Separating the minor components of T_2 ribonuclease digests on the two-dimensional thin-layer plates (Nishimura, 1972) we could not detect any difference in the minor base compositions of $tRNA^{Tyr}$, $tRNA^{Phe}$, $tRNA_2^{Ser}$, $tRNA_{UCG}^{Ser}$, $tRNA_{UUG}^{Leu}$ from ψ^+ and ψ^- strains (unpublished results). Unlinked nuclear mutations that modify suppressor function, 'antisuppressors', may affect levels and structures of suppressor tRNAs or may operate by altering components of the ribosome. In the fission yeast *Schizosaccharomyces pombe,* one antisuppressor mutation is known to cause gross undermodification of the adenosine to i^6A at the 3′ side of the anticodon of $tRNA^{Tyr}$ (F. Hubschmidt-Hanner, unpublished).

Conclusions

The investigations described above have succeeded in establishing the molecular identity of some of the nonsense suppressors that can be obtained in yeast whereas until recently the nature of such suppressors was known only in prokaryotes. The tyrosine-inserting and serine-inserting ochre and amber suppressors of *S. cerevisiae* can now be purified to a reasonable degree of homogeneity for use in screening tests for nonsense mutants in other eukaryotic organisms. By introducing yeast suppressor tRNAs into nonsense mutant mammalian cells it is possible to bring about a temporary correction of their genetic deficiency (Capecchi, 1977; Celis *et al.*, this volume). As yet the molecular nature of UGA suppressing tRNAs derived by single mutation in *S. cerevisiae* is unknown although for use in *in vivo* tests for suppression in mammalian cells it is possible to obtain known UGA suppressors by further mutation of characterised ochre suppressors. Also unknown are the identities of the leucine-inserting ochre and amber suppressors of yeast, and it should be possible to test whether these originate by mutation of UUG and UUA decoding leucine tRNAs using suppressor assays in combination with RNA fingerprinting since $tRNA^{Leu}_{UUG}$ and $tRNA^{Leu}_{UUA}$ are readily purified on two-dimensional gels (Piper and Wasserstein, 1977).

Acknowledgements

I thank Professor Kjeld Marcker who kindly provided laboratory facilities for much of this work, during which period I was supported by a long-term EMBO fellowship. I also thank Drs. Susan Liebman and Fred Sherman for supplying strains and Drs. F. Hubschmidt-Janner, S. Leibman and C. Waldron for permission to cite their unpublished work.

References

Anderson, K.W. and Smith, J.D. (1972). *J. Mol Biol.* **69**, 349.

Beckmann, J.S., Johnson, P.F. and Abelson, J. (1977). *Science* **196**, 205.

Brandriss, M.C., Soll, L. and Botstein, D. (1975). *Genetics* **79**, 551.

Brandriss, M.C., Stewart, J.W., Sherman, F. and Botstein, D. (1976). *J. Mol. Biol.* **102**, 467.

Capecchi, M.R., Hughes, S.H. and Wahl, G.M. (1975). *Cell* **6**, 269.

Celis, J.E., Coulondre, C. and Miller, J.H. (1976). *J. Mol. Biol.* **104**, 729.

Celis, J.E., Hooper, M.L. and Smith, J.D. (1973). *Nature New Biol.* **244**, 261.

Celis, J.E. (1977). *Brookhaven Symp. Biol.* **29**, 178.

Comer, MM., Foss, K. and McClain, W.H. (1975). *J. Mol. Biol.* **99**, 283.

Cox, B.S. (1971). *Heredity*, **26**, 211.

Gesteland, R.F., Wolfner, M., Grisafi, P., Fink, G., Botstein, D. and Roth, J.R. (1976). *Cell*, **7**, 381.

Goodman, H.M., Abelson, J., Landy, A., Brenner, S. and Smith, J.D. (1968). *Nature (London)* **217**, 1019.

Goodman, H.M., Olson, M.V. and Hall, B.D. (1977). *Proc. Nat. Acad. Sci.* **74**, 5453.

Gilmore, R.A., Stewart, J.W. and Sherman, F. (1971). *J. Mol. Biol.* **61**, 157.

Hawthorne, D.C. (1976). *Biochimie* **58**, 179.

Hawthorne, D.C. and Leupold, U. (1974). *Current Topics Microbiol. Immunol.* **64**, 1.

Kaltoft, K., Zeuthen, J., Engbaek, F., Piper, P.W. and Celis, J.E. (1976). *Proc. Nat. Acad. Sci.* **73**, 2793.

Liebman, S.W., Stewart, J.W., and Sherman, F. (1975). *J. Mol. Biol.* **94**, 595.
Liebman, S.W. and Sherman, F. (1976). *Genetics* **82**, 233.
Liebman, S.W., Sherman, F. and Stewart, J.W. (1976). *Genetics* **82**, 251.
Miller, C.G. and Roth, J.R. (1971). *J. Mol. Biol.* **59**, 63.
Nishimura, S. (1972). *Prog. Nucleic Acid Res. Mol. Biol.* **12**, 50.
Olson, M.V., Montgomery, D.L., Hopper, A.K., Page, G.S., Horodyski, F. and Hall, B.D. (1977). *Nature (London)* **267**, 639.
Piper, P.W., Wasserstein, M., Engbaek, F., Kaltoft, K., Celis, J.E., Zeuthen, J., Liebman, S. and Sherman, F. (1976). *Nature (London)* **262**, 757.
Piper, P.W. and Wasserstein, M. (1977). *Eur. J. Biochem.* **80**, 103.
Piper, P.W. (1978). *J. Mol. Biol.* **122**, 217.
Sherman, F., Liebman, S.W., Stewart, J.W. and Jackson, M. (1973). *J. Mol. Biol.* **78**, 157.
Soll, L. (1974). *J. Mol. Biol.* **86**, 233.
Soll, L. and Berg, P. (1969). *Proc. Nat. Acad. Sci.* **63**, 392.
Yaniv, M., Folk, W.R., Berg, P. and Soll, L. (1974). *J. Mol. Biol.* **86**, 245.
Zachau, H.G., Dutting, D. and Feldmann, H. (1966). *Z. Physiol. Chem.* **347**, 212.

BIOSYNTHESIS OF SUPPRESSOR TRANSFER RNA

S. ALTMAN

Department of Biology, Yale University,
New Haven, USA.

Introduction

Current understanding of tRNA gene organization and tRNA biosynthesis has come directly from the study of suppressor tRNAs (see Altman, 1975; McClain, 1977; Smith, 1976, for reviews; Smith, this volume). The study of tRNA biosynthesis has taken advantage of our ability to isolate mutants not only in the enzymes engaged in processing gene transcripts (Schedl and Primakoff, 1973; Sakano *et al.*, 1974) but also in the initial substrate of the pathway, the tRNA gene transcript itself (Altman, 1971; Altman and Smith, 1971; Guthrie, 1975; McClain *et al.*, 1975; McClain and Seidman, 1975; Seidman *et al.*, 1975). These latter mutants are of different varieties but all of them share certain common features: 1) they are in suppressor tRNA genes; and 2) they retard the action of certain processing enzymes on the mutant substrated when compared to normal rates with wild type substrates.

Mutations in tRNA Genes Which Alter Subsequent Steps in tRNA Biosynthesis

The E. coli *tRNA*^{Tyr} *Gene*

The construction of the su_3^+ system in *E. coli* and of ϕ80 transducing phages carrying the su_3^+ genes has already been described by J.D. Smith in this volume. Mutants of ϕ80 su_3^+ phage can be selected which have lost some or all of their suppressor activity. In many cases this appears to be due to a decrease in the production of mature tRNA in question (Abelson *et al.*, 1970; Anderson and Smith, 1972;

Smith, 1976; Smith *et al.*, 1970). In fact, the following statements can be made: i) mutations in the anticodon (G35→C35; C35→U35) do not alter the amounts of mature tRNA produced; ii) mutations near the termini of the molecule (G2→A2; C81→U81, etc.) reduce the amount of tRNA made by 80-90%; and iii) mutations elsewhere in the molecule (see Figure 1 and Table 1) decrease the amount of

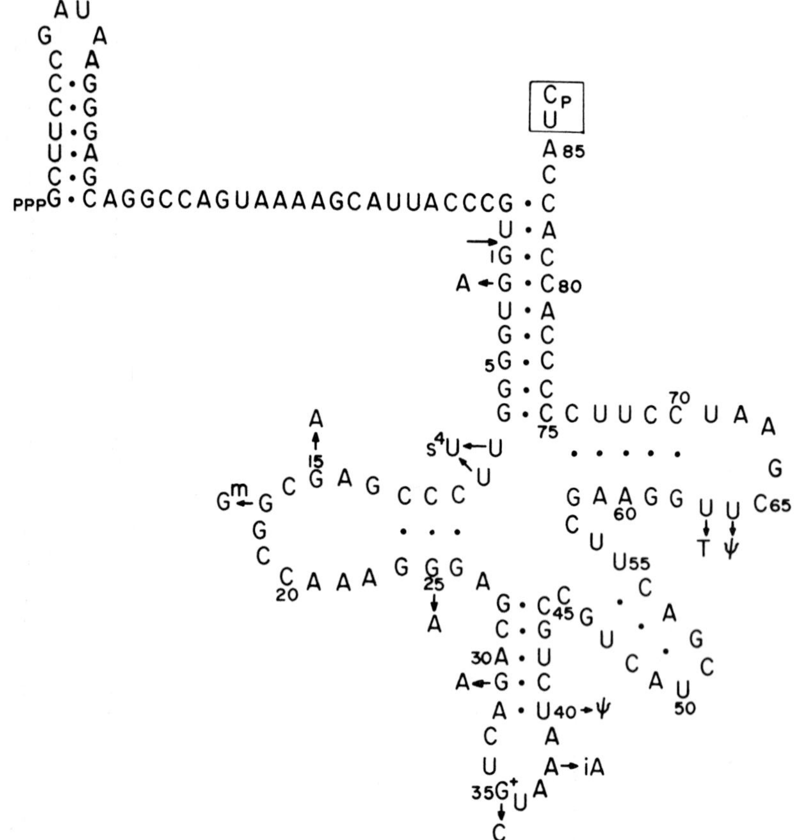

Fig. 1 Nucleotide sequence of precursor to *E. coli* tRNA$_1^{Tyr}$ su$_3$A25 (Altman and Smith, 1971). The arrow pointing towards the sequence indicates the site of RNase P cleavage, on the 5′ side of the G1 of the mature tRNA sequence. The boxed nucleotides are 'extra' nucleotides at the 3′ terminus. The arrows pointing away from the sequence show the nucleotide modifications found in the mature tRNA and the positions of some nucleotide substitutions which affect RNase P cleavage. For example, G15→A15, G25→A25 and G31→A31 all retard RNase P cleavage of the precursor molecule. G35→C35 is the su$_3^+$ mutation and does not affect processing events. See Table I for the locations and phenotypes of other mutations.

TABLE I

Location in precursor to E. coli *tRNA$_1^{Tyr}$ of nucleotide substitutions which affect processing events*

First mutation+	Second mutation	Affected function
G15→A15		RNase P cleavage
G25→A25		RNase P cleavage
G31→A31		RNase P cleavage
G31→U31		Probably RNase P cleavage
G46→A46		Probably RNase P cleavage
G1→A1		Not known: no precursor identified
G2→A2		Not known: no precursor identified
C(-4)→U(-4)*		Not known: enhanced tRNA production
G31→U31	C16→U16	Secondary mutation restores normal or almost normal tRNA biosynthesis in all these cases
	C45→U45	
	C41→A41	
G31→A31	C16→U16	
	C45→U45	
	C41→U41	
G25→A25	C11→U11	
G46→A46	C54→U54	
G1→A1	C81→U81	
G2→A2	C80→U80	

+See Figure 1 for location of mutations in tRNA$_1^{Tyr}$.

*This mutation is four nucleotides from the 5′-terminus of the mature tRNA sequence and is in the 5′-extra sequence.

Data for this table taken from Abelson *et al.*, 1970; Altman and Smith, 1971; Anderson and Smith, 1972; McClain, 1977; Smith *et al.*, 1970; and Smith and Celis, 1973.

mature tRNA produced to varying degrees but also show a short term accumulation of precursor tRNA molecules when RNA is extracted rapidly from the relevant cell types. Mutants having class i) or class ii) mutations show no precursor accumulation presumably because in i) processing occurs normally and efficiently, and in ii) because degradative enzymes compete with normal processing enzymes at the frayed ends of the mutant $tRNA^{Tyr}$ precursor molecules. Mutations of the kinds described can be induced by a variety of agents: among these are EMS and NG (nitrosoguanidine). Acridine half-mustards have also been used to induce su^- revertants in su_3^+ host cells (Altman, 1971). These latter cells can then be lysogenized with $\phi 80psu_3^+$. The mutant su^- gene can be picked up on transducing $\phi 80$ derivatives upon prophage induction. The resulting $\phi 80su^-$ derivatives (those which do have part of the $tRNA^{Tyr}$ gene system) appear to carry partial duplications of the su_3^+ gene which produce precursor tRNA but little or no mature tRNA *in vivo.*

The $\phi 80su_3^+$ system can also be manipulated in a fruitful fashion in a search for mutants of host cell enzymes which disallow the production of mature tRNA. This will be discussed below.

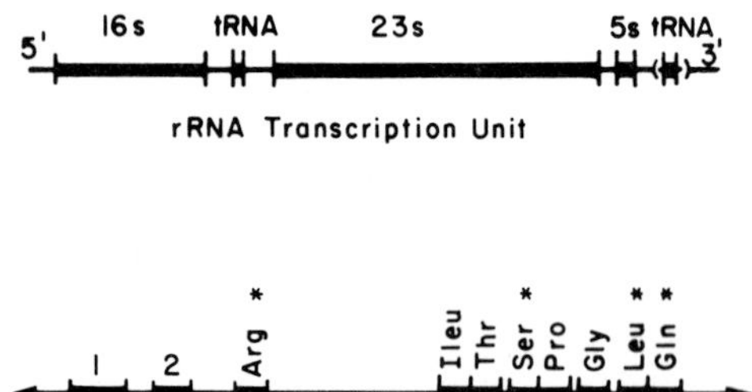

Fig. 2 Organization of rRNA transcription unit in *E. coli* and tRNA genes of bacteriophage T4. *Top,* the organization of an hypothetical *E. coli* rRNA transcription unit. tRNA hybridizes to a DNA restriction fragment in the spacer region between the 16S and 23S sequences and also near the 3' terminus. The total length of the transcription unit is about 5000 base pairs. Not all rRNA transcription units in *E. coli* necessarily have tRNA genes in both the indicated regions (adapted from Lund *et al.,* 1976, and Morgan *et al.,* 1978). *Bottom,* the organization of T4 tRNA genes as determined by hybridization mapping of the various tRNA species to DNA restriction fragments (adapted from Velten *et al.,* 1976). The abbreviations refer to the tRNA species found in this cluster; the relative size of the region can be judged from the size of the tRNA genes. In addition, there ar other RNA species of unknown function, bands 1 and 2, encoded in this region. The tRNA species marked by asterisks have been converted to nonsense suppressors in certain T4 derivatives (see text).

TABLE II

Location in T4 precursor tRNAs of nucleotide substitutions which affect processing events

Precursor tRNA	Mutation	Affected function
Pro-*Ser*+	G67→A67 C70→U70 C75→U75	Nucleotide modification; 3′-5′ - exonuclease action at the serine terminus in all these cases
	A12→G12 G17→U17 C29→A29 G30→A30 T68→C68 C70→U70 U77→C77	3′-5′ - exonuclease action at the serine terminus thereby also affecting RNase P cleavage in all these cases
	G1→U1 A3→03* U8→C8	Facilitate *in vivo* RNase P cleavage in all these cases
Gln-Leu	C11→U11 G40→A40 C62→U62	Nucleotide modification; partial reduction of RNase P cleavage and CCA repair activity at Gln moiety in all these cases
Gln-*Leu*	C72→U72	Nucleotide modification; RNase P cleavage reduced at both moieties

+Underlined moiety contains the mutation noted.
*03 indicates a deletion of A3.
Data for this table taken from Guthrie, 1975; McClain, 1977; McClain *et al.*, 1975; McClain and Seidman, 1975; Seidman *et al.*, 1975.

Bacteriophage T4 tRNAs

Bacteriophage T4 codes for eight tRNAs (Figure 2). Some of these tRNAs have been converted to suppressors by selection of phenotypic revertants of nonsense mutations in various T4 genes coding for proteins (McClain, 1970; Wilson and Kells, 1972). Once suppressors are available, full or partial revertants to wild type, non-suppressor phenotype phages are easily found as is the case also for $\phi 80su_3^+$. Many of these T4 revertants to non-syppressor function block the processing of T4 tRNA gene transcripts (Table II). McClain and his colleages have used this system to describe, for example, seven distinct steps in the maturation of T4 $tRNA^{Pro}$ and $tRNA^{Ser}$ from their dimeric precursor molecule (McClain, 1977; Seidman *et al.,* 1975).

Transducing phage carrying Missense Suppressor tRNA

In addition to the phages carrying the genes for nonsense suppressor tRNAs, a bacteriophage lambda-ϕ80 hybrid is available which carries a missense suppressor derived from *E. coli* $tRNA^{Gly}$ (Carbon *et al.,* 1974; Chang and Carbon, 1975). This phage, in fact, carries three tRNA genes. A tRNA gene cluster, which includes a $tRNA^{Gly}$ missense suppressor derivative, has been picked up in its entirety from the host cell chromosome. The precursor tRNAs can be isolated from a suitable host cell which makes their biosynthesis amenable to analysis. The general outline of the synthesis of these tRNAs is the same as for all others yet described.

Mutants Defective in Enzymes Necessary for tRNA Biosynthesis

Ribonuclease P

Ribonuclease P, an endoribonuclease which cleaves precursor tRNAs at the 5′-termini of the mature tRNA sequence (Altman and Smith, 1971; Robertson *et al.,* 1972) is essential for the biosynthesis of *E. coli* tRNAs. This conclusion is drawn from analysis of *E. coli* thermosensitive mutants in which no tRNA is made at restrictive temperatures. Schedl and Primakoff (1973) and Sakano *et al.,* (1974) showed that in these mutants RNase P was thermosensitive and precursors to most, if not all, tRNAs accumulated (Schedl *et al.,* 1974; Ikemure *et al.,* 1975) (Figure 3). T4 tRNAs also require this enzyme for their biosynthesis (McClain, 1977). The selection of RNase P ts mutants was the result of a search for *E. coli* derivatives which failed to suppress nonsense mutations, either in the host chromosome or in an infecting phage chromosome, at high temperatures (42°). These mutants were capable of suppression at 30° and synthesized wild type proteins normally at both temperatures.

Recently one of the *S. typhimurium* mutants which affects the biosynthesis of histidine and $tRNA^{His}$ has been shown to be defective in an endonucleolytic activity, possibly RNase P (Bossi and Cortese, 1977).

Enzyme Processing the 3′- Terminus

In addition to RNase P, a 3′ -5′ exonuclease which is absent in *E. coli* BN (McClain, 1977) and tRNA nucleotidyl transferase (CCA enzyme; Deutscher *et al.,* 1974) are necessary for the synthesis of some T4 tRNAs (Siedman *et al.,* 1975). These tRNAs (Pro, Ser, Gln) lack the 3′ -terminal C-C-A sequence in their gene transcripts. Thus, extra 3′ -terminal nucleotides must be first removed by an exonuclease and the remaining sequence repaired by the CCA enzyme to give the correct 3′ -terminal sequence of the mature tRNA. *E. coli* mutants with reduced levels of CCA enzyme (Deutscher *et al.,* 1974) are also blocked in the biosynthesis of these tRNAs.

It is true that *E. coli* tRNAs require some 3′ -5′ -exonuclease activity for their biosynthesis but to date no mutants have been found blocked in this particular step of *E. coli* tRNA biosynthesis. *E. coli* tRNAs have the sequence CCA encoded in their genes and apparently do not require the BN exonuclease or the CCA enzyme for their normal biosynthesis. However, some turnover of the 3′ -terminal nucleotides of tRNAs does occur normally. In strains with low levels of CCA enzyme growth is slower, presumably due to slow repair of the CCA sequence (Deutscher *et al.,* 1974).

It can be inferred from various kinds of *in vitro* studies (Bikoff and Gefter, 1975; Daniel *et al.,* 1976) that several RNases are needed for *E. coli* tRNA biosynthesis (see below) but mutants in RNase P function only have been isolated to date.

Nucleotide Modification Enzymes

E. coli mutants exist which are deficient in their ability to synthesize nucleotide modifications found in some mature tRNAs. These mutants, however, appear to have no effect on the processing of gene transcripts (Björk, 1975; Björk and Neidhardt, 1975; Colby *et al.,* 1976; Singer *et al.,* 1972).

Enzymes which Excise Intervening Sequences

In this volume Beckmann *et al.,* discuss a yeast mutant which fails to excise intervening sequences from the transcripts of certain yeast tRNA genes.

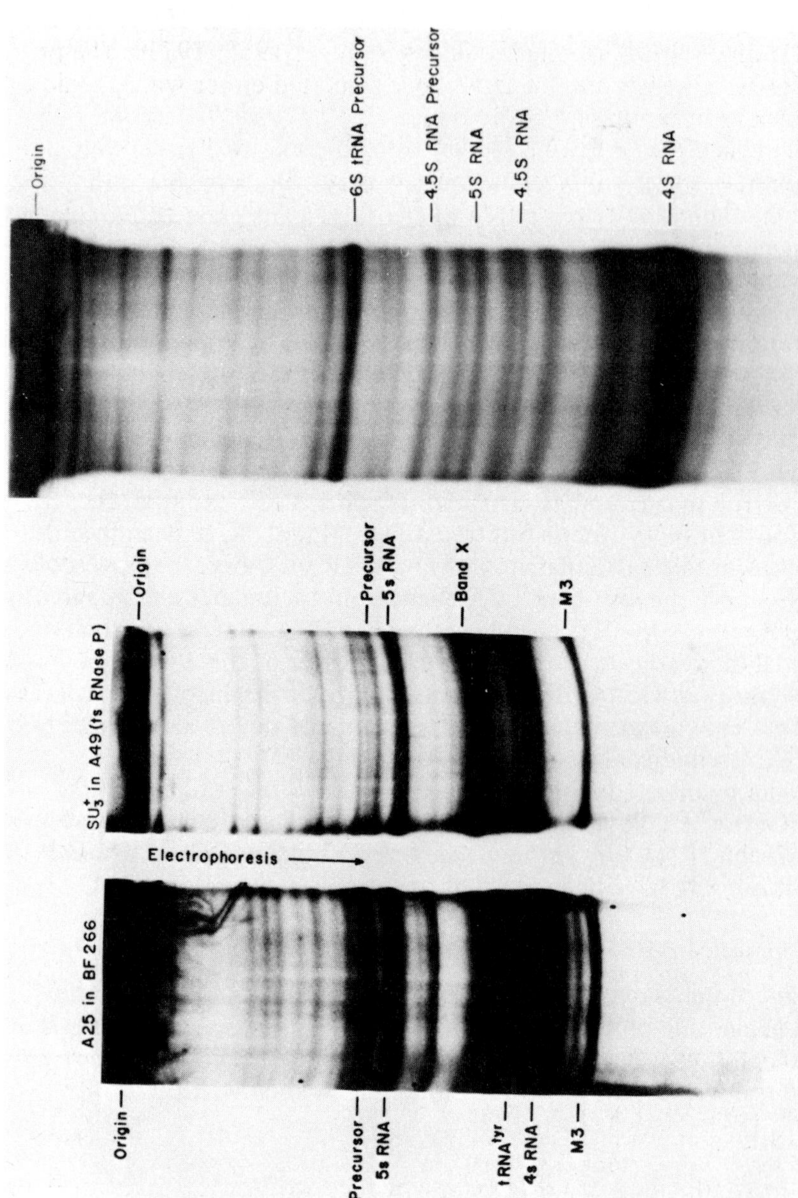
Origin
6S tRNA Precursor
4.5S RNA Precursor
5S RNA
4.5S RNA
4S RNA
SU$_3^+$ in A49 (ts RNase P)
Origin
Precursor
5s RNA
Band X
M3
Electrophoresis
A25 in BF 266
Origin
Precursor
5s RNA
tRNA^{Tyr}
4s RNA
M3

Precursor tRNA Sequences

The nucleotide sequences of several *E. coli* and T4 precursor tRNAs are known (Altman and Smith, 1971; Seidman *et al.*, 1975; Guthrie, 1975; Chang and Carbon, 1975; Vögeli *et al.*, 1975. 1977). The precursors themselves have generally been isolated either with the aid of mutations in suppressor tRNA genes which retard the action of processing enzymes or from cells with thermosensitive RNase P function. All these precursor molecules contain extra nucleotides at their 5′ - termini and many have extra nucleotides at their 3′ -termini also (see Altman, 1978, for review). RNase P cleaves all the precursor molecules at the 5′ termini of the mature tRNA sequences. Some precursors are isolated with less than their complete complement of nucleotide modifications but this seems to have little or no bearing (with the possible exception of the 2′-0-methyl guanosine modification) on the ability of processing enzymes to act on these molecules. Among the precursors studied to date, one finds great variability both in the numbers of extra 5′ - terminal nucleotides and the extent of 3′ - processing (recall that some T4 tRNAs do not have the CCA sequence in their transcripts). Precursor tRNAs which originally may have been part of multimeric gene transcripts, or part of transcription units which include non-tRNA segments, may be partly processed prior to isolation. (The T4 Pro-Ser and Gln-Leu precursors start with pXp as does the dimeric *E. coli* Gly-Thr precursor. The *E. coli* Glu_2 precursor starts with $_{OH}Cp$: all of these precursors are thought to be derived from larger transcripts. In particular, the Glu_2 precursor may be derived from a larger rRNA transcription unit (Figure 2).)

The isolation of precursor tRNA molecule up to 400 or more nucleotides long shows that certain tRNA genes can be clustered and transcribed together since some of these long precursor molecules contain several tRNA sequences. In some of these molecules the

Fig. 3 (see opposite) Autoradiographs of polyacrylamide gel separations of RNA species extracted from *E. coli* under different conditions of precursor tRNA preparation. *Left,* RNA extracted from a host normal in RNase P function infected with $\phi 80su_3A25$. The A25 mutation slows the action of RNase P on the precursor tRNA. *Center,* RNA extracted from host cells temperature sensitive for RNase P function and infected with $\phi 80su_3^+$ at the restrictive temperature. Note the underproduction of mature tRNA and the many bands in the *upper* portion of the gel, most of which represent multimeric precursor tRNAs. Band X is a shortened form of precursor to $tRNA^{Tyr}$ (see text). *Right,* RNA extracted from cells as shown in the center panel except that the cells are uninfected. Thus, there is no enrichment for any particular precursor tRNA species. Note again the many bands throughout the gel and which are mostly precursor tRNA species (taken from Altman, 1978).

sequence for the same isoaccepting tRNA may be repeated with few, if any, nucleotide changes from one iteration to another (Schedl *et al.*, 1974; Ozeki *et al.*, 1974; Ilgen *et al.*, 1975). In other clusters different accepting species are represented. Since suppressor tRNAs have been mapped at many positions around the *E. coli* genome (Smith, this volume) we know that not *all* tRNA genes are grouped together and transcribed in a small number of long transcripts (Smith 1977).

Recently some tRNA genes have been mapped by hybridization techniques to positions in rRNA cistrons (Lund *et al.*, 1976; Morgan *et al.*, 1978), between the sequences coding for the 16S and 23S rRNA segments and in some cases near the 3′ - terminus of the large transcript (Figure 2). If a tRNA located within such a transcript were to be processed first by an endoribonuclease different from RNase P, it would appear as a precursor tRNA with a few extra 5′ -nucleotides, possibly with a 5′ -hydroxyl group (see above for discussion of precursor to $tRNA_2^{Glu}$).

Detailed genetic deletion and restriction fragment mapping of T4 tRNA genes places them all in a cluster near the T4 lysozyme gene (Velten *et al.*, 1976). There is some evidence that all the genes are transcribed from one promoter (Kaplan and Nierlich, 1975; Goldfarb *et al.*, 1978) but no *in vivo* precursors larger than dimers have yet been isolated. It is likely, then, that an endoribonuclease that creates 5′ -phosphate groups has acted *in vivo* on the putative long transcript prior to action by RNase P.

In addition to those with completely determined sequences, several other *E. coli* precursor tRNAs have been partly characterized with respect to their 5′- and 3′ -terminal sequences. These latter data confirm the universal accuracy of RNase P cleavage and the presence of CCA sequence in all *E. coli* tRNA gene transcripts.

The isolation of precursor tRNA sequences which contain anticodon mutations conferring suppressor function on the resulting mature tRNAs again confirms that tRNA-mediated suppression arises from mutations in the tRNA genes themselves.

Ribonucleases in tRNA Biosynthesis

The requirement for RNase P in tRNA biosynthesis has already been discussed. This enzyme cannot be recognizing its cleavage sites through primary structure information because there is no homology in nucleotide sequence around the many RNase P cleavage sites so far characterized. The common features, however, which all tRNA precursor molecules have, is the conformation in solution of the tRNA moieties. (It is assumed that: a) tRNA moiety conformation in precursor tRNAs is similar to mature tRNA conformation; and b) that all tRNAs have roughly the same three-dimensional structure in solution.) In fact, those mutants of $tRNA^{Tyr}$ su_3^+ and T4 suppressor tRNAs which produce precursor molecules that are cleaved less rapidly *in vitro* by RNase P than the parent molecules, have nucleotide substitutions in parts of the tRNA moiety which can be expected to be important in determining conformation in solution. How RNase P recognizes a particular substrate structure is discussed below.

The absolute requirement in *E. coli* tRNA biosynthesis for endoribonucleases other than RNase P is not established. In heated extracts of *E. coli* RNase P ts mutants, other enzymatic activities have been identified (called RNase O or RNase P_2) (Sakano and Shimura, 1975; Schedl *et al.*, 1976) which can cleave precursor tRNAs in intercistronic regions. *In vitro,* some multimeric precursor molecules apparently cannot be processed to monomeric units by RNase P or RNases P_2 or O action alone but require a sequential order of action by both kinds of endoribonucleases to achieve this end. The relevance of this phenomenon to *in vivo* events has not been demonstrated. This is especially true considering the recent demonstration that in an RNase III^- strain, tRNA biosynthesis progresses as usual and no RNase O activity can be identified in extracts of these cells (H. Sakano, personal communication). Thus, RNase O can be identified with RNase III. It remains to be proved that RNase P_2 is unequivocally different from RNase III. Furthermore, as will be illustrated with some aspects of T4 tRNA biosynthesis, ordering enzymological events in a given pathway on the basis of *in vivo* studies alone seems as secure as walking on quicksand.

In vitro coupled transcription-processing systems have also provided useful data concerning tRNA biosynthesis (Bikoff and Gefter, 1975; Daniel *et al.*, 1976). These studies have been carried out primarily with $\phi 80psu_3^+$ DNA as template, to which has been added RNA polymerase and various other sub-cellular fractions. In this fashion, Bikoff and Gefter (1975) identified four ribonucleases involved in processing their *in vitro* tRNA gene transcript. Two of these RNases are very likely RNase P and RNase III and a third is a 3′-5′-exonuclease (but not RNase II). The fourth RNase is as yet uncharacterized. Similarly, Daniel and her coworkers (1976) identified the need for RNase P and a 3′-5′-exonuclease in their system. In general, these *in vitro* studies have tended to confirm the results with precursors isolated from cells and studies of mutant cell extracts.

Through the use of *E. coli* BN (Seidman *et al.*, 1975), CCA (Deutscher *et al.*, 1974) and RNase P ts mutants (Schedl and Primakoff, 1973; Sakano *et al.*, 1974) McClain and his coworkers have delineated a sequence of seven reactions needed to produce mature tRNA^{Pro} and tRNA^{Ser} from their dimeric precursor molecule. In studying the need for the CCA sequence as a recognition point for RNase P (since RNase P cleavage of the Pro-Ser dimeric precursor does not occur until the CCA sequence is present in the Ser moiety) *in vitro,* RNase P cleavage experiments were carried out with the Pro-Ser precursor encoded by bacteriophage T2, which has the CCA sequence in the tRNA genes. The results were compared with those from experiments using the T4 encoded Pro-Ser precursor (Schmidt and McClain, 1978; Schmid *et al.*, 1976). *In vitro,* RNase P is able to cleave the T4 precursor at *both* tRNA moieties. Furthermore, the rate of cleavage of the T4 precursor was much closer to that of the T2 Pro-Ser precursor than expected from *in vivo* studies. These results illustrate the difficulty in assigning physiological significance to data from *in vitro* studies with RNase P.

An interesting feature of T4 precursor tRNA molecules is their comparitively long lifetime *in vivo.* The precursor to tRNA^{Tyr}su$_3^+$ has a half-life of the order of seconds at 37^0 (Altman, 1971). The precursors transcribed from the mutant genes, su$_3^+$ A25, for example, have half-lives of the order of minutes but cannot be isolated unless the cells containing them are extracted rapidly with phenol. Wild type T4 precursor tRNAs, on the other hand, have half-lives of the order of minutes (Guthrie *et al.*, 1973). The features of T4 precursor tRNA molecules which lengthen their half-lives in comparison to other precursor tRNAs may be responsible for our inability to isolate any T4 precursors longer than dimeric ones. That is, the very long gene transcript, possibly containing eight tRNA sequences, may be surviving long enough inside cells to be cleaved by an enzymatic activity, like RNase III, which normally works on tRNA precursors with relatively slow kinetics. A similar explanation may be provided for the appearance of a fore-shortened precursor to tRNA^{Tyr} su$_3^+$ in RNase P ts mutan cells (see Figure 3, band X). The half-life of the precursor molecule in these latter cells is artificially lengthened by the absence of RNase P function at restrictive temperatures.

The exact nature of the 3′-5′-exonuclease(s) normally involved in the processing of *E. coli* tRNA precursors has not been described. Several candidate activities have been proposed but none has been shown, by genetic manipulation, to be absolutely required. It is quite possible that both an endoribonuclease and an exoribonuclease, or two exonucleases, may be needed for processing events at the 3′-termini of precursor tRNAs. For example, *in vitro* data indicate that the termination site for transcription of the tRNA^{Tyr} su$_3^+$ gene is tens, perhaps hundreds of nucleotides, from the end of the mature tRNA sequence (Daniel *et al.*, 1976; Küpper *et al.*, 1978). Yet precursor $tRNA^{Tyr}$ su_3^+ isolated from RNase P ts mutant cells has only three extra nucleotides. Perhaps one enzyme removes most of the extra transcribed nucleotides at the 3′-terminus but another is required for the final trimming (possibly after RNase P cleavage) of the last three nucleotides. Curiously, the BN nuclease, which is needed for final trimming of some T4 precursor tRNAs, is not needed for *E. coli* precursor tRNA maturation. One may speculate that the CCA sequence must already be in the gene transcript in order for the canonical *E. coli* 3′-5′-exonuclease to carry out the final 3′-terminal trimming.

Enzyme-substrate Recognition in tRNA Biosynthesis

RNase P, 3′-5′-exonucleases, nucleotide modification enzymes - all these must recognize certain aspects of precursor tRNA molecules with great accuracy. The case of the canonical exonuclease is perhaps easiest to deal with. We can suggest that this enzyme degrades the 3′-terminal extra nucleotides of a precursor molecule until it encounters secondary structure. If some of the CCA sequence has also been removed, no matter, the CCA enzyme can repair the damage. However, substrate recognition by this enzyme, as discussed above, probably involves more than just finding the 3′-termini of RNA molecules. If the case of the BN exonuclease can be used as an example, then several sites in the aminoacyl stem and the $T\psi C$ loop of the tRNA moiety of precursor tRNA molecules must be involved in enzyme-substrate recognition. Many single nucleotide changes have been shown to retard BN action on the T4 Pro-Ser precursor (Table II). Similarly, studies of both mutant $tRNA^{Tyr}$ and T4 precursor tRNA molecules have demonstrated the need for canonical tRNA moiety conformation for efficient action of several nucleotide modification enzymes. In the case of precursor to $tRNA^{Tyr}$ su_3 A25, the cleaved precursor could be modified at position 40 ($U40 \rightarrow \psi 40$) much more efficiently than uncleaved precursor (Schaefer *et al.*, 1973). On the other hand, precursor to $tRNA^{Tyr}$ su_3^+(G25) could modified efficiently at position 40 whether or not it was cleaved by RNase P (Ciampi *et al.*, 1977).

RNase P is the best-studied, from the point of view of both genetics and biochemistry, of the RNases involved in tRNA biosynthesis. Two genetic loci have been identified which can affect RNase P function (Schedl and Primakoff, 1973; Sakano *et al.*, 1974). This result suggests that the enzyme must have at least two subunits. Purified RNase P can cleave any *E. coli* or T4 precursor tRNA at the correct site, the beginning of the transcribed mature tRNA sequence(s) in the precursor molecule. As mentioned above, certain single nucleotide changes in the tRNA moieties of precursor molecules (see Table II) retard the action of RNase P on these substrates both *in vivo* and *in vitro*. How are these data to be understood in terms of enzyme-substrate recognition?

Recently it has been shown that highly purified RNase P has an essential RNA component (Stark *et al.*, 1978). This RNA molecule has a molecular weight of about 110,000 daltons (corresponding to about 350 nucleotides). In addition to the RNA moiety of the enzyme, there is one, possibly two, protein subunits of molecular weight about 20,000 daltons. The density of the enzyme complex in CsCl is 1.71g/ml. When RNase P is pretreated with other RNases it loses its ability to cleave precursor tRNA molecules. What role does the RNA subunit of the enzyme play in substrate recognition?

Since RNase P has to recognize RNA substrate, it seems reasonable to propose a model of enzyme function in which the enzyme's RNA subunit, rather than the protein moiety, governs the interaction with the RNA substrates. That is, the common feature of all precursor tRNA is their subset of invariant nucleotides (Rich and RajBhandary, 1976) which, we presume, performs a critical role in determining a common tRNA moiety solution conformation. The RNA subunit, acting through nucleotide-nucleotide interactions such as hydrogen bonding, may be able to position the enzyme complex on its substrates both accurately and rigidly. Once the complex is fixed in position, the nucleolytic subunit (protein) then cleaves the substrate always at a fixed point in space with respect to the positions of the invariant nucleotides of the tRNA moieties. One can envision the RNA subunit covering mainly the region of the bend in the tRNA moiety but not the anticodon loop (since mutations in the anticodon loop do not affect RNase P action), while the protein moiety lies mainly near the termini of the mature tRNA sequence of a precursor molecule. In this way a rough picture emerges of why certain nucleotide changes which disturb tRNA conformation interfere with RNase P action and why the enzyme may find having a RNA subunit particularly useful. It will be of interest to see if any of the other enzymes involved in tRNA biosynthesis also possess RNA moieties.

Initiation and Termination of Transcription

Initiation of transcription of the $\phi 80su_3^+$ gene occurs 41 nucleotides to the 5′-side of the start of the mature tRNA sequence (Altman and Smith, 1971). Khorana and his colleagues determined the sequence of another 50 nucleotides adjacent to this site and showed that there existed, close to the beginning of the transcript, a region of symmetry in the nucleotide sequence which is similar to other promoter elements (Loewen *et al.*, 1974). No further direct evidence is yet available to give proof that this symmetric element is the site of tight binding by RNA polymerase.

Various *in vitro* experiments indicated that transcription termination of the su_3^+ gene occurs 100 or more nucleotides past the CCA terminus of the mature tRNA gene sequence (Bikoff and Gefter, 1975; Daniel *et al.*, 1976; Küpper *et al.*, 1978). Recently Landy and his colleagues have identified a sequence of several nucleotides in the ϕ80psu$_3^+$ DNA which coincides with the end of the *in vitro* transcription product (Köpper *et al.*, 1978; Egan and Landy, 1978). This sequence is not identical to those, which are rich in U residues in the gene transcript, which terminate some *E. coli* mRNAs. Thus, the signals for termination (and possibly also for initiation) of transcription of tRNA genes may not be identical to those used for genes coding for non-tRNA gene products.

No data bearing on this problem are available for tRNA genes other that $tRNA^{Tyr}$.

tRNA Biosynthesis in Eukaryotes

Studies of the biosynthesis of suppressor tRNAs other than those of *E. coli* have been achieved only in yeast (Beckmann *et al.*, this volume). The main conclusions from this and other work with yeast are: a) tRNA genes appear to be transcribed as monomeric units; and b) some tRNA gene transcripts contain intervening sequences adjacent to the anticodons in the tRNA moieties. These intervening sequences must be excised during the maturation process. Aside from this latter phenomenon, we presume that other features of precursor tRNA processing are similar to those found in *E. coli.* Enzymatic activities resembling *E. coli* RNase P have been purified from *Bombyx mori* (silkworm), human KB tissue culture cells and chick embryonic tissue (Garber *et al.*, 1978; Koski *et al.*, 1976; Bowman and Altman, unpublished data). Individual precursor tRNAs have been identified from *B. mori* after two-dimensional polyacrylamide gel electrophoresis analysis of ^{32}P-labelled material extracted from posterior silkglands. Analysis of bulk precursor tRNA from *B. mori* and KB cells, as well as results from individual precursors from *B. mori,* indicate that precursor tRNAs from these sources have, on the average, about eight extra nucleotides at each end of the mature tRNA sequences (Garber *et al.*, 1978; Garber, 1977; Koski, 1978). As with yeast, transcripts contain only one tRNA sequence.

The ability of eukaryotic RNase P preparation to cleave *E. coli* precursor tRNAs has been conserved throughout evolution. KB cell RNase P, like *E. coli* RNase P, can be inactivated by pretreatment with micrococcal nuclease. Thus, it is likely that this RNase P also has an essential RNA component. This surprising feature of RNase P is, therefore, not unique to *E. coli* and may be the critical factor in preserving the substrate recognition properties of this enzyme through evolutionary time.

Acknowledgements

E.J. Bowman, R.L. Garber, R. Kole, R.A. Koski and B.C. Stark participated in the work reported here which was done in my laboratory. Research in my laboratory was supported by United States Public Health Service Grant GM-19422.

References

Abelson, J.N., Gefter, M.L., Barnet, L., Landy, A., Russell, R.L. and Smith, J.D. (1970). *J. Mol. Biol.* **47**, 15.
Altman, S. (1971). *Nature New Biol.* **229**, 19.
Altman, S. (1975). *Cell* **4**, 21.
Altman, S. (1978). International Review of Biochemistry, (ed. Clark, B.F.C.), 17, University Park Press, Baltimore (in press).
Altman, S. and Smith, J.D. (1971). *Nature New Biol.* **233**, 35.
Anderson, K.W. and Smith, J.D. (1972). *J. Mol. Biol.* **69**, 349.
Bikoff, E.G. and Gefter, M.L. (1975). *J. Biol. Chem.* **250**, 6240.
Björk, G. (1975). *J. Virol.* **16**, 741.
Björk, G.R. and Neidhardt, F.C. (1975). *J. Bacteriol.* **124**, 99.
Bossi, L. and Cortese, R. (1977). *Nucl. Acids Res.* **4**, 1945.
Carbon, J., Chang, S. and Kirk, L.L. (1974). *Brookhaven Symp. Biol.* **26**, 26.
Chang, S. and Carbon, J. (1975). *J. Biol. Chem.* **250,** 5542.
Ciampi, M.S., Arena, F., Cortese, R. and Daniel, V. (1977). *FEBS Letters* **77**, 75.
Colby, D.S., Schedl, P. and Guthrie, C. (1976). *Cell* **9**, 449.
Daniel, V., Beckman, J.S., Grimberg, J.I. and Zeevi, M. (1976). *In:* Control of Ribosome Synthesis, (eds. Kjeldgaard, N. and Maaloe, Ø.), 268. Munksgaard.
Deutscher, M.P., Foulds, J. and McClain, W.H. (1974). *J. Biol. Chem.* **249**, 6696.
Egan, J. and Landy, A. (1978). *J. Biol. Chem.* (in press).
Garber, R.L. (1977). Ph.D. Thesis, Vale University, New Haven.
Garber, R.L., Siddiqui, M.A.Q. and Altman, S. (1978). *Proc. Nat. Acad. Sci.* **75**, 635.
Goldfarb, A., Seaman, E. and Daniel, V. (1978). *Nature* **273**, 562.
Guthrie, C. (1975). *J. Mol. Biol.* **95**, 529.
Guthrie, C., Seidman, J.G., Altman, S., Barrell, B.G., Smith, J.D. and McClain, W.H. (1973). *Nature New Biol.* **246**, 6.
Ikemure, T., Shimura, Y., Sakano, H. and Ozeki, H. (1975). *J. Mol. Biol.* **96**, 69.
Ilgen, C., Kirk, L.L. and Carbon, J. (1976). *J. Biol. Chem.* **251**, 922.

Kaplan, D.A. and Nierlich, D.P. (1975). *J. Biol. Chem.* **250**, 934.
Koski, R.A. (1978). Ph.D. Thesis, Vale University, New Haven.
Koski, R.A., Bothwell, A.L.M. and Altman, S. (1976). *Cell* **9**, 101.
Köpper, H., Sekiya, T., Rosenberg, M., Egan, J. and Landy, A. (1978). *Nature* **272**, 423.
Loewen, P.C., Sekiya, T. and Khorana, H.G. (1974). *J. Biol. Chem.* **249**, 217.
Lubd, E., Dahlberg, J.E., Lindahl, L., Jaskunas, S.R., Dennis, P.P. and Nomura, M. (1976). *Cell* **7**, 165.
McClain, W.H. (1970). *FEBS Letters* **6**, 99.
McClain, W.H. (1977). *Accounts. Chem. Res.* **10**, 418.
McClain, W.H., Barrell, B.G. and Seidman, J.G. (1975). *J. Mol. Biol.* **99**, 717.
McClain, W.H. and Seidman, J.G. (1975). *Nature* **257**, 106.
Morgan, E.A., Ikemura, T., Lindahl, L., Fallon, A.M. and Nomura, M. (1978). *Cell* **13**, 335.
Ozeki, H., Sakano, H., Yamada, S., Ikemure, T. and Shimura, Y. (1974). *Brookhaven Symp. Biol.* **26**, 89.
Rich, A. and RajBhandary, U.L. (1976). *Ann. Rev. Biochem.* **45**, 805.
Robertson, H.D., Altman, S. and Smith, J.D. (1972). *J. Biol. Chem.* **247**, 5243.
Sakano, H. and Shimura, Y. (1975). *Proc. Nat. Acad. Sci.* **72**, 3369.
Sakano, H., Yamada, S., Ikemure, T., Shimura, Y. and Ozeki, H. (1974). *Nucl. Acids Res.* **1**, 355.
Schaefer, K., Altman, S. and Söll, D. (1973). *Proc. Nat. Acad. Sci.* **70**, 3626.
Schedl, P. and Primakoff, P. (1973). *Proc. Nat. Acad. Sci.* **70**, 2091.
Schedl, P., Primakoff, P. and Roberts, J. (1974). *Brookhaven Symp. Biol.* **26**, 53.
Schedl, P., Roberts, J. and Primakoff, P. (1976). *Cell* **8**, 531.
Schmidt, F.J. and McClain, W.H. (1978). *J. Biol. Chem.* (in press).
Schmidt, F.J., Seidman, J.G. and Bock, R.M. (1976). *J. Biol. Chem.* **251**, 2440.
Seidman, J.G., Barrell, B.G. and McClain, W.H. (1975). *J. Mol. Biol.* **99**, 733.
Seidman, J.G., Schmidt, F.J., Foss, K. and McClain, W.H. (1975). *Cell* **5**, 389.
Singer, C.E., Smith, G.R., Cortese, R. and Ames, B.N. (1972). *Nature New Biol.* **238**, 72.
Smith, J.D. (1972). *Ann. Rev. Genetics* **6**, 235.
Smith, J.D. and Celis, J.E. (1973). *Nature New Biol.* **243**, 66.
Smith, J.D. (1976). *Prog. Nucl. Acid Res. Mol. Biol.* **16**, 25.
Smith, J.D., Barnett, L., Brenner, S. and Russell, R.L. (1970). *J. Mol. Biol.* **54**, 1.
Stark, B.C., Kole, R., Bowman, E.J. and Altman, S. (1978). *Proc. Nat. Acad. Sci.* (in press).
Velten, J., Fukuda, K. and Abelson, J. (1976). *Gene,* **1**, 93.
Vögeli, G., Grosjean, H. and Söll, D. (1975). *Proc. Nat. Acad. Sci.* **72**, 4790.
Vögeli, G., Stewart, T.S., McCutchan, T. and Söll, D. (1977). *J. Biol. Chem.* **252**, 2311.
Wilson, J.H., and Kells, S. (1972). *J. Mol. Biol.* **69**, 39.

CONTROL MECHANISMS OF THE FORMATION OF RIBOSOMAL RNA AND TRANSFER RNA AND THE SYNTHESIS OF GUANOSINE TETRAPHOSPHATE

N.O. KJELDGAARD

Department of Molecular Biology,
University of Aarhus, Denmark.

Introduction

It is a basic feature of bacterial physiology that the cells, and in particular *Escherichia coli* cells, during growth adjust themselves to the most varied environmental conditions as to maintain constant chain growth rates for the polymerization of the macromolecules, DNA, RNA and proteins (Maaløe and Kjeldgaard, 1966). Thus regulatory mechanisms do exist which define a coupling between a potential for biosynthesis and the concentration of the catalytic complex performing the synthesis of such macromolecules. This is certainly true for the collections of molecules participating in protein biosynthesis.

Mechanisms of Control

The ribosomes of *E. coli* are composed of three species of mature rRNA molecules with a total nucleotide length of 5000 and around 55 protein molecules containing about 8500 amino acids.

Under all conditions of exponential growth there appears to be a coordinate regulation of the accumulation of all these ribosomal components and of several other components, proteins and tRNAs, involved in protein synthesis (Kjeldgaard and Gausing, 1974; Pedersen *et al.*, 1978). Similarly, during partial or complete amino acid starvation of a stringent strain, the biosynthesis of rRNA and tRNA as well as that of ribosomal proteins are strongly decreased to the same extend (Lazzarini and Dahlberg, 1971; Ikemura and Dahlberg, 1973; Dennis and Nomura, 1974).

Thus there might be some common regulatory features involved in the formation of all these components. The findings that there are little free rRNA (Lindahl, 1975) or free ribosomal proteins in the cells (Gausing, 1974) indicate that we either have a strong coordination of synthesis, or that scavenging mechanisms exist rapidly destroying any material not finding its place in a normal structural complex.

It is now clear that under certain conditions of slow growth an important breakdown of nascent rRNA molecules occurs. Indications in this direction was obtained for slow growing chemostat cultures (Norris and Koch, 1972) and for cultures growing on a pool carbon source (Pedersen, 1976). A more complete study involving *E. coli* cultures growing exponentially at 30° in several different media has been done by Gausing (1977). Here the rate of accumulation of mature rRNAs was compared with their rate of synthesis, as measured by hybridization of pulse labelled RNA to DNA from phage λ-dilv5 which carries a set of rRNA genes (Jørgensen and Fiil, 1976). The results are summarized in Figure 1, where the rates measured are shown as fractions of total rate of RNA synthesis. All rRNA loci contain the genetic information for the 16S, 23S and 5S rRNA and the primary transcript is a large molecule of about 6000 nucleotides. As the hybridization assay measures this large rDNA transcript, the values obtained for the rate of synthesis must be reduced to 5/6 to obtain values directly comparable to those of the rate of mature rRNA accumulation. The dashed line in Figure 1 represents these corrected values and the distance between this curve and that of mature rRNA represents newly synthesized material which never ends up in ribosomes and therefore must be degraded. At low growth rates this breakdown hits about 70% of the newly synthesized molecules, whereas it is not certain if the breakdown of about 10% shown at the faster growth rates is real on inherent to inaccuracies in the calculations. The amount of tRNA has been determined to 12% of total RNA at the medium to fast growth rates, increasing to about the double at low growth rates (Maaløe, 1978). This suggests that the mature tRNAs are the only entirely stable RNA molecules present in the cells, and under all conditions being transcribed in coordination with rRNA. The increase at low growth rates, therefore, reflects the breakdown of the nascent rRNA. The dotted area in Figure 1 represents at each growth rate the rate of synthesis of rRNAs calculated as 10% of the rate of the rDNA transcript.

It is clear from the results in Figure 1, that the breakdown of nascent rRNA cannot be the sole regulatory principle governing the formation of the mature rRNA. There is also a change in the distribution of RNA polymerases over the genes coding for rRNAs and tRNAs and those coding for mRNAs. Furthermore, during exponential growth conditions there is a strong increase in the total number of polymerase molecules involved in RNA synthesis with increasing growth rates, the overall rate of RNA biosynthesis per unit of protein

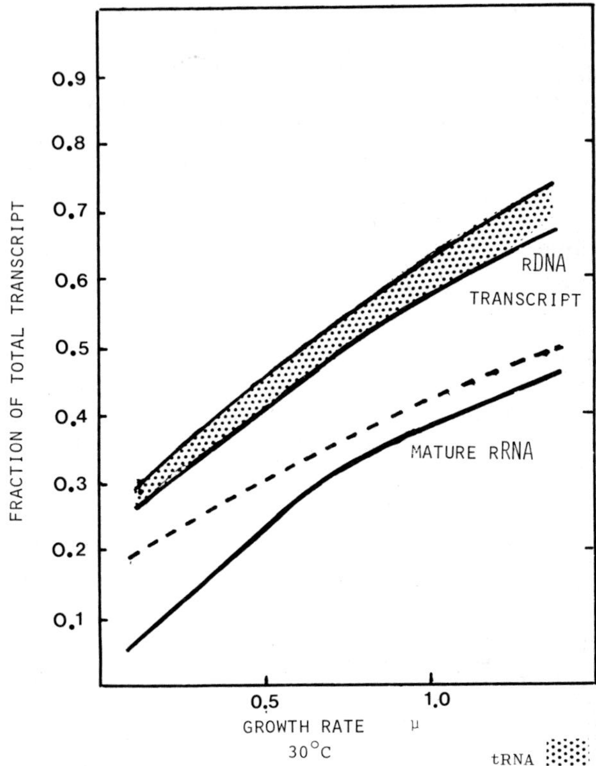

Fig. 1 Relative rates of RNA synthesis at different growth rates. The bottom curve represents the rate of accumulation of mature rRNA in ribosomes. The dashed curve shows the rate of synthesis of nascent nucleotide sequences (5000 nucleotides) corresponding to the mature rRNA. The distance between these two lines indicates the quantity of breakdown of nascent rRNA. The third curve shows the rate of synthesis of the 6000 nucleotide primary transcript of the *rrn* genes. The dotted area indicates the rate of tRNA synthesis. The top area of the figure represents the rate of mRNA synthesis. The figure is redrawn from Gausing (1977).

increasing from 0.002 min^{-1} at a generation time of 300 min at 30° to 0.012 min^{-1} at a generation time of 45 min (Gausing, 1977).

Transcription of rRNA and tRNAs

In the haploid genome of *E. coli* there appears to be seven loci for rRNA (Kiss *et al.*, 1977). This was shown by hybridization of rRNA to fragments of total *E. coli* DNA formed by a restriction enzyme (Bam H1) known not to cut within the rRNA genes. The fragments were separated by electrophoresis and a subsequent hybridization with 16S or 23S rRNA was found to have equal intensity in seven distinct bands. Six of these sets of genes have been isolated on bacteriophages or on plasmids and have been located on the *E. coli* genetic map (Nomura *et al.*, 1977). All loci are transcribed in the order 16S, 23S and 5S rRNA with a spacer region between the 16S gene and the 23S gene carrying the information for a few tRNA genes (Nomura *et al.*, 1977).

As all rRNA loci are situated close to the origin of replication, the gene dosage during exponential growth must be close to 10 rRNA loci per replicating genome (Figure 2).

Jarry and Rooset (1973a, 1973b) have identified a number of nucleotide sequence heterogenieties in the 5S RNA molecules of *E. Coli* K12 and MRE 600 and used these to map the genetic location of a number of rRNA genes. Measuring the frequency of occurrence of these heterogenieties in 5S RNA isolated from cells grown at widely different growth rates and during amino acid starvation of a relaxed strain, Morgan and Kaplan (1976) found evidence that the different 5S RNA genes are expressed with equal efficiency under all growth conditions. We can therefore assume that all seven rRNA loci always are transcribed at the same frequency.

From the work of Fiers and his colleagues on the nucleotide sequence of MS2 RNA and the amino acid sequences of the MS2 coded proteins it is known that all 61 possible codons can be translated in *E. coli* (cf. Fiers, 1975). Although not many species of tRNA have yet been identified in *E. coli* we can expect that all tRNA species responding to the 61 codons can be found.

A few tRNA genes corresponding to species mutated to translational suppressors have been mapped on the *E. coli* chromosome (Smith, 1976; Smith, chapter 6 this volume). The genes for tRNAs of AlalB, Ilel and Glu2 has been identified on the spacer regions of the rRNA genes and the tRNAs of Aspl and Trp have been found close to the rRN genes (Ikemura and Nomura, 1977). Recently a rather large number of additional tRNA genes has been mapped using F′ factors and λ transducing phages (Ikemura and Ozeki, 1977). As seen in Figure 2, where the tRNA gene positions are marked on the outside of the genetic map circle, these genes are widely scattered around the chromosome. In all cases using strains polyploid for a given tRNA gene, Ikemura and Ozeki (1977) found a gene dosage effect on the amount of the proper tRNA. It therefore seems possible that the distribution between major and minor isoaccepting tRNA species could be caused by different numbers of gene copies and different map locations of the tRNA genes.

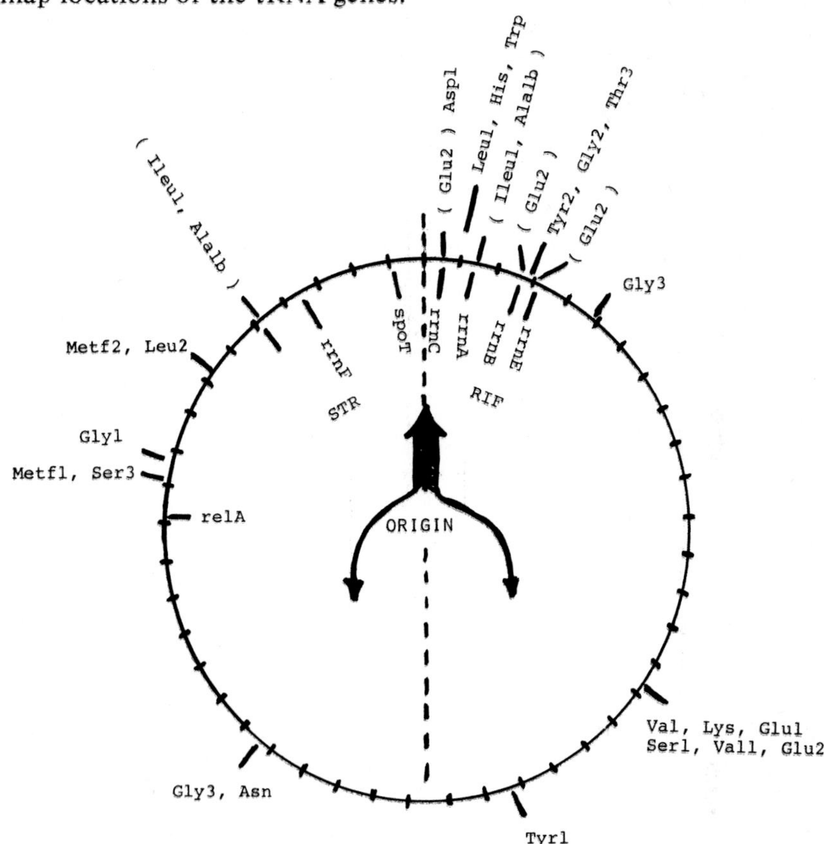

Fig. 2 Genetic map location of rRNA and tRNA genes of *E. coli* related to the origin of DNA replication. The rRNA genes are shown at the inside and the tRNA genes at the outside of the circle. The tRNA genes in brackets are those located in the spacer region between the 16S and the 23S of the rrn genes. The divisions on the circle are 100 kilobasepairs.

With 75 nucleotides per mature tRNA the total amount of genetic information coding for tRNA molecules must be 5000-6000 basepairs, or close to that of one rRNA gene. With a rate of synthesis for mature tRNA of about 1/10 of that of the rDNA transcript, it seems that the rate of transcription of the tRNA genes and the of the rRNA genes could be identical.

Exhaustive hybridization between RNA isolated from *E. coli* grown exponentially in glucose minimal medium to *E. coli* DNA indicated that one strand equivalent of the genome is transcribed, albeit with widely varying frequency for different parts of the genome (Hahn, *et al.,* 1977). Three major classes were recognized: A) a class comprising about 10% of the genome coding for 90-95% of the mRNA. The rRNA and the tRNA also belong to this group of busy genes; B) a class comprising about 55% of the genome and coding for about 5% of the mRNA. Calculated for an average length of the transcript of 1700 nucleotides, this corresponds to an abundance of these mRNA species of around one molecule per 30 cells; C) a class comprising about 30% of the genome, coding for mRNAs with an abundance of about one molecule per 1000 cells.

It is thus obvious that it is only about 10% of the genome or 4500 kilobases which are actively being transcribed in glucose grown cells, and for all practical general purposes of regulation the rest of the genome can be neglected.

The total genomic size for the rRNA and tRNA primary transcript is around 750 kilobases and as seen from Figure 1 these regions are responsible for the synthesis of about 50% of the RNA in glucose minimal medium ($\mu = 0.6$). The rRNA and tRNA genes must therefore accommodate about six times more polymerase molecules than the average busy gene.

In vitro Synthesis of rRNA

Studies of the synthesis of RNA by isolated *E. coli* nucleoids have shown that RNA chain elongation occurs by the endogenous RNA polymerases, that no new RNA chains are initiated and that the rRNA accounts for about 50% of the transcript (Pettijohn *et al.,* 1970; Murooka and Lazzarini, 1973; Muto, 1975). Addition of exogenous RNA polymerase leads to initiation of new polynucleotide chains of which only a few per cent has been found to be rRNA chains (van Ooyen *et al.,* 1975a). They reported, however, that the overall rate of rRNA synthesis was the same both for the endogenous and the exogenous RNA polymerase, whereas the addition of polymerase caused a strong increase of the synthesis of other RNA chains.

Similar low percentages of rRNA have been found repeatedly with *E. coli* DNA as template (Haseltine, 1972; Travers, 1973). In all these cases it is likely that it is not as much the rRNA synthesis which is decreased, as irregular initiations increase the synthesis of other RNA molecules not normally synthesized *in vivo.* DNA from transducing bacteriophages carrying one of the rrn genes, which accordingly accounts for a larger fraction of the DNA, are much better templates for the synthesis of rRNA, yielding about 25% of the transcript as rRNA (Jørgensen and Fiil, 1976; Travers and Baralle, 1976; Oostra, *et al.*, 1977). During amino acid starvation of a stringent strain, there is almost an arrest of rRNA synthesis, and simultaneously an accumulation of guanosinetetraphosphate (ppGpp). Using nucleoids isolated from an amino acid starved stringent strain Muto (1977) found a decrease in fraction of rRNA synthesized by the endogenous RNA polymerase molecules. This indicates that the decrease in rRNA synthesis is caused by a decrease frequency of initiation. No such decrease was observed in a starved relaxed strain. This is in agreement with the finding of Pedersen (1976) that there are no major breakdown of nascent rRNA during amino acid starvation of a stringent strain.

The effect of ppGpp on the *in vitro* synthesis of rRNA has been studied using different DNA templates. With nucleoids in the presence of added RNA polymerase as well as with isolated total *E. coli* DNA used as template, van Ooyen *et al.*, (1975b) found a specific inhibition of rRNA synthesis reaching about 70% at a ppGpp concentration of 1.4m*M*. Similar results were obtained by Reiness *et al.* (1975). Jørgensen and Fiil (1976) using DNA from λ*dilv*5 as template found a 75% inhibition of rRNA synthesis at a ppGpp concentration of 0.25m*M* and Travers (1976a) using ϕ80 *d3rrnB*$^+$ DNA as template found a K_i of 150 μM ppGpp for rRNA synthesis. It was shown that the effect of ppGpp was to decrease the initiation of new rRNA chains (van Ooyen *et al.*, (1976). It should, however, be emphasized that in several cases it has not been possible to demonstrate an effect of ppGpp (Haseltine, 1972; Travers, 1976b). When using DNA carrying the $tRNA^{Tyr}$ gene as template the effect of ppGpp has been doubtfull (Manley *et al.*, 1973; Reiness *et al.*, 1975), but recently Debenham and Travers (1977) have found the same system to be sensitive with a K_i of 10 μM ppGpp.

The Pool of ppGpp

During amino acid starvation of stringent strains of *E. coli* the internal concentration of ppGpp rapidly increases to 3-4m*M*, or close to the normal concentration of GTP (Table I). It is thus possible that the accumulation of ppGpp is a direct cause of the strong decrease of rRNA biosynthesis. In some cases it has been found that the basal level of ppGpp during exponential growth vary inversely with the growth rate around a concentration of 0.2m*M*. It is not at all certain whether these variations by themselves explain the changes with the growth conditions of the distribution of the RNA polymerase molecules over the genes for rRNA, tRNAs and the other genes to be

TABLE I

Concentrations in Average Glucose Grown E. coli *Cell*

	m*M*
GTP	3.2
ATP	7.4
UTP	3.7
CTP	1.8
ppGpp Basal level	0.2
ppGpp, maximum at amino acid starvation	3.3

Nucleotide concentrations in an average glucose grown *E. coli* cell.

transcribed. The effect of ppGpp, however, is not limited to that of rRNA and tRNA, but with several coupled transcriptional-translational systems a stimulation or an inhibition is seen upon ppGpp addition (Gallant and Lazzarini, 1976; Yang *et al.*, 1974; Nomura, 1977).

The pool of ppGpp in *E. coli* cells is established by an interplay of a biosynthetic and a degradative pathway (Figure 3). Any variation in the rate of synthesis, or the rate of degradation, will lead to changes in the pool size.

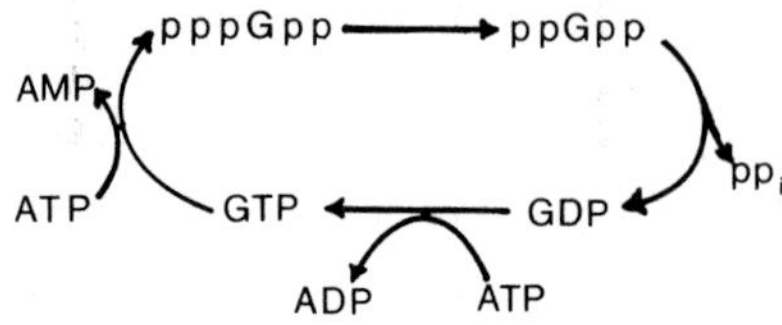

Fig. 3 Turnover pathway of the guanosinepolyphosphates

The biosynthesis of ppGpp from GTP and ATP is catalyzed by the product of the *rel*A gene, the 'stringent factor' or ATP:GTP 3′ pyrophosphotransferase (Haseltine *et al.*, 1972; Friensen *et al.*, 1976). The pppGpp is hydrolyzed into ppGpp, but although most GTPases will catalyze the conversion, it is not with certainty known if a specific enzyme or enzyme complex is primarily responsible for the reaction. The further breakdown of ppGpp is undertaken by the gene product of the *spo*T locus (Just Justesen, personal communication; Sy, 1977), the immediate breakdown products being GDP and pyrophosphate (Justesen, 1978).

ATP:GTP 3′ Pyrophosphotransferase

This enzyme of a molecular weight of 77000 has been isolated in a pure and a highly active state from a strain carrying a *λdrelA pyrG* bacteriophage (Pedersen and Kjeldgaard, 1977). The enzyme is highly unstable during the preparation, which therefore has to be carried through without undue delays. By itself the enzyme has very little activity, but it is highly activated by 70S ribosomes, a mRNA and an uncharged tRNA corresponding to the codon occupying the A-site on the ribosome (Haseltine and Block, 1974). In the absence of ribosomes, the pyrophosphotransferase can be activated by the addition of methanol (Sy *et al.*, 1973) or ethanol at concentrations around 20%. The turnover numbers for the ribosome stimulated enzyme at 37° were found to be about 10000 molecules of pppGpp per enzyme molecule per minute, and for the ethanol activated enzyme at 25° to around 300. The $S_{0.5}$ (K_m) for the ribosome stimulated assay was found to be 0.53 m*M* for GTP, and for the ethanol activated assay to be 0.44 m*M* and 0.48 m*M*, respectively (Justesen, 1978). Whereas impure enzyme preparations form both pppGpp and ppGpp when GTP and ATP are used as substrates, the purified enzyme is devoid of GTPases and has pppGpp as the reaction product. The pyrophosphotransferase concentration in the *E. coli* cells is very low. By immunological methods (Pedersen and Kjeldgaard, 1977) and by two dimensional gel electrophoresis (Friesen *et al.*, 1976) the amount of enzyme was estimated to 0.005% of the proteins in wild type strains. This corresponds to about 1 molecule of enzyme per 200 ribosomes in cells grown in glucose minimal medium. The concentration of the enzyme show very little variation with the growth rate of the culture (Lund, 1978). This obviously means that the number of enzyme molecules per ribosome must vary from about one per 500 ribosomes in fast growing cells to about one per 100 ribosomes in slow growing cells. The suggestion of Schmalf *et al.*, (1978) that the stringent factor is present in equimolar amount to the ribosomes, is based on their use of a rather inactive enzyme as a reference.

Although low in concentration, the high activity of the pyrophosphotransferase can easily explain the strong synthesis of ppGpp in stringent cells during amino acid starvation, where uncharged tRNA molecules corresponding to the missing amino acid accumulate. However, since only few of the ribosomes will be located at a codon corresponding to the missing amino acid, it seems clear that the pyrophosphotransferase has to find its place on those blocked ribosomes. Therefore, the enzyme cannot be too strongly bound to the ribosomes.

The Role of tRNA

As stated above, uncharged tRNA is essential for the activity of the pyrophosphotransferase when attached to the ribosomes (Pedersen *et al.*, 1973). Charging of the tRNA completely abolishes the effect of the enzyme-ribosome complex. Other changes at the 3′-end of the tRNA, such as peroxidation and removal of the terminal adenosine residue also destroys the stimulating activity (Lund *et al.*, 1973). Sprinzl and Richter (1976) have tested different 3′-adenosine modifications of $tRNA^{Phe}$ and found no stimulating activity for the 3′-deox and the 3′-amino modifications. The 2′-deoxy-adenosine modification, however, was found to be fully active (Table II).

TABLE II

Effect of tRNA Modifications on the Activity of ATP:GTP 3′-pyrophosphotransferase

	Specific activity U/MG	
	Ribosome activated	Ethanol activated
No tRNA	7.1	3.7
E. coli bulk tRNA	56.0	5.9
Yeast $tRNA^{Phe}$	94.7	5.5
- -CC-3′NH_2 A	17.8	3.8
- -CC-3′ dA	14.2	4.0
- -CC-2′dA	70.2	2.1
- -CC-OH	38.1	3.9
- -CC-2′ dA + $tRNA^{Phe}$	-	5.4

Ribosome activated system + 0.1 mg/ml tRNA
22% ethanol activated system + 0.33 mg/ml tRNA
The activity is given in units/mg enzyme. One unit corresponds to the formation of 1 μmole pppGpp per min. The assays were performed in the presence of ribosomes (1 mg/ml), polyU (0.1 mg/m1), ATP, GTP and tRNAs at 0.1 mg/ml. The ribosome activated assay was done at 37°. The ethanol activated assay was done as described in Figure 4, with 22% ethanol and tRNAs at 0.33 mg/ml.

These experiments were performed with yeast tRNA which can fully replace *E. coli* tRNA. All other changes in the tRNA molecule has been found to have no influence on the stimulating activity of the pyrophosphotransferase (Ofengand and Liou, 1978). The earlier reported lack of stimulating activity of tRNA from a yeast mutant strain impaired in the synthesis of the m_2^2G residue (Lund *et al.*,

1973) cannot be repeated (Kjellin-Stråby, personal communication). Other submethylated tRNA species have been found active in stimulating ppGpp formation (Lund *et al.*, 1973).

Richter (1976) suggested that the pyrophosphotransferase is of importance for the binding of the uncharged tRNA to the ribosomes. It is thus a question if the stimulating effect of uncharged tRNA is a direct action on the enzyme, or an effect mediated through the ribosomes. Using the highly purified enzyme, and the activation of the catalytic activity by ethanol, we have recently found that uncharged tRNA does stimulate the formation of pppGpp (Figure 4), although at rather high tRNA concentrations.

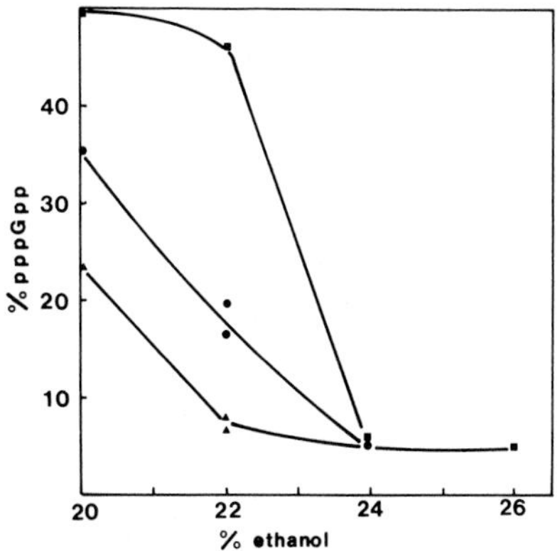

Fig. 4 Effect of uncharged tRNA on the activity of the ATP:GTP pyrophosphotransferase at different ethanol concentrations. The assay contained in 30 μl volumes the enzyme (0.7 μg/ml), ATP (3.3 m*M*), GTP (0.4 m*M*) without tRNA (▲—▲) or with bulk *E. coli* tRNA at (■—■) 0.67 mg/ml or (●—●) 0.033 mg/ml. Ethanol was present at the indicated concentrations. It should, however, be pointed out that it is extremely difficult to establish a very precise ethanol concentration in an assay mixture of 30 μl. The synthesis of pppGpp after 60 min at 25° was measured as % conversion of the GTP added. From Justesen (1978).

The effect which most likely is a protection of the enzyme against the rather high ethanol concentrations (20-22%), is specific for the uncharged tRNA. Modifications of the 3′ -end of tRNA abolishes the stimulation (Table II). This is also true for the 2′ -deoxy-adenosine derivative found to stimulate the ribosome coupled system. Amino-acylation of tRNA even results in an inhibition of the synthesis of pppGpp (Figure 5) (Justesen, 1978). It thus seems clear that the pyrophosphotransferase contains a site specific for tRNA molecules, besides a site of attachment to the ribosomes, as well as catalytic sites for GTP and ATP.

Richter *et al.*, (1974) reported that the tetranucleotide fragment of tRNA, TpΨpCpGp stimulated the ribosome coupled activity of the pyrophosphotransferase. We have tested this fragment in the ethanol stimulated assay system and found no effect, just as we were unable to repeat the results for the ribosome coupled system.

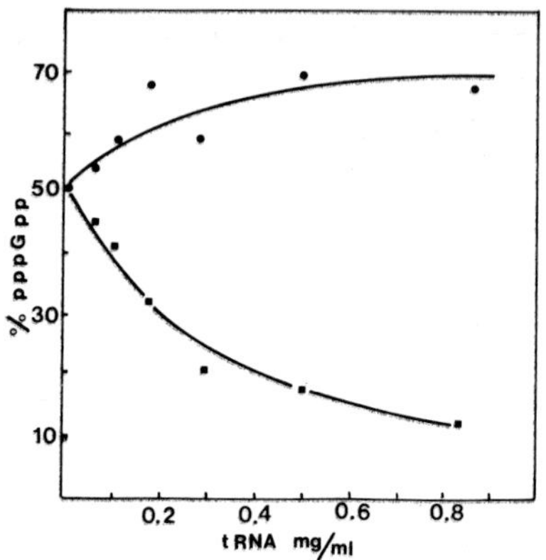

Fig. 5 Effect of uncharged $tRNA^{Val}$ and val-$tRNA^{Val}$ on the activity of ATP: GTP pyrophosphotransferase. The assays were performed as in Figure 4 with 20% ethanol. Uncharged $tRNA^{Val}$ (● —●) and val-$tRNA^{Val}$ (■ —■) were added at the indicated concentrations. From Justesen (1978).

Conclusion

Through the ppGpp biosynthesis, the cells possess a mechanism which is able to monitor the concentration of amino acids and to amplify that signal. As the protein synthesis is the major effort in cell duplication such a monitoring device can be of major importance for the establishment of a balance between the biosynthetic potential and the supply of amino acids.

The pleiotropic effect of ppGpp makes this nucleotide a good candidate for the fulfilment of such a general regulatory function. It is clear from the results of the *in vitro* transcription of rRNA genes that the ppGpp has a direct effect on the RNA polymerase. Travers (1976) has proposed that the extremely large RNA polymerase complex exists in a number of different conformational states, with varying affinity towards different types of promoter regions. The ppGpp is one of several effectors which might result in such a change in specificity, lowering the number of initiations in certain classes of promoters and increasing the initiations in others.

It has always been an enigma how the cells secure the strict coordination in ribosome synthesis between the RNA components which enters directly into the particles and the ribosomal proteins, which are formed in a copy number of 20-30 per mRNA molecule transcribed. Ideally this would require a 20 fold decreased frequency of initiations of the ribosomal protein messengers compared to that of the rRNAs. It is clear from the observed degradation of nascent rRNA that such a constant ratio in initiation does not exist under all conditions. It might therefore be that we intuitively expect a refinement of the regulatory functions of the cells, which are not met and that scavengering mechanisms might be more widespread than hitherto imagined.

Acknowledgement

I want to thank my collaborators Kirsten Gausing, Just Justesen and Torben Lund for their permission to use and redraw their data.

References

Block, R. and Heseltine, W.A. (1975). *In:* Ribosomes (eds. Nomura, M., Tissieres, A. and Lengyel, P.), 747, Cold Spring Harbor Laboratory, Cold Spring Harbor, New York.

Debenham, P. and Travers, A. (1977). *Eur. J. Biochem.* **72**, 515.

Dennis, P.P. and Nomura, M. (1974). *Proc. Nat. Acad. Sci.* **71**, 3819.

Fiers, W. (1975). *In:* RNA Phages (ed. Zinder, N.), 353, Cold Spring Harbor Laboratory, Cold Spring Harbor, New York.

Friesen, J.D., Parker, J., Watson, R.J., Fiil, N.P., Pedersen, S. and Pedersen, F.S. (1976). *J. Bacteriol.* **127**, 917.

Gallant, J. and Lazzarini, R.A. (1976). *In:* Protein Synthesis: A Series of Advances (ed. McConkey, E.H.) 309, Dekker, New York.

Gausing, K. (1974). *Mol. Gen. Genet.* **129**, 61.

Gausing, K. (1977). *J. Mol. Biol.* **115**, 335.

Hahn, W.E., Pettijohn, D.E. and Van Ness, J. (1977). *Science* **197**, 582.

Haseltine, W.A. (1972). *Nature (London)* **235**, 329.

Haseltine, W.A., Block, R., Gilbert, W. and Weber, K. (1972). *Nature (London)* **238**, 381.

Ikemura, T. and Dahlberg, J.F. (1973). *J. Biol. Chem.* **248**, 5033.

Ikemura, T. and Nomura, M. (1977). *Cell* **11**, 779.

Ikemura, T. and Ozeki, H. (1977). *J. Mol. Biol.* **117**, 419.

Jarry, B. and Rosset, R. (1973a). *Mol. Gen. Genet.* **121**, 151.

Jarry, B. and Rosset, R. (1973b). *Mol. Gen. Genet.* **126**, 29.

Justensen, J. (1978). Ph.D. thesis, Aarhus University, Aarhus, Denmark.

Jørgensen, P. and Fiil, N.P. (1976). *In:* Control of Ribosome Synthesis (eds. Kjeldgaard, N.O. and Maaløe, O.) Alfred Benzon Symposium IX, Munksgaard, Copenhagen.

Kiss, A., Sain, B. and Venetianer, P. (1977). *FEBS Letters* **79**, 77.

Kjeldgaard, N.O. and Gausing, K. (1974). *In:* Ribosomes (eds. Nomura, M.,

Tissieres, A. and Lengyel, P.), 369, Cold Spring Harbor Laboratory, Cold Spring Harbor, New York.
Lazzarini, R.A. and Dahlberg, A.E. (1971). *J. Biol. Chem.* **246**, 420.
Lindahl, L. (1975). *J. Mol. Biol.* **92**, 15.
Lund, E., Pedersen, F.S. and Kjeldgaard, N.O. (1973). *In:* Ribosome and RNA Metabolism, 307, Slovak Academy of Sciences, Bratislava.
Lund, T. (1978). Ph.D. thesis, Aarhus University, Aarhus, Denmark.
Maaløe, O. (1978). *In:* Biological Regulation and Development (ed. Goldberger, R.F.), Plenum, New York. (in press).
Maaløe, O. and Kjeldgaard, N.O. (1966). Control of Macromolecular Synthesis, W.A. Benjamin, New York.
Manley, J., Reiness, G., Zubay, G. and Gefter, M.L. (1973). *Arch. Biochem. Biophys.* **157**, 50.
Morgan, E.A. and Kaplan, S. (1976). *Biochem. Biophys. Res. Commun.* **68**, 969.
Murooka, Y. and Lazzarini, R.A. (1973). *J. Biol. Chem.* **248**, 6248.
Muto, A. (1975). *Mol. Gen. Genet.* **138**, 1.
Muto, A. (1977). *Mol. Gen. Genet.* **152**, 153.
Nomura, M. (1977). *In:* Nucleic Acid - Protein Recognition (ed. Vogel, H.J.), 443, Academic Press, New York.
Nomura, M., Morgan, E.A. and Jaskunas, S.R. (1977). *Annu. Rev. Genet.* **11**, 297.
Norris, T.E. and Koch, A.L. (1972). *J. Mol. Biol.* **64**, 633.
Ofengand, J. and Lion, R. (1978). *Nucl. Acids Res.* **5**, 1325.
Oostra, B.A., van Ooyen, A.J.J. and Gruber, M. (1977). *Mol. Gen. Genet.* **152**, 1.
Pedersen, F.S. and Kjeldgaard, N.O. (1977). *Eur. J. Biochem.* **76**, 91.
Pedersen, F.S., Lund, E. and Kjeldgaard, N.O. (1973). *Nature New Biol.* **243**, 13.
Pedersen, S. (1976). *In:* Control of Ribosome Synthesis (eds. Kjeldgaard, N.O. and Maaløe, O.). 345, Alfred Benzon Symposium IX, Munksgaard, Copenhagen.
Pedersen, S., Bloch, P.L., Reeh, S. and Neidhardt, F.C. (1978). *Cell* **14**, 179.
Pettijoh, D.E., Clarkson, K., Kossman, C.R. and Stonington, O.G. (1970). *J. Mol. Biol.* **52**, 281.
Reiness, G., Yang, H.L., Zubay, G. and Cashel, M. (1975). *Proc. Nat. Acad. Sci.* **72**, 2881.
Richter, D. (1976). *Proc. Nat. Acad. Sci.* **73**, 707.
Richter, D., Erdmann, V.A. and Sprinzl, M. (1974). *Proc. Nat. Acad. Sci.* **71**, 3226.
Schmalf, H., Fehr, S. and Richter, D. (1978). *Hoppezeyler's Z. Physiol. Chem.* **359**, 125.
Smith, J.D. (1976). *In:* Progress in Nucleic Acid Research and Molecular Biology (ed. Cohn, W.E.), 25, Academic Press, New York.
Sprinzl, M. and Richter, D. (1976). *Eur. J. Biochem.* **71**, 171.
Sy, J. (1977). *Proc. Nat. Acad. Sci.* **74**, 5529.
Sy, J., Ogawa, Y. and Lipman, F. (1973). *Proc. Nat. Acad. Sci.* **70**, 2145.
Travers, A. (1973). *Nature (London)* **244**, 15.
Travers, A. (1976a). *Mol. Gen. Genet.* **147**, 225.
Travers, A. (1976b). *FEBS Letters* **69**, 195.
Travers, A. (1976c). *Nature (London)* **263**, 641.
Travers, A. and Baralle, F.E. (1976). *In:* Control of Ribosome Synthesis (eds.

Kjeldgaard, N.O. and Maaløe, O.), 241, Alfred Benzon Symposium IX, Munksgaard, Copenhagen.
Van Ooyen, A.J.J., Gruber, M. and Jørgensen, P. (1976). *Cell* **8**, 123.
Van Ooyen, A.J.J., De Boer, H.A., Ab, G. and Gruber, M. (1975a). *Biochim. Biophys. Acta* **395**, 128.
Van Ooyen, A.J.J., De Boer, H.A., Ab, G. and Gruber, M. (1975b). *Nature (London)* **254**, 530.
Yang, H.-L., Zubay, G., Urm, E., Reiness, G. and Cashel, M. (1974). *Proc. Nat. Acad. Sci.* **71**, 63.

ORGANIZATION AND EXPRESSION OF YEAST tRNA GENES

J.S. BECKMANN, P.E. JOHNSON, G. KNAPP, H. SAKANO, S.A. FUHRMAN, R.C. OGDEN AND J. ABELSON

Departments of Chemistry and Biology, University of California, San Diego, California, USA.

Introduction

Recombinant DNA technology provides an approach to study the fine structure and organization of eukaryotic genomes with a precision previously unattainable. Moreover, it allows us to ask questions regarding the control and mechanism of gene expression. The yeast *Saccharomyces cerevisiae* has been chosen for study by many laboratories because it is a eukaryotic system well suited for detailed analysis of its genetic organization. For example, the ribosomal RNA genes, reiterated some 140 times per haploid genome (Schweizer *et al.*, 1969; Rubin and Sulston, 1973; Feldman, 1976), are tandemly repeated (Cramer *et al.*, 1977) and it has been suggested that they might be present on only one chromosome (Petes *et al.*, 1977). The transfer RNA genes, of which there seem to be about 360 copies (Schweizer *et al.*, 1969; Rubin *et al.*, 1973), are located on several chromosomes (Capecchi *et al.*, 1975; Piper *et al.*, 1976; Brandiss *et al.*, 1976; Gesteland *et al.*, 1976). The genes coding for $tRNA^{Tyr}$ have been mapped and they are located at eight different loci on several chromosomes (Gilmore *et al.*, 1971).

The construction of a bank of *Escherichia coli* clones containing yeast DNA inserted into the plasmid pBR313 (Bolivar *et al.*, 1977) and the subsequent identification of those clones containing rRNA or tRNA genes are reported. Data previously described in part (Beckmann *et al.*, 1977) are discussed here with regard to the distribution of several specific tRNA genes among these clones. The data suggest that the yeast tRNA genes are neither generally present as tandemly repeated units nor are they significantly clustered, although limited clustering does exist. Some of these tRNA genes have been further characterized. One clone carrying sequences coding for $tRNA_3^{Arg}$,

$tRNA^{Asp}$ and $tRNA^{Tyr}$ was selected for detailed restriction analysis. The nucleotide sequence of the region containing $tRNA_3^{Arg}$ and $tRNA^{Asp}$ has been determined. These data, though still preliminary, will be presented here.

In addition, a study of the *in vivo* expression of yeast tRNA genes has been undertaken. Precursors, larger than 4S RNA, for at least 5 tRNAs ($tRNA^{Tyr}$, $tRNA^{Phe}$, $tRNA_3^{Leu}$, $tRNA^{Ser}_{UCG}$ and $tRNA^{Trp}$) were isolated and identified. It was shown that the precursors for $tRNA^{Tyr}$, $tRNA^{Phe}$, $tRNA_3^{Leu}$ and $tRNA^{Trp}$ have the same 5′- and 3′- ends as the mature tRNAs. Their increased size is solely due to the presence of intervening sequences (Knapp *et al.*, 1978, and unpublished data). The availability of these RNA precursors as substrates allowed us to demonstrate that yeast contains an activity that can excise the intervening sequences and religate the ends.

Cloning Experiments

A large collection of *E. coli* clones carrying yeast DNA fragments was established which could subsequently be screened for those containing sequences coding for the 4S, 5S and 5.8S RNAs. The vehicle used for these experiments was the plasmid pBR313 (Bolivar *et al.*, 1977) which carries antibiotic resistance genes for tetracycline (tet^r) and ampicillin (amp^r). Foreign DNA can be inserted by ligation into this plasmid at a number of single restriction sites. In our experiments we used endo R· *Hind* III and *Bam*H I restriction endonucleases to cleave the plasmid DNA. Both these enzymes, recognizing respectively the DNA sequences A*AGCTT and G*GATCC (Roberts, 1976), cleave the circular DNA at a single site in the tet^r gene yielding linear molecules with 5′ -single-stranded ends. Yeast DNA was purified from *S. cerevisiae* strain X2180-1A (this strain, originally from Dr. Mortimer, was described by Duntze *et al.*, 1970) and cleaved with endo R· *Hind* III or *Bgl* II. The latter enzyme recognizes and cuts an A*GATCT sequence and produces the same sticky ends as endo R· *Bam*H I (Roberts, 1976). The yeast DNA fragments were mixed in three-fold molar excess with cleaved plasmid DNA and the mixture was ligated (Velten *et al.*, 1976) to produce a random collection of recombinant plasmids. These plasmids were introduced into *E. coli* strain C600 SF8 (Struhl *et al.*, 1976) by transformation (Velten *et al.*, 1976). After addition of the DNA, the transformation mixture was divided into a number of separate growth tubes in order to minimize the chances of obtaining sibling clones. From 33 independent transformations, about 4000 ampicillin resistant and tetracycline sensitive clones were isolated.

Screening for Yeast rRNA and tRNA Genes

This collection of clones containing yeast DNA was screened by the colony hybridization technique (Grunstein and Hogness, 1975; Beckmann *et al.*, 1977) to detect those clones with DNA sequences complementary to yeast 4S, 5S and 5.8S RNA. *In vivo* labeled ^{32}P-RNAs were utilized as probes in these experiments.

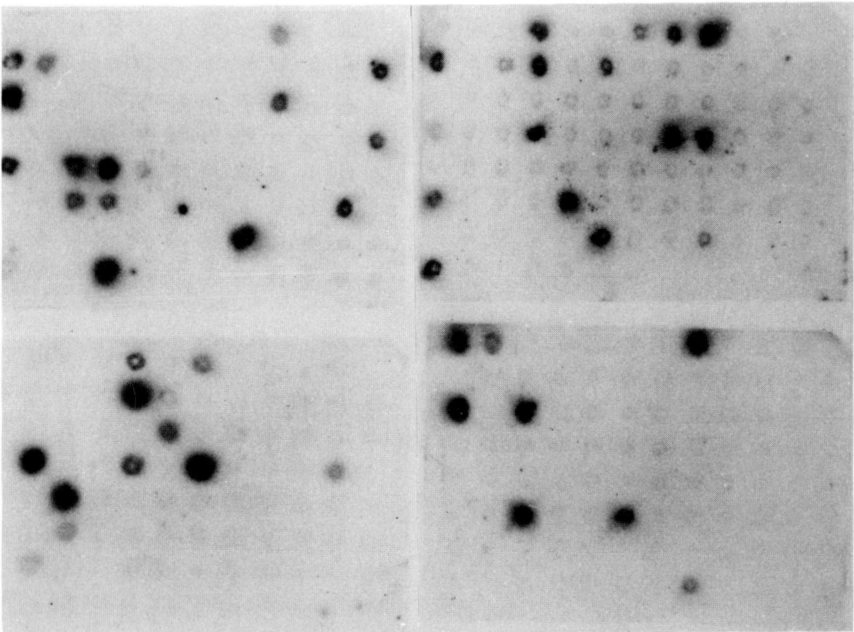

Fig. 1 Screening for recombinant clones carrying yeast tRNA genes. Three hundred and eighty clones carrying fragments of the yeast genome inserted into pBR313 are shown after having been assayed for the presence of yeast tRNA genes by hybridization with *in vivo* ^{32}P-labeled 4S RNA.

One hundred eighty-seven clones hybridized specifically with 4S RNA (Figure 1 illustrates the detection of clones hybridizing with 4S RNA); 127 and 123 clones hybridized with 5S and 5.8S RNAs, respectively. Only two of these clones contained both 5S and 5.8S RNA genes. The 5S and 5.8S clones also hybridized with 4S RNA, but this hybridization was shown to be due to the presence of contaminating ribosomal RNA sequences in the 4S RNA and was not attributable to tRNA. Five of these clones, however, exhibited true hybridization to tRNAs, but it has not been determined whether this hybridization represents true linkage between certain tRNA genes and rRNA genes or whether this linkage was artificially generated by the ligation of multiple fragments with the plasmid vectors.

Screening for Individual tRNA Genes

The collection of 187 clones hybridizing to 4S RNA was screened with purified tRNAs. Prior to the final screening process each clone was purified and retested for hybridization to 4S RNA. Two grids were formed containing the entire 4S collection. The results of screening these clones with 18 different tRNA species have been previously reported (Beckmann *et al.*, 1977). This screening has now been extended to a total of 26 non-overlapping tRNAs (Table I).

Pure labeled tRNAs were obtained in two ways: by *in vivo* labeling or 5′-end labeling of individual unlabeled tRNAs (Beckmann *et al.*, 1977). The purity and in some cases the identity of the individual *in vivo* labeled tRNAs were ascertained by T1 ribonuclease fingerprinting. These labeled tRNAs were then hybridized to the tRNA clone bank. The results of hybridization with 26 pure tRNA probes are given in Tables, I, II and III. One hundred thirty-four of the 187 clones in the 4S collection hybridize to at least one of the tRNAs (71% of our collection). Only 22 of the 4S clones hybridize to more than one tRNA species. Six of these clones are found to hybridize to three different tRNAs and can be subdivided into three groups according to the tRNAs to which they hybridize (Table III). It is interesting to note that a possible cluster of Arg_3, Asp and R9 tRNA genes occurs four times in our collection.

In these Tables only clones that were shown to hybridize to a defined tRNA are listed. It is worth noting that this collection was also screened for the presence of $tRNA_f^{Met}$ and $tRNA_m^{Met}$ (provided by P. Sigler), and $tRNA_{suf5}^{Gly}$ (M. Culbertson, unpublished data) and was found not to contain copies of these genes.

The distribution of various yeast tRNA genes among the recombinant clones suggests some tentative conclusions regarding the organization of these genes in the yeast genome. The average molecular weight of an endo R· *Hind* III or *Bgl* II fragment is 2.5×10^6 daltons. Thus, each fragment represents one four thousandth of the yeast haploid genome (ca. 10^{10} daltons) (Bhargava and Halvorson, 1971). We have isolated 4000 clones at random. If the 360 tRNA genes contained in a haploid genome (Schweizer *et al.*, 1969, Petes *et al.*, 1977) were widely spaced we would have expected to isolate by chance about 360 clones each containing a gene for only one tRNA species (although in our collection the chance of obtaining a *particular* tRNA gene as determined by Poisson distribution is $1\text{-}e^{-1}$ or 0.63). On the other hand, if tRNA genes are tightly clustered far fewer 4S clones would have been detected. The size of the collection (187 4S clones) suggests that the tRNA genes are not tightly clustered.

TABLE I

Catalog of 4S Clones

Clones	tRNA species
1-i	Thr
2-b	Val$_2$(R12),R18
c	Asp
e	Thr
f	Thr
j	
k	Leu$_3$(R5)
m	Leu$_3$(R5)
p	Thr,R9
r	Val$_2$(R12)
3-b	
d	Gly,T9
h	
i	Val$_2$(R12)
p	R8 (Ala)
q	T15
4-a	Leu$_3$(R5),R1
d	R7
e	
g	
i	
j	
k	R8(Ala)
l	T3(Ser$_2$)
o	R13
5-a	Asp
e	Phe(R11)
f	Gly
i	R8(Ala),T16
j	
6-a	Leu$_3$(R5)
c	
g	Cys
h	Tyr(R10)
l	Gly
7-j	T2(Leu$_{UUA}$)
m	Gly
8-a	Thr
c	Thr
e	T2(Leu$_{UUA}$),T9
8-f	Asp
j	
k	R18
l	R13
n	
p	Asp
q	
s	T8
u	Leu$_3$(R5)
9-d	T9
e	T9
g	T3(Ser$_2$)
k	T15
l	R18
n	
10-b	R13
e	
f	
g	
h	Tyr(R10)
i	
l	R7
m	Arg$_3$(R19),Asp,R9
p	Arg$_3$(R19),R9
q	T2(Leu$_{UUA}$)
11-d	Thr
f	Leu$_3$(R5)
g	T3(Ser$_2$)
h	R7
j	
k	R18
l	
n	R18
o	T3(Ser$_2$)
p	
q	T8
s	Phe(R11),R8(Ala)
12-a	R17
d	Leu$_3$(R5),R1
e	Asp
o	
p	Cys

TABLE I cont'd.

Clones	tRNA species
12-q	Asp
13-a	
14-c	T9,R15
d	
e	T15,R17
g	R4(Ser_{UCG})
k	R8(Ala)
15-a	
c	
d	
g	T2(Leu_{UUA})
h	T9
16-h	
17-a	
c	Leu_3(R5)
d	R15
f	Asp
h	T9
p	T9
18-c	Cys
f	R15
g	R4(Ser_{UCG})
k	R9
l	R7
n	R18
q	Val_2(R12)
s	
u	Arg_3(R19),Tyr (R10),Asp
19-b	
f	Arg_3(R19),Asp,R9
g	T9
i	T15
k	T15
m	Leu_3(R5),R1
n	T8
r	R18
t	
20-b	Trp
f	T9
g	Cys
20-m	Trp
n	T3(Ser_2)
o	Cys
21-d	Arg_3(R19),Asp,R9
k	R7
n	Arg_3(R19),Asp,R9
22-c	
f	
g	
h	
i	R7
23-a	T9
f	Leu_3(R5)
g	
h	Gly
k	Tyr(R10)
l	R4(Ser_{UCG})
24-b	Asp
c	T15,R17
d	
k	
o	
25-l	R18
p	Val_2(R12)
26-d	R8(Ala)
f	Val_2(R12)
h	T9
27-a	
b	T3(Ser_2)
d	R9,Cys
g	
j	
28-b	R13
c	T9
e	
i	Asp
29-e	
i	T15,R17
l	Val_2(R12),T16
n	Arg_3(R19),R9
p	R7

TABLE I cont'd.

Clones	tRNA species	Clones	tRNA species
29-q	T2(Leu_{UUA})	141	
r	Arg_3(R19),T9,R9	142	
30-e		143	T3(Ser_2)
g	T15	144	Tyr(R10)
h	Gly	241	Arg_3(R19)
i	R6	242	Leu_3(R5)
l	T9	243	Val_2(R12)
		244	
31-1	Asp	341	T3(Ser_2)
32-e		342	
f	T16	541	R6
g	T16	542	Leu_3(R5)
h	T16		

All 187 clones containing 4S genes obtained by a series of 33 independent transformations are listed. The identity of the various purified tRNA species found to hybridize with each clone is indicated. Screening for the various tRNA species was done using either individual 4S spots obtained by 2-dimensional separation of *in vivo* labeled tRNAs (T1-T16; R1-R19) or by incubating purified tRNA species with bacteriophage T4 polynucleotide kinase and γ-labeled [^{32}P]-ATP. Spots T2, T3, R4, R5, R8, R10 and R11 were respectively shown by fingerprint analyses to be tRNAs specific for Leu_{UUA}, Ser_2, Ser_{UCG}, Leu_3, Ala, Tyr and Phe; T9 was not fingerprinted but did cohybridize with the same clones as $tRNA^{Lys}$. Abbreviations for the tRNA species are as follows: Asp, aspartic acid; Arg, arginine; Cys, cysteine; Gly, glycine; Leu, leucine; Lys, lysine; Phe, phenylalanine; Ser, serine; Thr, threonine; Trp, tryptophan; Tyr, tyrosine.

These results are in good agreement with data from Craig Chinault and John Carbon (personal communication). If the collection of 187 clones contains 360 tRNA genes, on the average there will be two genes per clone. The possibility that yeast tRNA genes are widely spaced in the genome is consistent with the genetic mapping of tyrosine-inserting suppressors. These suppressors are unlinked and often map on separate chromosomes (Gilmore *et al.,* 1971). Clarkson and Birnstiel, 1973 have studied the organization of tRNA genes in *Xenopus.* Their data suggest a model in which identical tRNA genes are arranged in tandem, interspersed by spacer DNA. The results presented here, though not excluding the possibility of tandem repetition of genes for identical tRNA species, do argue in favour of a different general principle of organization in yeast.

Our data also suggest that the frequency of occurrence of the

TABLE II

Frequency of Occurrence of the Different tRNA Clones

tRNA species[a,c]	Number of clones found[b]	Minimum independent number
Ala(R8) (G. Keith) (1)	6	6
Asp (G. Keith; P. Bolton and D. Kearns) (2)	15	12
Arg_3(R19) (J. Weissenbach and G. Keith) (3)	9	8
Cys (N. Holness) (4)	6	5
Gly (P. Bolton and D. Kearns)	6	6
Leu_{UUA}(T2) (5)	5	5
Leu_3(R5) (S.H. Chang; G. Pixa and G. Keith) (6)	12	10
Phe(R11) (B. Reid) (7)	2	2
Ser_2(T3) (H. Zachau) (8)	8*	6
Ser_{UCG}(R4)	3	3
Thr (J. Weissenbach and G. Keith)	7	4
Trp (G. Keith; B. Reid) (9)	2	1
Tyr(R10) (B. Reid; P. Bolton and D. Kearns) (10)	5	5
Val_2(R12) (S. Montasser and G. Keith) (11)	8	7
T8	3	3
T9	14	12
T15	8	7
T16	5	3
R1	3	3
R6	2	2
R7	7	7
R9	10	7
R13	4	4
R15	3	3
R17	4	4
R18	8	7

The data reported in Table I are summarized here as follows:
[a]The total number of clones hybridizing to each tRNA.
[b]Minimum number of independently obtained clones for each tRNA, computed by subtracting any possible siblings.
[c]We gratefully acknowledge the various donors of the different purified tRNA species as indicated for each tRNA.

(1) Holley *et al.*, 1965. (2) Gangloff *et al.*, 1972. (3) Kuntzel *et al.*, 1972. (4) Holness and Arfield, 1976. (5) Piper and Wasserstein, 1977. (6) Chang *et al.*, 1973. (7) RajBhandary *et al.*, 1967. (8) Zachau *et al.*, 1966. (9) Keith *et al.*, 1971. (10) Madison *et al.*, 1966. (11) Axelrod *et al.*, 1974.

TABLE III

Possible Clusters of Two or More tRNA Genes[a]

tRNA species	Number of clones found	Minimum independent number
Gly, T9	1	1
Ala(R8), T16	1	1
Leu_{UUA}(T2), T9	1	1
Ala(R8), Phe	1	1
Arg_3, Asp. R9	4	3
Arg_3, Asp, Tyr	1	1
Arg_3, T9, R9	1	1
Arg_3, R9	2	2
Val_2, R18	1	1
Thr, R9	1	1
Cys, R9	1	1
Leu_3, R1	3	3
T9, R15	1	1
T15, R17	3	3

[a]The clones which react with more than one of the tRNAs tested and their frequency of occurrence are listed as described under Table II.

different tRNA genes on the yeast genome may vary extensively from one tRNA species to another. Table II shows that whereas ten clones hybridize with $tRNA_3^{Leu}$, only one hybridizes with $tRNA^{Trp}$. It is obviously of importance to analyze and compare the DNA from various of our clones. In doing so we hope to answer the following questions: for a set of clones carrying genes for a particular tRNA, what is the conserved sequence - is it confined to the structural gene or does it contain auxiliary information? Do iso- or hetero-clusters of tRNA genes exist and if so, what is their organization and mode of transcription? Can one by comparison of the sequences adjacent to the structural genes recognize any special features which could serve as regulatory signals for transcription?

Restriction Mapping of $tRNA_3^{Arg}$ Clones

A first step in our attempt to answer these questions was undertaken by a comparison of the plasmid restriction maps of seven clones which hybridize to $tRNA_3^{Arg}$, three of which also contained a gene for $tRNA^{Asp}$. The restriction patterns for each of these plasmids were found to be unique, indicating that each clone chosen for this analysis carried a different segment of the yeast genome.

The $tRNA_3^{Arg}$ clones were also analyzed by comparing the sizes of the DNA restriction fragments containing the $tRNA_3^{Arg}$ coding sequences, using the DNA blotting technique (Southern, 1975). The size of the DNA fragments containing the tRNA genes, generated by cleavage with either endo R· *Hind* III, endo R· *Eco* RI, endo R·*Hae* III (Figure 2) or endo R·*Hha* I were found to differ in all cases. Therefore, it appears that the flanking sequences of the $tRNA_3^{Arg}$ genes are not conserved, at least at this gross level of analysis.

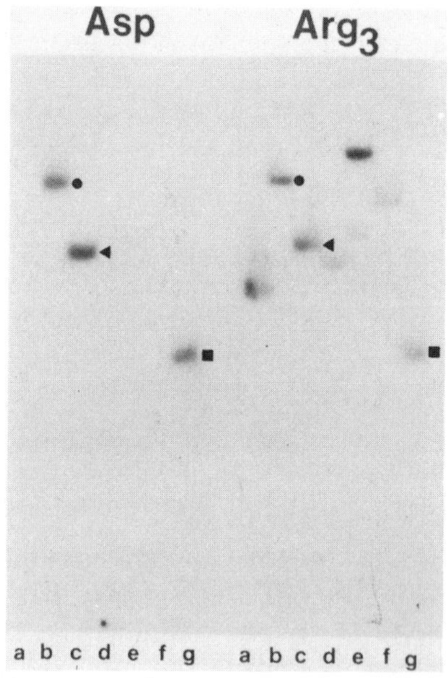

Fig. 2 Hybridization with $tRNA^{Asp}$ and $tRNA_3^{Arg}$ to a restriction digest of seven $tRNA_3^{Arg}$ containing plasmids. The plasmid DNAs were cleaved with endo R· *Hae* III, separated by agarose gel electrophoresis and transferred to nitrocellulose filter paper (Southern, 1975). The filters were then hybridized to either labeled $tRNA^{Asp}$ or $tRNA_3^{Arg}$ and the labeled bands revealed by autoradiography. It can be seen that each plasmid contains a different endo R·*Hae* III fragment which hybridizes to $tRNA_3^{Arg}$. The symbols indicate that identical fragments hybridize to both $tRNA_3^{Arg}$ and $tRNA^{Asp}$ (lanes b, c and g). The identities of the plasmid DNAs are as follows: 1, 10p; b, 10m; c, 18u; d, 29n, e, 29r; f, 241; g, 19f.

The high frequency of coincidence of $tRNA_3^{Arg}$ and $tRNA^{Asp}$ cistrons suggests a close linkage between these genes. Our restriction mapping data for three Arg_3/Asp clones support this conclusion. In each case, the two genes are found within single endo R· *Hind* III and *Hae* III fragments. A more detailed restriction map for one such clone (18μ, see below) confirms this linkage.

Characterization of 18μ DNA

A more detailed analysis of the plasmid DNA of clone 18μ was undertaken. This clone contains genes for $tRNA_3^{Arg}$, $tRNA^{Asp}$, and $tRNA^{Tyr}$. The gene for $tRNA^{Tyr}$ is separated from those for $tRNA_3^{Arg}$ and $tRNA^{Asp}$ by approximately one kilobase, in which there is a single site for endo R· *Eco* RI. (Figure 3 shows a restriction map of a 2.2 kilobase segment of the 18μ plasmid). It can be seen that the $tRNA_3^{Arg}$ and $tRNA^{Asp}$ sequences are nearly contiguous. In order to examine their close relationship in more detail, the endo R· *Hind* II/III fragment of approximately 430 base pairs was cleaved with endo R· *Taq* I; the 3′ -ends were labeled by 'filling in' (the details of this procedure will be published elsewhere) by Klenow enzyme in the presence of α-[^{32}P]-labeled dGTP and dCTP. The two labeled fragments were separated and sequenced according to the method of Maxam and Gilbert (Maxam and Gilbert, 1977).

While the results of these analyses are still preliminary, several features of the tRNA gene organization are worth noting. The two tRNA genes are tightly linked and in the same orientation; only nine nucleotides, AAACAAAGA, separate the coded 3′ -end of the $tRNA_3^{Arg}$ gene from the 5′ -end of the $tRNA^{Asp}$ gene. Furthermore, the 3′ -end of the $tRNA_3^{Arg}$ gene does not encode the last two nucleotides of the 3′ -CCA terminus of the mature $tRNA_3^{Arg}$. These data also reveal that both tRNA genes are uninterrupted. This observation is significant given the demonstration by Goodman *et al.*, (1977) and by Valenzuela *et al.*, (1978) for the yeast $tRNA^{Tyr}$ and $tRNA^{Phe}$ genes, respectively, of the existence of small intervening sequences within the structural genes. The fact that the tRNA genes for $tRNA^{Asp}$ and $tRNA_3^{Arg}$ are not interrupted excludes the possibility that all yeast tRNA genes possess intervening sequences.

Isolation of *in vivo* Precursors of Yeast tRNAs

As mentioned above, intervening sequences have been reported in two species of tRNA genes: $tRNA^{Tyr}$ (Goodman *et al.*, 1977) and $tRNA^{Phe}$ (Valenzuela *et al.*, 1978). A primary concern, therefore, was to establish the mechanism by which the cell eliminates these sequences during the formation of the mature tRNAs. One possible mechanism is that both the structural and intervening sequences are transcribed producing precursors which contain extra internal nucleotides not found in the mature tRNAs. These additional nucleotides would subsequently be removed enzymatically. Previously, Blatt and Feldman (1973) had observed 4.5S sized RNAs in pulse-labeled wild type yeast. They showed *in vitro* that these RNAs produced 4S-sized material on incu-

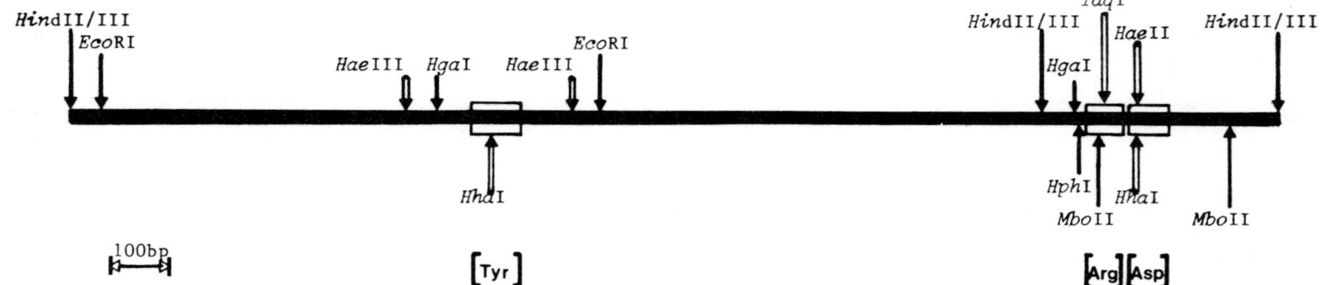

Fig. 3 Partial restriction map of a segment of plasmid 18μ coding for $tRNA^{Tyr}$, $tRNA_3^{Arg}$ and $tRNA^{Asp}$. 18μ DNA was analysed by restriction enzyme digestion and the tRNA genes mapped by hybridization. For reference, the endo R· *Hind* II/III fragment containing the $tRNA^{Asp}$ and $tRNA_3^{Arg}$ genes is approximately 430 bp, while the endo R· *Eco* RI fragment containing the $tRNA^{Tyr}$ gene is about 900 bp.

bation with a crude yeast extract.

Tye isolation of yeast tRNA precursors was greatly facilitated by the observation of Hopper *et al.*, (1978) that tRNA precursors accumulate at the non-permissive temperature in the yeast mutant ts-136. This temperature-sensitive mutant (in the *rna*1 locus of *S. cerevisiae* (Hutchison *et al.*, 1969)) accumulates RNA in the nucleus at the non-permissive temperature. The defect in the mutant is thought to be in transport of RNA nucleus to cytoplasm. Figure 4 shows the accumulation at the non-permissive temperature of 4.5S RNA in a diploid strain (M304 (Knapp *et al.*, 1978)) that is homozygous for the *rna*1 locus.

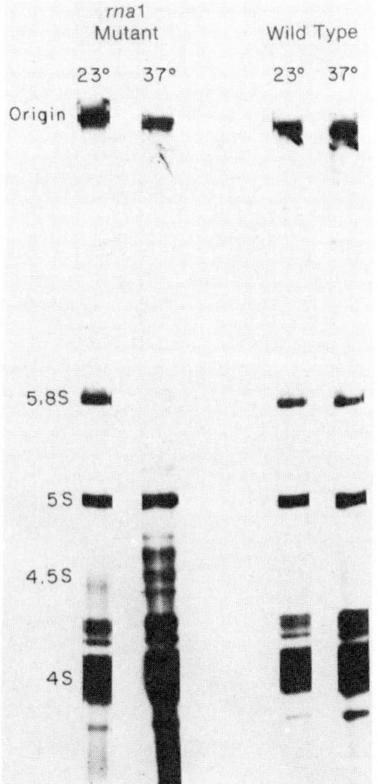

Fig. 4 Polyacrylamide gel electrophoresis of ^{32}P-RNA from the temperature-sensitive *rna*1 mutant (M304) and wild type (GX-124). The two diploid yeast strains were grown in 25 ml cultures and labeled for 30 minutes at the indicated temperatures with 1 mCi of [^{32}P]-orthophosphate. The ^{32}P-RNA was extracted and polyphosphate was removed.

Figure 5 shows the two-dimensional electrophoretic separation of ^{32}P-RNA from M304 labeled at 37°. The availability of the bank of clones carrying yeast tRNA genes allowed the identification of the individual tRNA precursors by hybridization. The *rna*1 mutant accumulates precursors for $tRNA^{Phe}$, $tRNA^{Tyr}$, $tRNA_3^{Leu}$, $tRNA^{Trp}$ and

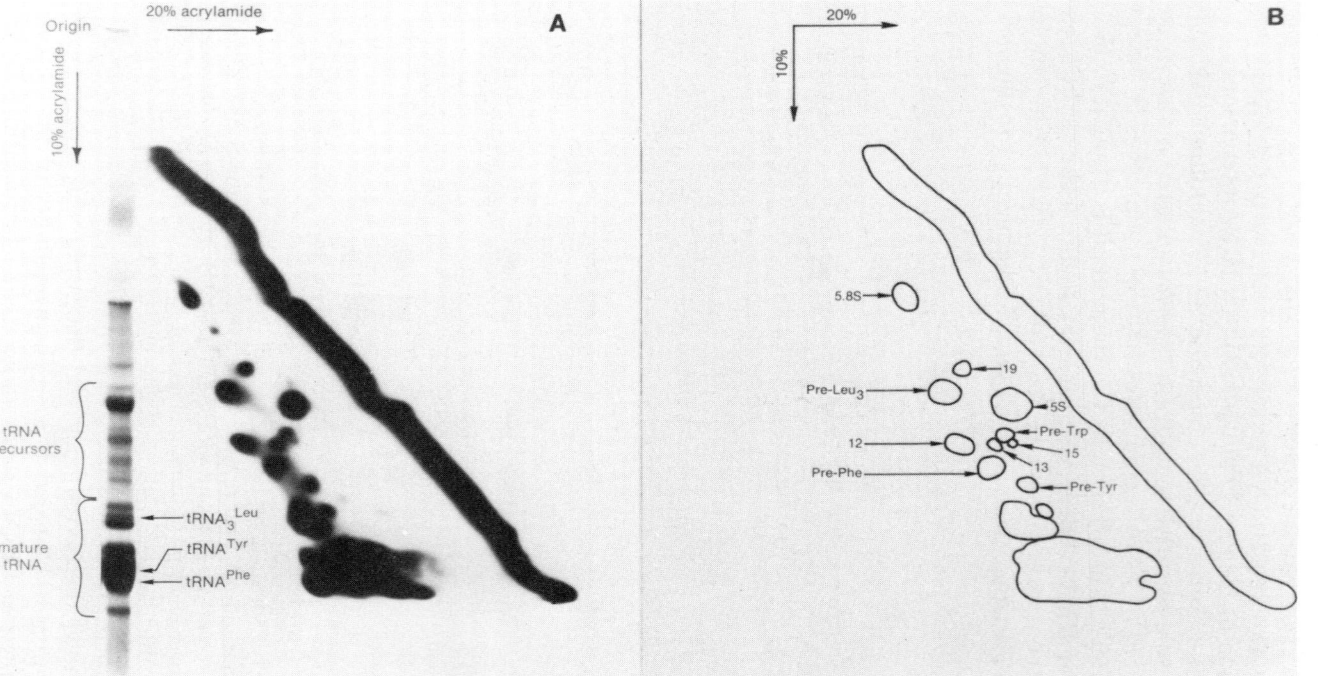

Fig. 5 Two-dimensional polyacrylamide gel electrophoresis of ^{32}P-RNA isolated from M304. ^{32}P-RNA, extracted from cells of the temperature-sensitive mutant (M304), was purified by electrophoresis in a 10% polyacrylamide gel followed by electrophoresis in a second dimension using a 20% polyacrylamide gel. Autoradiography revealed the positions of the different pre-tRNAs. The dark, diagonal band is presumably heterogeneous polyphosphate. (A) Autoradiogram of the polyacrylamide gel separations. (B) Diagrammatic representation and identification of the ^{32}P-pre-tRNAs. For brevity, here and in the subsequent Figures and Tables, pre-Phe, pre-Tyr, pre-Trp and pre-Leu$_3$ are used to represent, respectively, pre-tRNA^{Phe}, pre-tRNA^{Tyr}, pre-tRNA^{Trp}

$tRNA^{Ser}_{UCG}$ (spot 15) in large amounts. Spots 11 and 12 hybridize to a unique set of clones containing as yet unidentified tRNA genes. Spot 19 also hybridizes to clones which hybridize pure yeast $tRNA^{Leu}_{3}$. It is striking that only a limited number of tRNA precursors have been detected. Thus, a very select set of precursors accumulate in this mutant.

Characterization of tRNA Precursors

It is possible to estimate the size of the tRNA precursors from their electrophoretic mobilities on 10% polyacrylamide gels (Figure 5). The estimated size of pre-$tRNA^{Tyr}$ is 91±5 nucleotides and that of pre-$tRNA^{Phe}$ is 97±5 nucleotides. If pre-$tRNA^{Tyr}$ includes the intervening sequences, it must be at least 92 nucleotides in length (mature tRNA plus insert). The minimum size of a pre-$tRNA^{Phe}$ containing intervening sequences is 94 nucleotides. Thus, the presence of intervening sequences could entirely account for the observed size of these precursors. Their presence can be determined by sequencing the tRNA precursors. We have not yet completed the sequence analysis, but the data prove conclusively that pre-$tRNA^{Tyr}$ and pre-$tRNA^{Phe}$ contain the intervening sequences, that they do not have extra sequences at their 5′-ends and that their 3′-ends contain the uncoded CCA termini. Similarly, the size of the other precursors can be esimated as follows: pre-$tRNA^{Trp}$, 113±5 nucleotides; pre-$tRNA^{Ser}_{UCG}$, 113±5 nucleotides; pre-$tRNA^{Leu}_{3}$, 121±5 nucleotides; and 19, 135±5 nucleotides. These compare with the mature sizes of 75 ($tRNA^{Trp}$), 85 ($tRNA^{Ser}_{UCG}$) and 85 ($tRNA^{Leu}_{3}$). If these precursors also have mature 5′- and 3′-ends, these size predictions would suggest that their intervening sequences are much larger than those in pre-$tRNA^{Tyr}$ and pre-$tRNA^{Phe}$.

After additional purification by hybridization to plasmid DNA containing the appropriate yeast tRNA gene, the pre-tRNAs were digested with either RNAase T1 or pancreatic RNAase and oligonucleotides were separated by two-dimensional paper electrophoresis. Figures 6 and 7 give the fingerprints of the T1 and pancreatic ribonuclease digestion products of pre-$tRNA^{Tyr}$. Each oligonucleotide was eluted and further characterized as described previously (Knapp *et al.*, 1978). Table IV lists the oligonucleotides that would result from T1 or pancreatic RNAase digestion of the anticodon loop region of $tRNA^{Tyr}$ in the presence or the absence of the intervening sequence.

Analysis of pre-$tRNA^{Tyr}$ oligonucleotides revealed that the intervening sequences are indeed present. In the RNAase T1 oligonucleotide map, we have confirmed the presence of (but have not yet

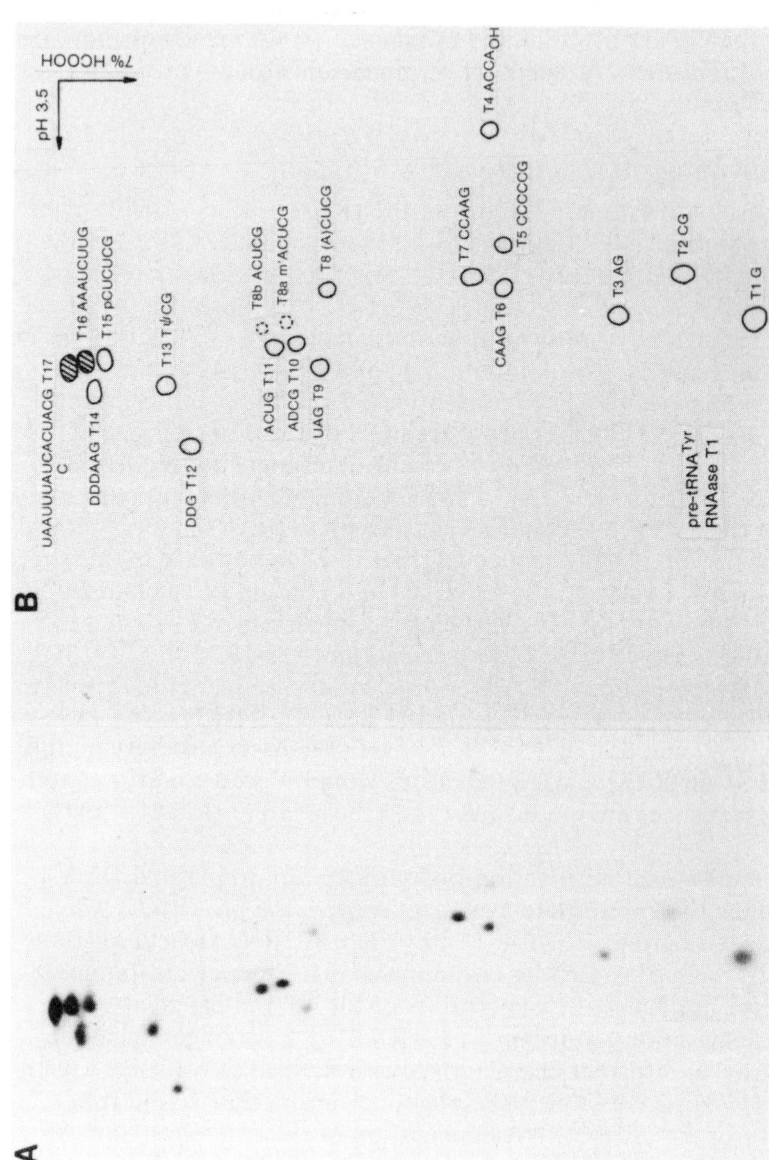
A
B
pH 3.5
7% HCOOH
UAAUUUAUCACUACG T17
C
T16 AAAUCUUG
T15 pCUCUCG
DDDAAG T14
T13 TψCG
DDG T12
ACUG T11
T8b ACUCG
T8a m¹ACUCG
ADCG T10
UAG T9
T8 (A)CUCG
T7 CCAAG
CAAG T6
T5 CCCCCG
T4 ACCA$_{OH}$
T3 AG
T2 CG
T1 G
pre-tRNA^{Tyr}
RNAase T1

sequenced) the long oligonucleotide T17, that comprises most of the insert. Oligonucleotide T17 is heterogeneous; it can be resolved into two spots of equal intensity by treating the pre-tRNA^{Tyr} with RNA-ase T1 and alkaline phosphatase (data not shown). Both oligonucleotides were further characterized by digestion with pancreatic RNAase. One of them contained an AU among its pancreatic RNAase digestion products and the other did not. Goodman *et al.*, (1977) observed two different intervening sequences in the three tRNA^{Tyr} genes that they examined. The predicted RNA sequences in Table IV show that one of the T17 oligonucleotides should have an AU pancreatic RNAase product and the other should not. Thus, both intervening sequences are present in pre-tRNA^{Tyr}.

The DNA sequence predicts the existence in the pancreatic RNAase fingerprint of two 'overlap' oligonucleotides, AAU and GAAAU, which span the junctions between the intervening sequence and the mature tRNA. It can be seen in Figure 7 that both of these products are present (P8 and P15) while the sequence AAAU that is present in the mature tRNA is clearly absent (see Table IVB).

The RNAase T1 fingerprint of pre-tRNA^{Tyr} (Figure 6) shows the presence of the mature 5′ - (T15) and 3′- (T4) ends. Other oligonucleotides corresponding to possible sequences at the ends of the precursor are not present leading to the conclusion that the only extra sequences contained in pre-tRNA^{Tyr} are the predicted intervening sequences.

Similarly, pre-tRNA^{Phe} has been shown (Knapp *et al.*, 1977) to contain the two intervening sequences detected by Valenzuela *et al.*, (1978). Oligonucleotides observed in the T1 and pancreatic RNAase

Fig. 6 See opposite. RNAase T1 digestion of pre-tRNA^{Tyr}. (A) Autoradiogram of the digestion products. Purified ^{32}P-pre-tRNA^{Tyr} was digested with RNAase T1 and the products were separated by two-dimensional paper electrophoresis (Shiokawa and Pogo, 1974). For comparison, a RNAase T1 fingerprint of mature tRNA^{Tyr} may be found in Piper *et al.*, (1976). (B) Schematic representation of the positions of the oligonucleotides. Open symbols indicate the oligonucleotides which are common to both the pre-tRNA and the tRNA. Closed symbols denote those oligonucleotides that are uniquely derived from the intervening sequence. Hatched symbols designate the locate of 'overlap' oligonucleotides which span the junction between the tRNA and the intervening sequence. Speckled symbols represent oligonucleotides which are derived from the tRNA sequence but are generated only when the intervening sequence is present. Spots T8, T8a and T8b originate from a common sequence in the pre-tRNA. They differ in the extent of modification of the m^{1}A residue: spot T8b has an unmodified A, spot T8a contains m^{1}A and spot T8 results from depurination or chemical modification of m^{1}A.

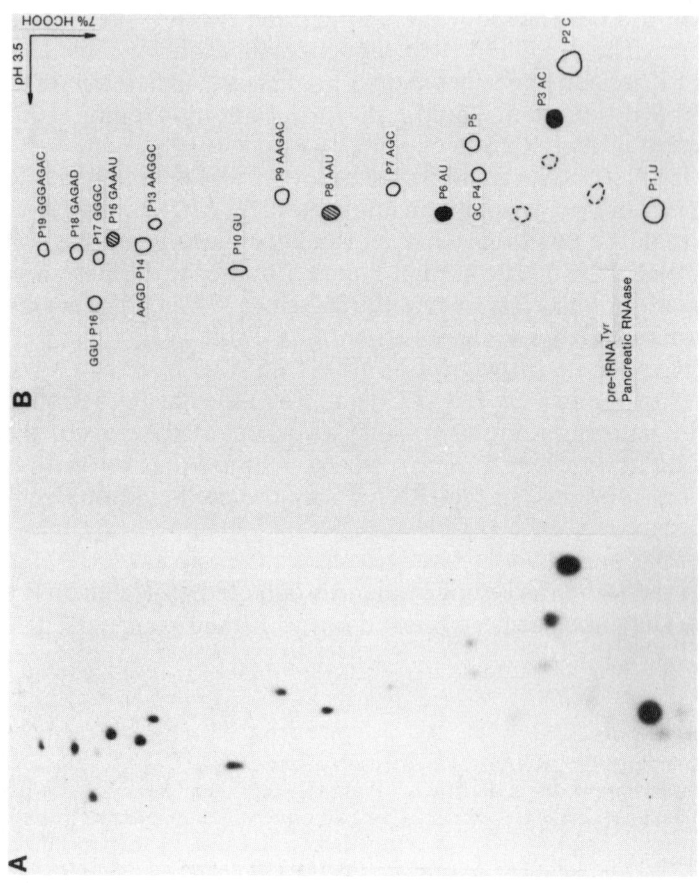
A
B
pH 3.5
7% HCOOH
P19 GGGAGAC
P18 GAGAD
P17 GGGC
P15 GAAAU
GGU P16
AAGD P14
P13 AAGGC
P10 GU
P9 AAGAC
P8 AAU
P7 AGC
P6 AU
GC P4
P5
P3 AC
P2 C
P1 U
pre-tRNA Tyr
Pancreatic RNAase

fingerprints also confirm the presence of the mature 5′ - and 3′ - ends of pre-$tRNA^{Phe}$. Preliminary data (not shown) from fingerprints of pre-$tRNA^{Ser}_{UCG}$, pre-$tRNA^{Trp}$ and pre-$tRNA^{Leu}_{3}$ suggest that these precursors also contain mature 5′ - and 3′ -ends and intervening sequences. Further analysis of pre-$tRNA^{Ser}_{UCG}$ by M.C. Etcheverry, D. Colby and C. Guthrie (private communication) confirms the presence of mature 5′ - and 3′ - ends and that the intervening sequence is located at the 3′ -end of the anticodon loop.

Thus, transcription of the intervening sequences observed in tRNA genes has been proven by analyses of the sequences of pre-$tRNA^{Tyr}$ and pre-$tRNA^{Phe}$. The preliminary analyses of pre-$tRNA^{Trp}$ and pre-$tRNA^{Leu}_{3}$ also support this observation. Furthermore, the analyses of pre-$tRNA^{Tyr}$, pre-$tRNA^{Phe}$, pre-$tRNA^{Trp}$ and pre-$tRNA^{Leu}_{3}$ indicate that these 4.5S precursors are not primary transcripts: preceding events in tRNA biosynthesis have produced mature 5′ - and 3′ -ends of the molecules and, in addition, some modified nucleosides have already been introduced in the pre-tRNAs.

Analysis of Modified Nucleosides in tRNA Precursors

Although a careful cataloguing of the modified nucleosides in the precursors has not been completed, it is clear some are present in good yield while others are totally absent. Analysis of the base composition of the four pre-tRNAs was by two dimensional cellulose thin layer chromatography (Mazzara *et al.*, 1977). The four pre-tRNAs contain excellent yields of ψ, T and D. For example, in pre-$tRNA^{Tyr}$ analysis of RNAase T1 oligonucleotides shows that T12 (DDG) and T13 (TψCG) occur primarily with the modifications found in the mature tRNA. Oligonucleotide T14(DDDAAG) contains approximately equivalent amounts of U and D. Other minor bases are present in the precursors in lesser yields as follows: m^2_2G (pre-$tRNA^{Phe}$ and probably pre-$tRNA^{Leu}_{3}$), m^7G (pre-$tRNA^{Phe}$), m^5C (pre-$tRNA^{Phe}$ and pre-$tRNA^{Leu}_{3}$), m^1A (pre-$tRNA^{Phe}$, pre-$tRNA^{Tyr}$ and pre-$tRNA^{Trp}$). No 2′ -0-methylation of G or C has been detected in any of the precursors and the hypermodified bases, notably the Y base and i^6A, are absent in pre-$tRNA^{Phe}$ and pre-$tRNA^{Tyr}$, respectively..

Fig. 7 See opposite. Pancreatic RNAase digestion of pre-$tRNA^{Tyr}$. (A) Autoradiogram of the oligonucleotides separated by two-dimensional paper electrophoresis. (B) Schematic representation of the positions of the oligonucleotides. Symbols are as described in Figure 6B.

TABLE IV

A. Predicted RNAase T1 Oligonucleotides from the Anticodon Loop of tRNA^Tyr and pre-tRNA^Tyr a

tRNA^Tyr	common to both	pre-tRNA^Tyr
	Cm_2^2GCAAGp	
	(Cm_2^2Gp plus CAAGp)	
	ACUGp	
ψAi^{6}AAψCUUGp		UAAUUUA^U_CCACUACGp
		AAAψCUUGp

B. Predicted Pancreatic RNAase Oligonucleotides from the Anticodon Loop of tRNA^Tyr and pre tRNA^Tyr

tRNA^Tyr	common to both	pre-tRNA^Tyr
	AAGACp	
Ai^{6}AAψp		AAUp
		ACp
		AUp
		GAAAψp

[a]Underlined sequences indicate the intervening sequence; An oligonucleotide which is partially underlined is defined as an 'overlap' oligonucleotide.

Excision of the Intervening Sequences

If these RNA molecules are to serve as precursors to tRNAs, there must be an activity which removes the intervening sequence. The possibility exists that more than one enzyme constitutes this activity. The tRNA precursors were used as substrates to search for an activity which can excise the intervening sequences and religate the ends to form the anticodon loop. Yeast cells were lysed in an Eaton press and the cell-free extract was fractionated to search for the RNA splicing enzyme ('splicase'). Details of the initial fractionation procedure may be found in Knapp *et al.*, (1978). A high salt ribosomal wash of wild-type yeast extract contains this activity. Figure 8 shows that the RNA splicing enzyme converts pre-tRNA^{Tyr}, pre-tRNA^{Phe}, pre-tRNA^{Trp} and pre-tRNA$_3^{Leu}$ to 4S RNAs with mobilities identical to those of the mature tRNAs (illustrated here for tRNA^{Trp} and tRNA$_3^{Leu}$).

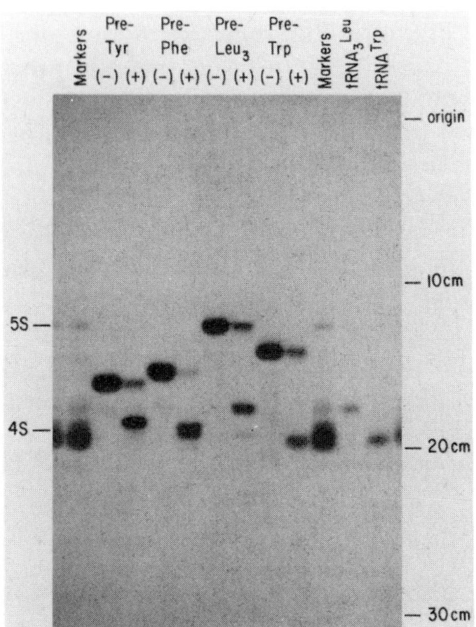

Fig. 8 Processing of yeast tRNA-precursors isolated from M304 to 4S-size molecules. Individual pre-tRNAs were incubated in the presence (+) or absence (-) of a ribosomal wash prepared from the wild type diploid GX-124. Conditions for the reaction are: 50 m*M* Tris-HC1, pH 8.0, 10 m*M* $MgC1_2$, 100 m*M* KC1, 2 m*M* dithiothreitol, 0.1 m*M* EDTA, 5% glycerol, 2 m*M* ATP, 50 μg/ml poly(U), ^{32}P-tRNA precursor and 5 A_{280} units/ml of the ribosomal wash fraction in a total volume of 100 μl. The reactions were incubated for 1 hour at 30°. The RNAs were analyzed on a 10% polyacrylamide gel. Marker 4S tRNA^{Trp} and tRNA$_3^{Leu}$ were obtained by DNA filter hybridization.

While the ribosome wash fraction and reaction conditions used for *in vitro* processing produce good yields of 4S RNA, we wish to reserve judgement on the intracellular localization of the 'splicase'. The yeast cells are lysed in a medium low ionic strength and this could lead to the fortuitous association of the processing activity with the ribosomes. The high salt ribosomal wash has been fractionated further using two chromatographic methods and the activity still processes at least three different pre-tRNAs. Thus far, the purification suggests that there will not be a separate enzyme for each precursor but that the activity might be a single RNA splicing enzyme.

The tRNA precursors are not processed to 4S RNA in the absence of Mg^{+2} suggesting an absolute Mg^{+2} (or divalent cation) requirement (unpublished data). Reasonable ionic strength (0.05-0.10*M*) is also a requirement. More precise requirements of the RNA splicing enzyme cannot be accurately determined at this level of purification. However, ATP is included in the reaction because there is a ligation activity which may require ATP. Poly(U) is included to protect the ^{32}P-pre-tRNA from degradation by random nucleases. Mixed yeast tRNAs could not be used for this purpose since inclusion of tRNA inhibits processing of precursors to 4S-sized RNA and results in the formation of partial reaction products.

The 4S products of *in vitro* processing were analyzed. For example Figure 9 shows the map of oligonucleotides produced by digestion of processed pre-$tRNA^{Tyr}$ (4S) with RNAase T1. The two overlap oligonucleotides T16 (AAAUCUUG) and T17 (UAAUUUA$^{U}_{C}$CACUACG) are now absent but the oligonucleotide ψAAAψCUUG, which is diagnostic for the mature $tRNA^{Tyr}$, has appeared (indicated in Figure 9). This oligonucleotide can only be produced by the covalent rejoining of the polynucleotide chain to produce the mature tRNA. The processed pre-$tRNA^{Tyr}$ (4S) has not been fully matured by the ribosomal wash fraction, since certain oligonucleotides are generated in the RNAase T1 digest which would occur only in the absence of certain methylations of G residues. For example, the fingerprint of fully mature $tRNA^{Tyr}$ contains Cm^2_2GCAAG and DDGmG whereas in Figure 9 CAAG and DDG are observed instead.

Secondary Structures of pre-$tRNA^{Tyr}$ and pre-$tRNA^{Phe}$

In Figure 10, we propose secondary structures for pre-$tRNA^{Tyr}$ and pre-$tRNA^{Phe}$. These structures have maximized base pairing which was determined by the calculation of the most favorable free energy

Fig. 9 See opposite. Oligonucleotide map of the 4S-size RNA produced by incubation of pre-$tRNA^{Tyr}$ with the GX-124 ribosomal wash. Pre-$tRNA^{Tyr}$ was incubated as described and the products were separated by 10% polyacrylamide gel electrophoresis. The 4S RNA band was eluted from the gel, digested with T1 RNAase and the oligonucleotides were separated by two-dimensional paper electrophoresis. (A) Autoradiogram of the oligonucleotides separated by two-dimensional paper electrophoresis. This electrophoretic separation was performed under slightly altered conditions from that of the RNAase T1 digest of pre-$tRNA^{Ty}$ (see Figure 4); therefore, the relative mobilities of certain identical oligonucleotides are different in the two fingerprints. (B) Schematic representation of the positions of the oligonucleotides. The cross-hatched symbol represents the oligonucleotide ψAAAψCUUG which is generated by the removal of the intervening sequence.

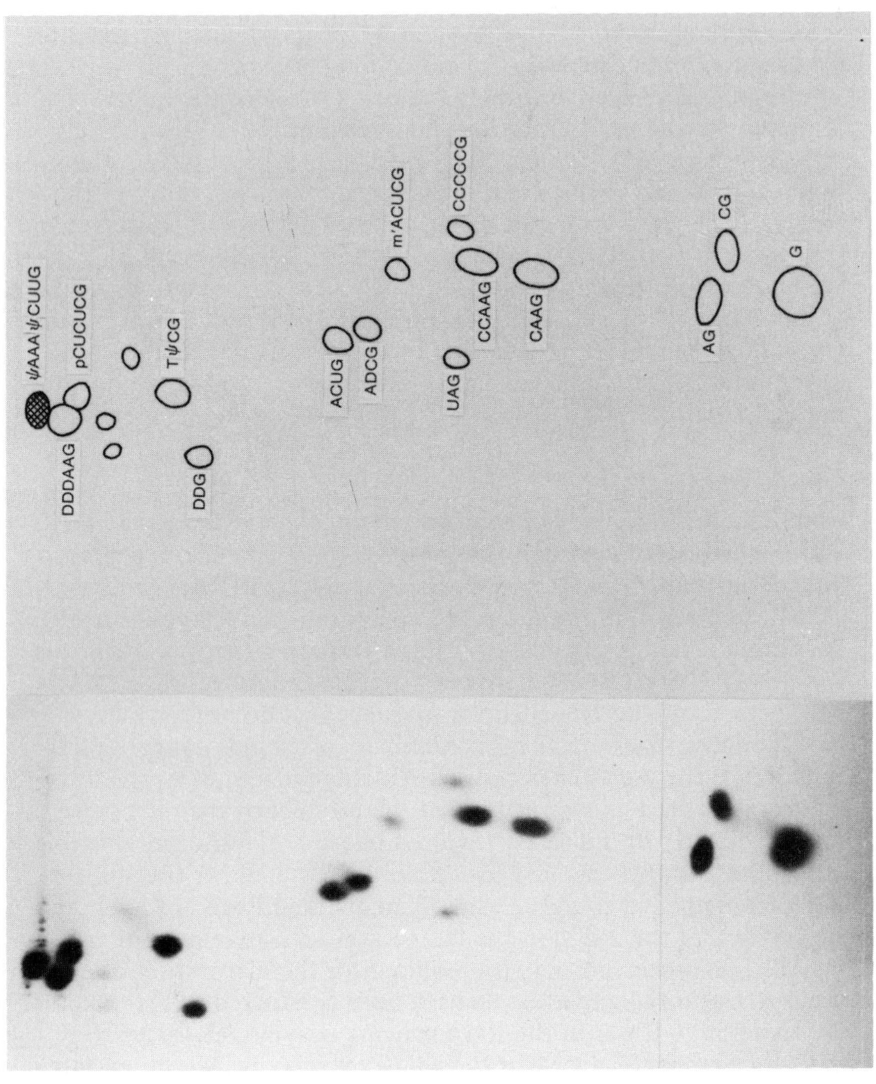
ψAAAψCUUG
pCUCUCG
DDDAAG
TψCG
DDG
ACUG
ADCG
m1ACUCG
UAG
CCAAG
CCCCCG
CAAG
AG
CG
G

using the rules derived by Tinoco *et al.*, (1973) and Borer *et al.*, (1974). Both structures retain the principal features of the tRNA cloverleaf representation. The D and TψC loops and helical stems are retained in the model. The loops contain the modified residues in good yield. The proposed structures diverge from the cloverleaf representation in the anticodon stem which is extended by additional base pairing. In each case the anticodon pairs specifically with the intervening sequences. Another common feature of the intervening sequences of the pre-tRNAs thus far characterized is that they all appear to be A rich. The fact that the 'splicase' processes several precursors raises the question of the recognition specificity of the enzyme. There are no sequence homologies between the intervening sequences of pre-$tRNA^{Tyr}$ and pre-$tRNA^{Phe}$. The base pairing of the anticodon with the intervening sequence places the excision points within loops. The two structures in Figure 10 do not, however, suggest a single model for recognition of the precursors and excision of the intervening sequences. Three-dimensional models may prove more instructive.

Conclusions on the Transcription of Intervening Sequences

Only a limited number of precursors accumulate in the ts-136 diploid. We have demonstrated that most of these precursors contain intervening sequences. It may be that the selected tRNA precursors accumulating in this mutant at the non-permissive temperature all contain intervening sequences. It is worth noting that no precursors for $tRNA_3^{Arg}$ or $tRNA^{Asp}$ have been detected. Both of these tRNAs have been shown to be coded for by genes that do not have intervening sequences. These facts raise interesting questions concerning the function of the *rna*1 gene product. Although incomplete, the information presented here suggests that the product of the *rna*1 gene may be directly or indirectly involved in the removal of intervening sequences. This processing event could be required for transport of RNA from nucleus to cytoplasm. Shiokawa and Pogo (1974) have suggested that the lesion in the temperature-sensitive mutant ts-136 may lie in a processing activity required for the maturation of nuclear pre-mRNA. Initial experiments have been conducted to test whether the lesion in ts-136 is in the RNA splicing enzyme. Unfortunately, the data were inconclusive as the wild-type enzyme is temperature

Fig. 10 See opposite. Possible secondary structures for pre-$tRNA^{Tyr}$ and pre-$tRNA^{Phe}$. The most stable structures were derived according to the criteria given in Borer *et al.*, (1974).and Tinoco *et al.*, (1973). The arrows indicate the location of the intervening sequences.

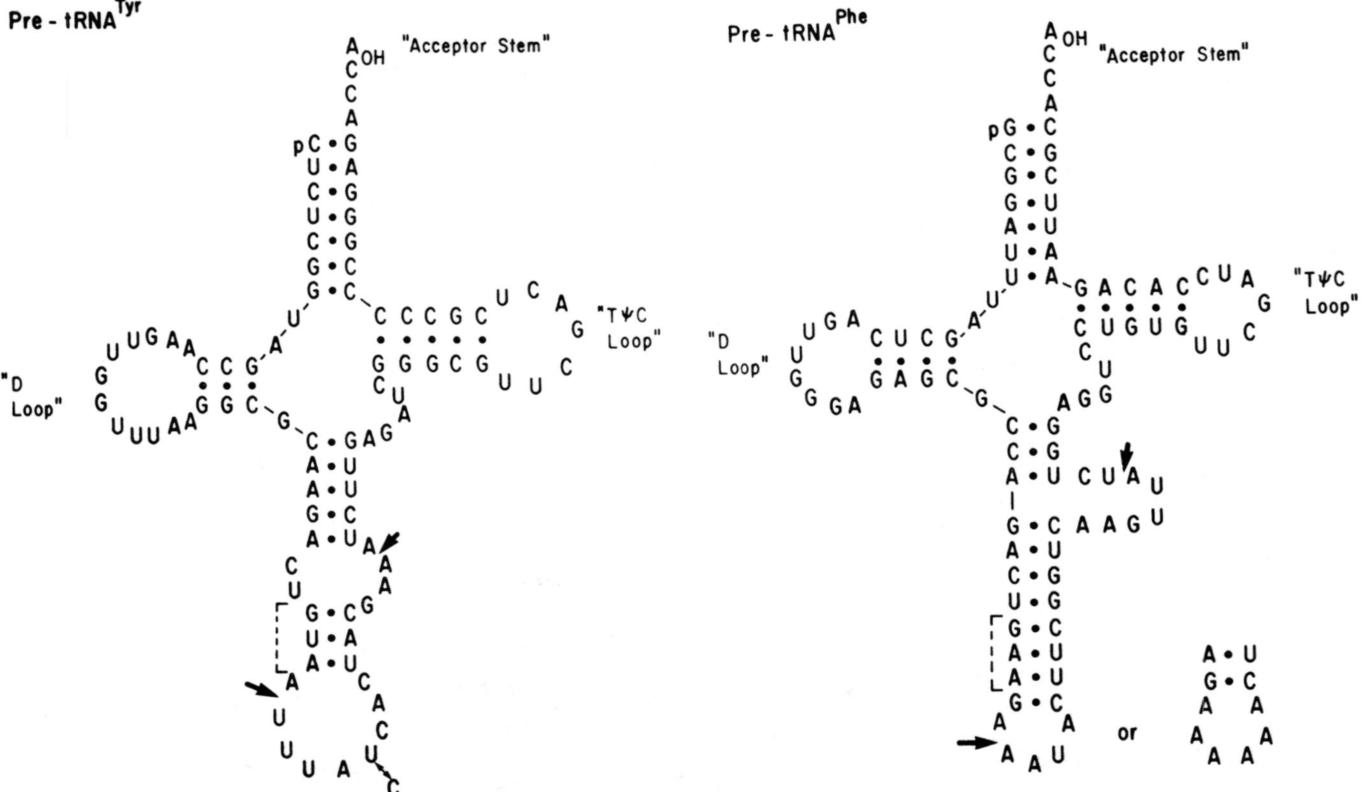

Pre - tRNA^{Tyr}
"Acceptor Stem"
"TψC Loop"
"D Loop"
Pre - tRNA^{Phe}
"Acceptor Stem"
"TψC Loop"
"D Loop"
or

sensitive *in vitro*. The temperature sensitivities of the activity from mutant and wild-type cells are similar.

Alternatively, the *rna*1 gene product may directly mediate RNA transport from nucleus to cytoplasm for final processing. A precise localization of the RNA splicing and of the tRNA precursors and 4S tRNAs accumulating in the mutant may help to differentiate between these two hypotheses.

Summary

The construction of a large bank of *E. coli* clones carrying different fragments of yeast DNA enabled us to prospect within a significant fraction of the yeast genome for the distribution and arrangement of specific tRNA genes. The data indicated that yeast tRNA genes are, in general, not highly clustered. In addition it was observed that the reiteration numbers of different tRNA genes vary extensively. The study of the molecular arrangement of some tRNA genes was undertaken by restriction and sequence analysis. These data revealed that the $tRNA_3^{Arg}$ and $tRNA^{Asp}$ genes carried on one of the clones are tightly linked; the structural genes are separated by only nine base pairs. In addition, the DNA sequence data demonstrated that both genes were entirely colinear with their products. In contradistinction, three $tRNA^{Tyr}$ genes (Goodman *et al.*, 1977) and three $tRNA^{Phe}$ genes (Valenzuela *et al.*, 1978) were shown to exhibit non-colinearity with their products. In the middle of each of the sequenced $tRNA^{Tyr}$ genes there is a 14-base pair sequence not present in the mature tRNA. In three $tRNA^{Phe}$ genes intervening sequences of 18 and 19 base pairs were found. In both sets of genes the insertion occurs at a position near the anticodon. The intervening sequences are closely related within each set but the tyrosine and phenylalanine specific sequences are dissimilar (Knapp *et al.*, 1978).

Finally, in the study of the *in vivo* expression of the yeast tRNA genes, precursors to $tRNA_3^{Arg}$ and $tRNA^{Asp}$ were not detected. However, precursors were detected and isolated for $tRNA^{Tyr}$ and $tRNA^{Phe}$, as well as for $tRNA_{UCG}^{Ser}$, $tRNA^{Trp}$, $tRNA_3^{Leu}$, and a few as yet unidentified tRNA species. It is possible that all the tRNAs for which precursors accumulate in the *rna*1 mutant will share one common feature; they all will be found to contain intervening sequences. This was fully demonstrated for the $tRNA^{Tyr}$ and $tRNA^{Phe}$ precursors (Knapp *et al.*, 1978) and our present data on precursors for $tRNA_{UCG}^{Ser}$, $tRNA^{Trp}$ and $tRNA_3^{Leu}$ agree with this. It was also shown that all of these precursors have the same 5′ - and 3′ -ends as the mature tRNAs. The availability of the precursors as substrates allowed us to demonstrate that yeast contains an activity that can excise the intervening sequences and splice the ends. Characterization of this activity is in progress.

Acknowledgements

We thank our colleages listed in Table II for sending us numerous pure species of tRNA. Drs. P. Piper and H. Feldman supplies us with valuable information used in identifying yeast tRNA fingerprints and Dr. A.K. Hopper provided us with her data prior to publication. We acknowledge Mrs. Esther Smith for her assistance and encouragement; Dr. T. Friedman for advice and facilities for the Maxam-Gilbert DNA sequencing; Dr. E.P. Geiduschek for his continuous interest and stimulating discussions; and I. Tinoco, Jr. for the computer analysis of the pre-tRNA sequences.

These investigations were supported by a grant from the National Cancer Institute, CA 10984. G.K. and S.A.F. were recipients of post-doctoral fellowships from the NIH and the American Cancer Society, respectively.

References

Axelrod, V.D., Kryubkov, V.M., Isaenko, S.N. and Bayev, A.A. (1974). *FEBS Letters* **45**, 333.

Beckmann, J.S., Johnson, P.F. and Abelson, J. (1977a). *Science* **196**, 205.

Beckmann, J.S., Johnson, P.F., Abelson, J. and Fuhrman, S.A. (1977b).Molecular Approaches to Eukaryotic Systems, (eds. G. Wilcox, J. Abelson, and C.F. Fox). 213. Academic Press, New York.

Bhargava, M.M. and Halvorson, H.O. (1971). *J. Cell Biol.* **49**, 423.

Blatt, B. and Feldmann, H. (1973). *FEBS Letters* **37**, 129.

Bolivar, F., Rodriguez, R.L., Betlach, M.C. and Boyer, H.W. (1977). *Gene* **2**, 75.

Borer, P.N., Dengler, B., Tinoco, I., Jr. and Uhlenbeck, O.C. (1974). *J. Mol. Biol.* **86**, 843.

Brandiss, M.C., Stewart, J.W., Sherman, F. and Botstein, D. (1976). *J. Mol. Biol.* **102**, 467.

Capecchi, M.R., Hughes, S.H. and Wahl, G.M. (1975). *Cell* **6**, 269.

Chang, S.H., Kuo, S., Hawkins, E. and Miller, N.R. (1973). *Biochem. Biophys. Res. Comm.* **51**, 951.

Clarkson, S.G. and Birnstiel, M.L. (1973). *Cold Spring Harb. Symp. Quant. Biol.* **38**, 451.

Cramer, J.H., Farrelly, F.W., Barnitz, J.T. and Rownd, R.H. (1977). Molecular Approaches to Eukaryotic Systems, (Eds. G. Wilcox, J. Abelson and C.F. Fox). 227. Academic Press, New York.

Duntze, S., Mackay, V. and Manney, T.R. (1970). *Science* **168**, 1472.

Feldmann, H. (1976). *Nucleic Acids Research* **3**, 2379.

Gangloff, J., Keith, G., Ebel, J.P. and Dirheimer, G. (1972). *Biochem. Biophys. Acta.* **259**, 198.

Gesteland, R.F., Wolfner, M., Grisafi, P., Fink, G., Botstein, D. and Roth, J.R. (1976). *Cell* **7**, 381.
Gilmore, R.A., Stewart, J.W. and Sherman, F. (1971). *J. Mol. Biol.* **61**, 157.
Goodman, H.M., Olson, M.V. and Hall, B.D. (1977). *Proc. Nat. Acad. Sci.* **74**, 5453.
Grunstein, M. and Hogness, D.S. (1975). *Proc. Nat. Acad. Sci.* **72**, 3961.
Holley, R.W., Apgar, J., Everett, G.A., Madison, J.T., Marquisee, M., Merrill, S.H., Penswick, J.R. and Zamir, A. (1965). *Science* **147**, 1462.
Holness, J.N. and Atfield, G. (1976). *Biochem. J.* **153**, 447.
Hopper, A.K., Banks, F. and Evangelidis, V. (1978). *Cell,* **17**, 211.
Hutchison, H.T., Hartwell, L.H. and McLaughlin, C.S. (1969). *J. Bacteriol.* **99**, 807.
Keith, G., Roy, A., Ebel, J.P. and Dirheimer, G. (1971). *FEBS Letters* **17**, 306.
Knapp, G., Beckmann, J.S., Johnson, P.F., Fuhrman, S.A. and Abelson, J. (1978). *Cell,* **14**, 221.
Kuntzel, B., Weissenbach, J. and Dirheimer, G. (1972). *FEBS Letters* **25**, 189.
Madison, J.T., Everett, G.A. and Kung, H. (1966). *Science* **153**, 531.
Maxam, A.M. and Gilbert, W. (1977). *Proc. Nat. Acad. Sci.* **74**, 560.
Mazzara, G.P., Seidman, J.G., McClain, W.H., Yesian, H., Abelson, J. and Guthrie, C. (1977). *J. Biol. Chem.* **252**, 8245.
Petes, T.D., Hereford, L.M. and Botstein, D. (1977). *In* Molecular Approaches to Eukaryotic Systems, (eds. G. Wilcox, J. Abelson and C.F. Fox), 239. Academic Press, New York.
Pinkerton, T.C., Paddock, G. and Abelson, J. (1973). *J. Biol. Chem.* **248**, 6348.
Piper, P.W. and Wasserstein, M. (1977). *Europ. J. Biochem.* **80**, 103.
Piper, P.W., Wasserstein, M., Engbaek, F., Kaltoft, K., Celis, J.E., Zeuthen, J., Liebman, S. and Sherman, F. (1976). *Nature (London)* **262**, 757.
RajBhandary, U.L., Chang, S.H., Stuart, A., Faulkner, R.D., Hoskinson, R.M. and Khorana, H.G. (1967). *Proc. Nat. Acad. Sci.* **57**, 751.
Roberts, R.J. (1976). *Crit. Rev. Biochemistry* **3**, 123.
Rubin, G.M. and Sulston, J.E. (1973). *J. Mol. Biol.* **79**, 521.
Schweizer, E., MacKechnie, C. and Halvorson, H.O. (1969). *J. Mol. Biol.* **40**, 261.
Shiokawa, K. and Pogo, A.O. (1974). *Proc. Nat. Acad. Sci.* **71**, 2658.
Southern, E.M. (1975). *J. Mol. Biol.* **98**, 503.
Struhl, K., Cameron, J.R. and Davis, R.W. (1976). *Proc. Nat. Acad. Sci.* **73**, 1471.
Tinoco, I., Jr., Borer, P.N., Dengler, B., Levine, M.D., Uhlenbeck, O.C., Crothers, D.M. and Gralla, J. (1973). *Nature New Biol.* **246**, 40.
Valenzuela, P., Venegas, A., Weinberg, F., Bishop, R. and Rutter, W.J. (1978). *Proc. Nat. Acad. Sci.* **75**, 190.
Velten, J., Fukada, K. and Abelson, J. (1976). *Gene,* **1**, 93.
Zachau, H.G., Dutting, D. and Feldmann, H. (1966). *Angew. Chem.* (Int. Ed. Engl.) **5**, 422.

CHARACTERIZATION OF PREMATURE CHAIN TERMINATION MUTANTS OF HYPOXANTHINE GUANINE PHOSPHORIBOSYL TRANSFERASE AND THEIR REVERTANTS IN CHINESE HAMSTER CELLS

C.T. CASKEY, R.G. FENWICK, G. KRUH AND D. KONECKI

The Howard Hughes Laboratory for the Study of Genetic Disorders, Departments of Medicine and Biochemistry, Baylor College of Medicine, Houston, Texas, USA.

Introduction

Mutations have been classified in prokaryotic genes by a variety of biochemical and genetic methods. Study of the mechanisms by which phenotypic correction occurs has led to a molecular understanding of reversion and suppression. Intergenic or suppressor mechanisms of phenotypic mutant correction are well understood at the molecular level through the efforts of investigators studying both viral and cellular mutations (Garen, 1968; Roth, 1974). Suppressor genes for premature chain termination (PCT) (Smith, 1973), frameshift (Yourno and Kohno, 1972), and missense mutations (Hill, 1975) can occur on the basis of anticodon mutations in tRNA genes (Goodman *et al.*, 1968, 1977) enabling correction of mutations at translation. Genetic studies of yeast PCT mutants indicate mutations can be phenotypically corrected by either reversion (Sherman *et al.*, 1970) or suppression (Sherman *et al.*, 1973). *In vitro* study of tRNAs isolated from clones carrying suppressed PCT mutations indicate these suppressors are also translational suppressors (Gesteland *et al.*, 1976; Capecchi *et al.*, 1975). Thus both prokaryotic and eukaryotic cells have the potential for mutant correction via suppressor mutations.

The following studies are directed toward the investigation of animal cell mutants. The goals are to describe at a molecular level the mechanisms of mutation and their correction. The male Chinese hamster lung fibroblast (RJKO) was chosen for study (Ford and Yerganian, 1958). These cells are extremely useful for the study of mutagenesis since genetic studies suggest they are pseudodiploid (Siminovitch, 1976). Our laboratory and others have successfully used Chinese hamster cells for isolation of both recessive and dominant mutants. A number of laboratories have selected Chinese hamster mutants in protein synthesis (Gupta and Siminovitch, 1976), amino acid auxotrophy (Kao and Puck, 1967), mitochondrial regulation (Soderberg *et al.*, 1977) and drug and amino acid analogue resistance (Astrin and Caskey, 1976; Wasmuth and Caskey, 1978). For these studies of mutation and suppression we are investigating the genetic locus for the enzyme hypoxanthine-guanine phosphoribosyl transferase (HGPRT, E.C.2.4.2.8.). The HGPRT gene is X-linked in man (Henderson *et al.*, 1969), mouse (Chapman and Shows, 1976), and hamster (Chasin and Urlaub, 1975) and thus selection of mutants is not complicated by diploidy. Both selective and counter selective (Szybalski and Smith, 1959) media are available for facilitating HGPRT$^-$and HGPRT$^+$ mutant clone isolation. While earlier studies questioned the usefulness of this locus for mutational study (Harris, 1971), the overwhelming body of data from numerous laboratories clearly establishes teh ability to obtain HGPRT mutants (Siminovitch, 1976). These studies do not rule out the occurrence of epigenetic changes affecting HGPRT expression but place the burden of proof on investigators who propose that epigenetic changes occur. We have applied genetic, biochemical and immunologic methods toward the identification and classification of HGPRT structural gene mutants, and have attempted to separate inter- and intragenic reversion events.

HGPRT mutants

We have utilized a number of mutagenic agents to obtain clones of Chinese hamster cells that are resistant to 8-azaguanine (Gillin *et al.*, 1972). Since our prominent goal was to isolate strains carrying PCT mutations, we were primarily interested in mutants that arose after treatment with the point mutagens N-methyl-N′-nitrosoguanidine (NG) of ethyl methanesulphonate (EMS). As shown in Table I, we have also isolated clones that arose spontaneously as well as ones induced by ultraviolet light or the frameshift mutagen ICR-191. The various mutagens increased the frequency of 8-azaguanine resistance from 8- to 70-fold and, as will be described below, they also influenced the distribution of mutant phenotypes.

We initially examined the isolates to determine whether they had detectable *in vivo* or *in vitro* HGPRT activity. Although we have demonstrated that we can detect HGPRT activity in extracts of some of the mutant cells (Gillin *et al.*, 1972), experiences has shown us that our most sensitive assay is the measurement of cellular incorporation of [^{14}C] hypoxanthine into nucleic acids, especially when *de novo* synthesis of purine nucleotides is blocked by adding aminopterin to the medium. Under these conditions we have observed levels of incorporation that are comparable to those of wild-type cells by cell

TABLE I

Isolation and Characterization of Chinese Hamster Cells Resistant to 8-Azaguanine

Mutagen	Frequency	Number of Isolates			
		Selected	$HGPRT^+$	$HGPRT^+$ and/or CRM^+	$HGPRT^-$ and CRM^-
None	1×10^{-5}	10	0	0	10
NG	7×10^{-4}	23	14	16	7
EMS	7×10^{-4}	18	5	7	11
UV-light	2×10^{-4}	10	1	1	9
ICR-191	8 to 10×10^{-5}	18	1	2	16
TOTAL		79	21	26	53

lines that have very low levels of HGPRT activity when assayed *in vitro*. In Table I we demonstrate that that the type of mutagenic treatment used in the isolation of our resistant strains affected the proportion of the clones that retained HGPRT activity. Such strains are most prominent among the populations isolated after treatment with the point mutagens NG and EMS but they are less frequent in the groups generated with UV light and ICR-191, which are mutagens that cause more extensive damage to DNA or the products of affected genes. It is also of interest that none of the spontaneous mutants that were isolated have detectable enzyme activity.

It was of particular interest to us that several of the isolates which incorporate normal amounts of [^{14}C] hypoxanthine in the presence of aminopterin show little or no incorporation without the drug (Gillin *et al.*, 1972). The incorporation of guanine but not adenine by these isolates is also dependent on the presence of aminopterin (Caskey *et al.*, 1975) but the drug has little effect on the incorporation of any of these purines by wild-type cells. As shown in Figure 1, the

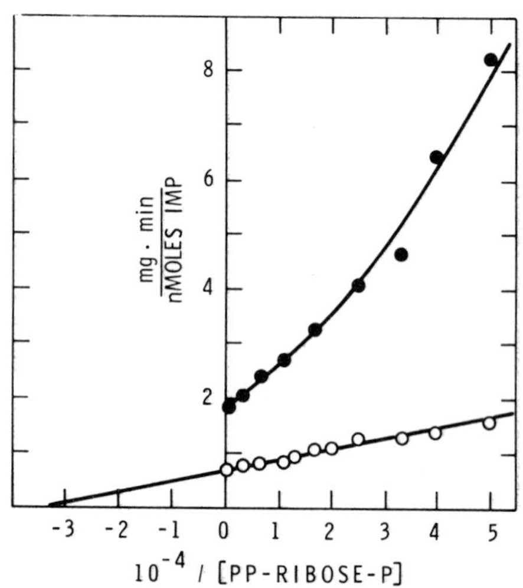

Fig. 1 Kinetic characteristics of PRPP binding for wild-type (RJKO,○) and mutant (RJK3,●) HGPRT.

kinetic parameters of HGPRT from wild type (RJKO) and mutant (RJK3) (Fenwick *et al.*, 1977) were found to differ. Although not shown here, RJK44 and RJK47 were similarly altered. The Km for 5-phosphoribosyl-1-pyrophosphate (PRPP) is increased for each of the mutant enzymes and, unlike the wild-type enzyme, they are activated by that substrate. Thus, they are less active than wild-type enzymes at low PRPP concentrations. Since PRPP is also a substrate for the first enzyme in the *de novo* pathway of purine biosynthesis, one might predict that the defect of the mutant enzymes would put HGPRT activity at a kinetic disadvantage to the *de novo* pathway. However, we have recently found that Virazole stimulates incorporation of guanine but not hypoxanthine by the mutants (Fenwick, unpublished observation). As an inhibitor of IMP dehydrogenase (Lowe *et al.*, 1977), Virazole would be expected to reduce the intracellular pool of GMP but not that of IMP, whereas the *de novo* purine pathway inhibitor, aminopterin, will reduce both. Since HGPRT of the mutant cells is activated *in vivo* for both GMP and IMP formation by *de novo* pathway inhibition and only for GMP formation by IMP dehydrogenase inhibition, it would appear that the mutant enzymes are more sensitive to product inhibition than the wild-type enzyme. Whether cellular alterations in PRPP can also influence the *in vivo* regulation demonstrable *in vitro* is less clear. Although the phenotypes of RJK3, 44, and 47 are similar, they are not caused by the same mutational alteration because HGPRT from RJK3 can be distinguished from wild-type enzyme by its electrophoretic mobility and other properties to be described below while the enzymes from RJK44 and 47 cannot (Fenwick *et al.*, 1977). We have not analyzed all of our clones that are resistant to 8-azaguanine and retain HGPRT activity but our results with RJK3, 44 and 47 indicate that such lines produce altered HGPRT. Thus, we have concluded that most of these clones carry mutations, probably of the missense class, in the structural gene for the enzyme.

Two-thirds of the 8-azaguanine isolates listed in Table I do not have any measurable HGPRT activity. To further categorize this group we examined them for the presence of HGPRT protein. This was done by using antiserum against HGPRT purified from Chinese hamster brain (Beaudet *et al.*, 1973). Material that cros reacts with wild-type HGPRT was found in two mutants induced with NG (RJK5 and 15) and one induced with EMS (RJK39) by using cell extracts of these clones to displace the wild-type enzyme from an immunoprecipitation complex. We have also coupled the immunoglobulin fraction of the immune serum to Sepharose beads and have used this material to purify HGPRT protein from extracts of cells that have been labelled with radioactive amino acids (Fenwick *et al.*, 1977). Material bound to the immunoabsorbent is subjected to SDS-polyacrylamide electrophoresis and the presence of HGPRT protein in the radioactive cell extract is determined by autoradiography of fluorography. By this method most of our HGPRT$^-$ clones do not produce a detectable HGPRT protein. The additional CRM$^+$

clones identified by this method are RJK45 and RJK63 which were isolated after mutagenesis with EMS and ICR-191, respectively. In a third immunological approach, these results were confirmed with a micro complement fixation assay (Chiang, 1977).

These immunological studies did not detect a very interesting class of mutants because of limitations imposed by the immune sera that were used. Evidence indicates that Chinese hamster HGPRT is composed of identical protein subunits (Olsen and Milman, 1974). Thus, mutations that greatly alter the structure of the subunit protein, especially PCT mutations, will probably prevent assembly of the subunits into the native enzyme conformation. When we have used a variety of chemical or physical agents to denature HGPRT to the point that it no longer has enzyme activity, the preparations were no longer found to have immunological activity by the assays described above. Thus, we do not feel enzyme subunits or fragments of the HGPRT peptide could be detected with our original immune sera. For this reason we have used purified HGPRT that has been denatured with urea as an antigen to raise antibodies that will recognize enzyme that is not in the native conformation. A comparison of antibody developed in response to native HGPRT immunizations (Ab_n) and denatured HGPRT immunizations (Ab_d) is given in Figure 2. These studies used the radioimmunologic method previously described for identifying *in vivo* labelled HGPRT. The Ab_n does not recognize HGPRT from urea denatured extracts while Ab_d does. Although not clearly demonstrated here, Ab_d also has weak recognition of native HGPRT. Ab_n only recognized native HGPRT. These results have been confirmed using complement fixation techniques (Chiang, 1977). These data indicate that HGPRT denaturation exposes additional antigenic sites recognized by Ab_d and that native conformation is required for Ab_n recognition.

PCT mutations have predictable consequences for the structure of mutant proteins. They lead to loss of peptide sequence beyond the PCT codon toward the carboxyl terminus and the protein is reduced in size. UGA mutations frequently have been reported to be leaky and can lead to synthesis of both reduced and normal molecular weight proteins. Detection of protein fragments is complicated by their lability and the sensitivity and specificity of the detection method. We have therefore used both Ab_d and Ab_n immunoadsorbent Sepharose beads in our efforts to screen radioactive mutant extracts for the presence of immunologically cross reactive material (CRM). While one must always reserve the possibility that CRM^- clones produce peptide fragments at levels below detection or that the available Ab_d does not recognize amino terminal fragments, the positive results are informative. As discussed earlier, we have studied 8-AG^r clones and found only 5 of the 53 HGPRT$^-$ clones CRM$^+$ against Ab_n. There were 15 additional HGPRT$^-$ clones determined

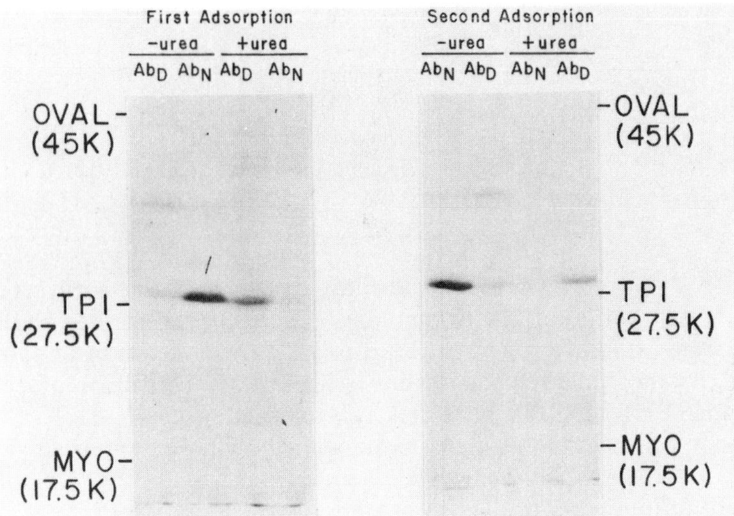

Fig. 2 Specificity of immune sera for native and denatured HGPRT.

CRM$^+$ against Ab_d. Thus a significant percentage (37%) of the HGPRT$^-$ mutants could be shown to be structural HGPRT mutants which did not have the capacity to form detectable native HGPRT molecules. We are continuing to develop the technology of use of Ab_d toward the purpose of identifying specific regions of the HGPRT protein.

We have carried out reversion analysis of a significant number of our HGPRT$^-$, CRM$^-$ mutants in an effort to clarify the nature of their phenotype. Spontaneous or mutagen induced revertants have been isolated from 23 such mutants to date by selecting subclones that will grow in HAT medium. For 2 out of 17 mutants extensively studied the frequency of spontaneous revertants in the populations is less than 10^{-7} or undetectable with the number of cells examined in typical experiments. The mutagens used to induce reversion included EMS, NG, and ultraviolet irradiation. As shown in Table II, heterogeneity among our mutants is suggested by their reversion frequencies and mutagen responsiveness. Our data is not adequate at this time to allow us to draw conclusions about the nature of the original mutation based upon mutagen specificity of reversion. RJK36, an EMS induced HGPRT$^-$, CRM$^-$ mutant, has been used

TABLE II

Mutagen Effect for Mutant Reversion to HGPRT$^+$

Mutant	(Mutagen)	Reversion Frequency x 10^5		
		Spontaneous	EMS	Ultraviolet
RJK10	(NG)	0.17	1.4	9.5
RJK30	(EMS)	0.02	0.03	>2.3
RJK51	(U.V.)	0.3	>7.5	>12.0
RJK76	(Spon)	<0.08	7.4	37.0

extensively in our laboratory and those of others both for reversion and somatic cell hybridization studies, and never reverted to HGPRT$^+$. Its extreme stability suggests that it might carry a deletion or some other chromosomal rearrangement that prevents restitution of a functional HGPRT gene. We have closely examined other HGPRT$^-$, CRM$^-$ clones which reverted at high frequency following mutagenesis since these mutants probably carry point mutations, some of which will be PCT mutations.

It is difficult to be sure such clones carry mutations in the structural or regulation elements of the HGPRT gene. To identify HGPRT$^-$, CRM$^-$ mutants with structural gene alterations we have examined *in vitro* the HGPRT of their revertants for alterations. Revertants of a regulatory mutant would not be expected to possess an altered HGPRT while structural gene mutants would have a high probability of differing from the wild-type. One of our standard tests of the HGPRT from revertant clones is to examine its immunoprecipitation from cell extracts (Fenwick *et al.*, 1977) and kinetic parameters. The enzymes from RJK159 and all other revertants of RJK10 are unique in that it takes about three times more antiserum to precipitate a given amount of these enzymes than is required to precipitate a comparable amount of wild-type enzyme. This is illustrated in Figure 3 which also shows that a mixture of RJK159 and wild-type gives a biphasic immuno precipitation pattern. Thus, the antiserum can distinguish the two enzymes and precipitate each independent of the other. This indicates that HGPRT from the revertants of RJK10 lacks one or more antigenic determinants that are present on the wild-type enzyme and is presumptive evidence that RJK10 does carry a structural gene mutation. The genetic changes responsible for reversion of RJK10 compensate for but do not actually reverse the original damage. Revertants of the HGPRT$^-$, CRM$^-$ mutant RJK37 have also been found to produce an altered but functional HGPRT (Fenwick *et al.*, 1977).

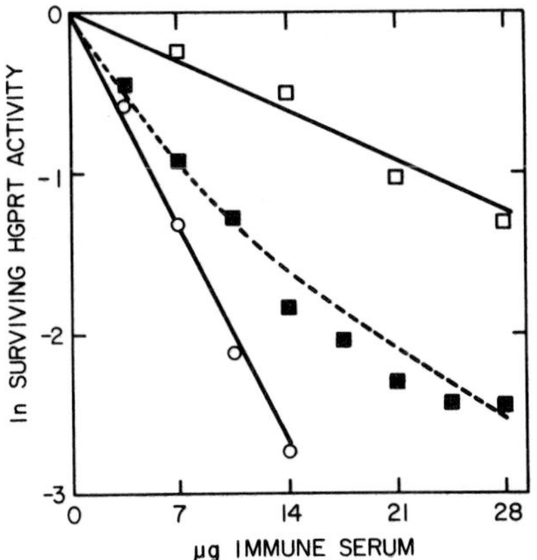

Fig. 3 Immunoprecipitation of wild-type (RJKO,○) revertant (RJK159,□) and mixed HGPRT (■).

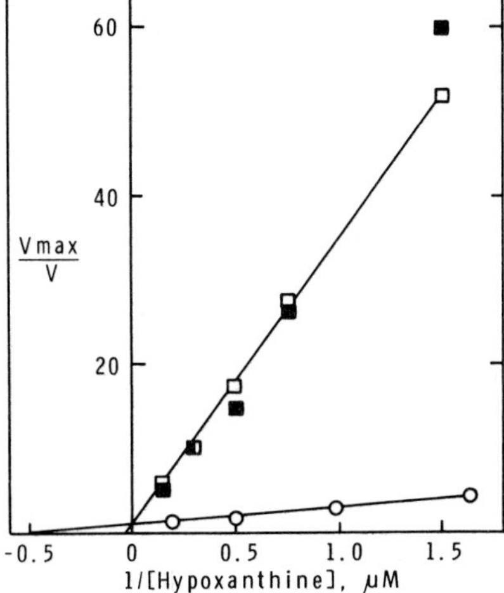

Fig. 4 Kinetic characteristics of Hx binding for wild-type (RJKO,○) and revertant (RJK255, □, and RJK256, ■) HGPRT.

Kinetic analysis of RJK37 revertants, RJK255 and 256, (Figure 4) indicate the Km for hypoxanthine is strikingly higher than that of wild-type enzyme. We therefore conclude both RJK10 and 37, mutants of the HGPRT⁻, CRM⁻ class, possess structural gene mutations. This large class of hamster mutants has similar features to the alkaline phosphatase PCT mutants of *E. coli* and can be anticipated to contain PCT mutants. More direct evidence for HGPRT PCT mutants is provided by our identification of molecular weight alterations. The

RJK3 and RJK39 mutants have been most extensively studied. The RJK3 mutant, selected following EMS mutagenesis, possesses HGPRT activity but is altered in its PRPP Km.

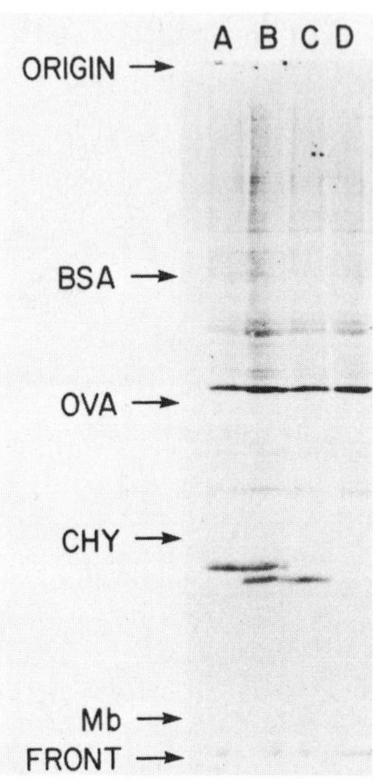

Fig. 5 Radioimmuno identification of wild-type (RJKO, A), mixed (RJKO + RJK3, B), and mutants (RJK3, C and RJK36, D).

The RJK39 mutant was selected following NG mutagenesis and is HGPRT$^-$. The molecular weight reductions of RJK3 and RJK39 are 10,000 and 5,000 daltons respectively. It is unlikely both are PCT mutations since the smaller protein is enzymically active.

In an effort to try to clarify these mutants we have analyzed their tryptic peptides using high pressure ion exchange chromatography (Milman *et al.*, 1977). *In vivo* radiolabeled HGPRT was purified by immunoabsorption to Sepharose bound Ab_n and SDS PAGE electrophoresis prior to trypsinization and peptide analysis. We have conducated peptide analysis of HGPRT labeled *in vivo* with [^{35}S] methionine, [^{3}H]lysine, [^{3}H]arginine, and [^{3}H]leucine following trypsine cleavage. Thus far no alterations in RJK3 have been identified. The peptide composition of RJK39 differs from wild type HGPRT. As shown in Figure 7, [^{3}H]lysine labeled HGPRT from RJK39 differs from wild type (RJKO) in the number of lysine peptides. The RJK39 mutant lacks a late eluting lysine peptide. No peptide differences were observed when [^{3}H]arginine was used. As shown in Figure 8

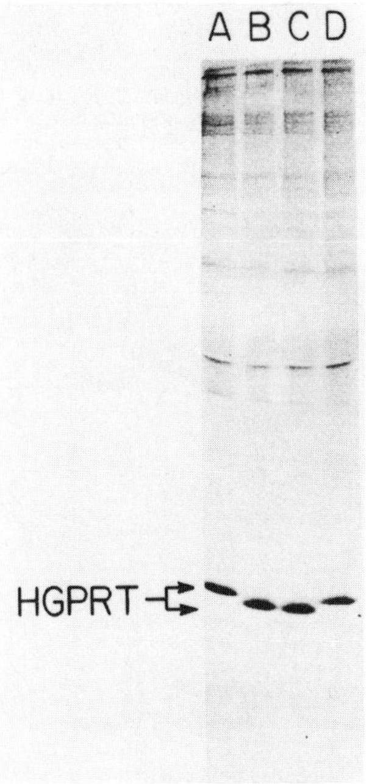

Fig. 6 Radioimmuno identification of wild type (RJKO, A) and mutant (RJK39, B and C) and RJK39 revertant (RJK472, D).

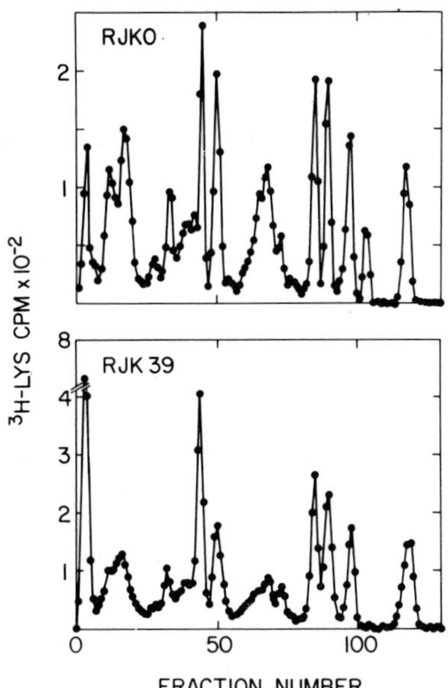

Fig. 7 Tryptic peptides labeled *in vivo* with [^{3}H]lysine and fractionated by HPLC.

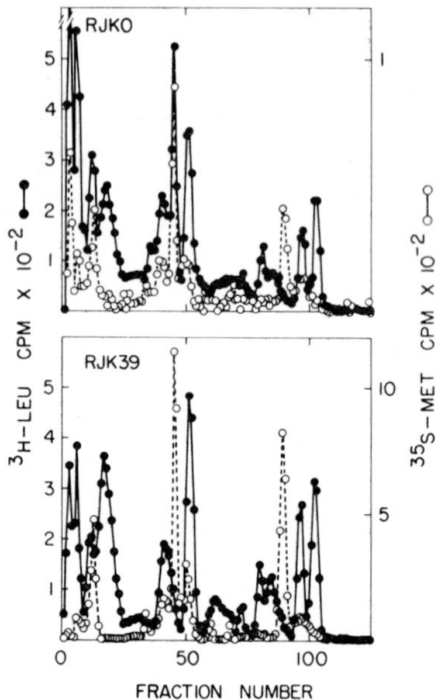

Fig. 8 Tryptic peptides labeled *in vivo* with [^{35}S] methionine (○) and [^{3}H] leucine followed by HPLC fractionation.

which uses HGPRT doubly labeled with [^{35}S] methionine and [^{3}H] leucine, RJK39 is also missing [^{3}H] leucine labeling in the second [^{35}S] methionine peak. Thus the RJK39 mutation has led to an alteration of at least two peptides.

These studies indicate RJK39 is reduced in molecular weight on the basis of structural gene mutation and not a post-translational modification defect. Since at the present time we have not identified the carboxyl peptide of HGPRT with certainty we can not conclude RJK39 is a PCT mutant. Celis and his associates have reported in this volume preliminary experiments indicating *in vivo* suppression of HGPRT by microinjection of RJK39 with yeast amber suppressor tRNA. Thus the peptide and microinjection data are compatible with RJK39 being an amber mutant. Further study of RJK39 is needed to be certain of this interpretation.

Since our immunologic and biochemical studies have indicated some of our HGPRT mutants have features of PCT mutations, we have extended our studies to biochemical and genetic analysis of their revertants in an effort to identify clones with suppressor mutations. Revertants derived from a given HGPRT$^-$ mutant differ widely in their *in vitro* HGPRT activity (Table III). The specific activity for a cloned revertant does not vary but remains a property of that clone.

TABLE III

HGPRT in vitro *Activity of RJK10 Revertants*

Isolate	Mutagen	Specific Activity
	Wild type	
RJKO	------	1.18
	$8AG^r$	
RJK10	NG	0.00
	HAT^r	
RJK159	Spontaneous	0.76
RJK160	Spontaneous	0.29
RJK169	NG	0.14
RJK173	Ultraviolet	0.10

These studies suggest the genetic mechanisms of their phenotypic reversion is heterogeneous. The stability of revertants for the $HGPRT^+$ character is of particular interest. The frequency at which revertant cell clones were found to survive in media containing 6-thioguanine (6-TG) has also been found to be a property of individual revertants as illustrated in Table IV.

TABLE IV

Stability of $HGPRT^+$ Revertants

Mutant	Specific Activity HGPRT	Frequency of $6\text{-}TG^r$
	$HGPRT^-$ Mutant	
RJK10	0.00	1.0
	$HGPRT^+$ Revertants	
RJK159	0.76	4.4×10^{-3}
RJK160	0.29	4.5×10^{-4}
RJK167	0.10	1.5×10^{-4}
RJK169	0.14	2.9×10^{-5}

In each case the revertant and $6\text{-}TG^r$ clone were studied biochemically to exclude the possibility it was not the consequence of aminopterin resistance or an HGPRT mutant capable of surviving in both media. The $HGPRT^+$ character of some revertants is 'unstable' since it is lost at a frequency higher than expected for mutation (i.e., 10^{-3}).

Since the HGPRT$^-$ clones derived from unstable revertants can revert to HGPRT$^+$ by further mutagenesis and selection in HAT media, they retain the structural gene for HGPRT. The genetic basis of this unstable character has been of interest to our laboratory since we have considered the possibility that it may reflect intergenic suppression. We are currently investigating this possibility by means of cellular hybridization (Chasin and Urlaub, 1975). Preliminary data indicates that a gene not syntenic with Glucose-6-phosphate dehydrogenase (G-6PD, X-linked) is responsible for the HGPRT$^+$ character of the unstable mutant RJK159. This genetic study suggests the HGPRT$^+$ activity depends upon a non X-linked function and therefore would provide evidence for a 'suppressor' gene activity. Additional study is required before a definitive statement of the molecular basis of this nonsyntenic HGPRT$^+$ character can be determined. A simple explanation not related to suppression could be that the X chromosome of these clones is fragmented and thus synteny with G-6PD lost. Cytogenetic and biochemical study of other X-linked markers should resolve these possibilities.

The structural analysis of revertant clones has been facilitated by the available immunologic methods developed for HGPRT. Using both complement fixation analysis and direct immune precipitation demonstrable differences from the wild-type HGPRT are found for mutants RJK10, RJK76, RJK53 and RJK37 (HGPRT$^-$, CRM$^-$ mutants) The molecular weight of some revertants (determined by SDS PAGE) also identified revertants differing from wild type. The revertants with altered molecular weight were also derived from HGPRT$^-$, CRM$^-$, mutants. It appears likely from such studies that both point and frameshift mutations can lead to an HGPRT$^-$, CRM$^-$ phenotype for the 8-AG^r mutants.

We have had particular interest in the revertants of RJK39, the HGPRT$^-$, CRM$^+$ mutant with reduced molecular weight whose HGPRT phenotype was suppressed when the cells were microinjected with amber tRNA. The SDS PAGE analysis of a number of HAT^r clones derived from RJK39 are shown in Figure 9. The clone (RJK633) which exhibits the m.wt. of RJK39 has been proven to be aminopterin resistant and not HGPRT$^+$. The revertant clone, RJK630, is of special interest since it expresses HGPRT with both mutant and wild-type m.wt. Suppressed PCT prokaryotic mutants have been shown to express both the mutant and suppressed proteins. It is quite tempting, taking into account all information available on RJK39, to interpret the data as providing evidence for a PCT suppressor gene in RJK630. There is, however, a simple explanation which must be considered. The revertant clone expressing both HGPRT bands is known to con-

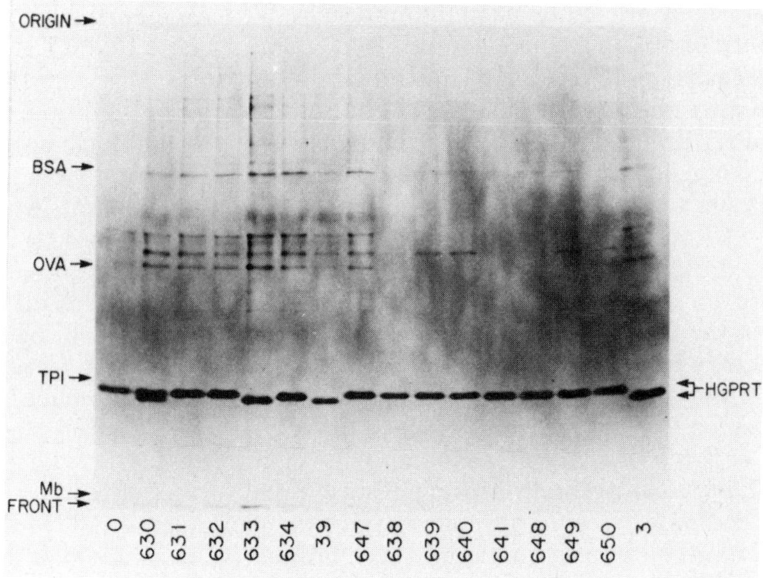

Fig. 9 Radioimmune detection of HGPRT from mutant and revertant clones.

tain more than the diploid number of chromosomes. Thus, the possibility the clone contains two HGPRT structural genes, one mutant, the other reverted, can not be excluded at the time this manuscript was prepared.

Discussion

The mammalian cell is known to terminate protein synthesis upon recognition of the codons UAA, UAG, and UGA (Goldstein, *et al.*, 1970). Therefore point mutations which alter codons corresponding to amino acids to one of these 3 codons will lead to premature chain termination. This expectation has been realized in the case of PCT mutations in the *Herpes* Thymidine kinase (Summers *et al.*, 1975), Adenovirus-2-SV40 hybrid virus (Gesteland *et al.*, 1977) and mouse immunoglobulin mutants (Adetugbo *et al.*, 1977). In each of these cases the convincing evidence for PCT mutations was provided by the *in vitro* suppression of the respective mRNA translation products by yeast suppressor tRNA. We are not able to carry out *in vitro* translation of HGPRT mRNA from our mutant clones at this time due to the low cellular level of mRNA. There is little doubt at this time that the structural gene for HGPRT in hamster (Beaudet *et al.*, 1973), mouse (Wahl *et al.*, 1975) and human (Sato *et al.*, 1972) cultured cells can be altered by chemical mutagens. We have reported here the ability to obtain Chinese hamster 8-AG^r clones whose HGPRT structural genes are altered as a consequence of mutation. Many are HGPRT and have normal molecular weight but differ from the wild type HGPRT in thermal lability, kinetic parameters, immunologic properties, or gel migration. The simplest interpretation of this data is that they are altered by missense mutation. The HGPRT$^-$ mutants which have CRM$^+$ products of normal molecular weight recognized by either Ab_n or Ab_d are also likely the consequence of mis-sense mutation.

We have identified revertants with molecular weights which differ from the wild-type and on the basis of this information have concluded they represent mutations of small deletion or insertion type. Large deletions or insertions may account for mutants which have not been observed to revert.

Our study of HGPRT$^-$, CRM$^-$ mutants indicate that structural gene mutations can account for the phenotype and furthermore that this is a common category of mutant. Since the molecular weight of their revertants have most commonly been normal, we conclude that point mutations account for many. Since PCT mutations have been common point mutations for extreme negative phenotypes in prokaryotic genes (Garen, 1968; Suzuki and Garen, 1969; Fowler and Zabin, 1968) and furthermore since such mutants were frequently CRM$^-$, we anticipate PCT mutations are to be found in this mutant set. The biochemical proof for such is difficult and depends upon the development of new genetic and biochemical probes.

The opportunity for establishing that mutants with reduced molecular weight are PCT mutations is greater. Our laboratory as well as that of Capecchi (1977) and Milman (1977) have used high pressure ion exchange chromatography to demonstrate peptide alterations in HGPRT mutants. We have had the opportunity to study mutants with reduced molecular weight and, as reported here, have demonstrated peptide alterations. The mutant RJK39 has at least two peptide alterations. This mutant was found to be suppressed by the injection of yeast amber suppressor tRNA and on these grounds has properties of a PCT amber mutant (Celis *et al.*, this volume). Capecchi (1977) has previously reported a mouse HGPRT$^-$, CRM$^+$ mutant which has an apparent normal molecular weight but an altered position of migration for an [^{35}S] methionine-containing peptide. This methionine-containing peptide is felt to represent the carboxyl peptide of mouse HGPRT and thus the mutant was concluded to be a PCT mutation. Using red blood cell ghost which have entrapped amber tRNA Capecchi reported suppression of HGPRT upon fusion of the rbc's to the mouse mutant. In both cases the evidence for PCT mutation rests heavily upon the *in vivo* HGPRT suppression and the potential for possible false positive results must be considered. With these reservations in mind the studies indicate PCT mutations have been identified at the HGPRT locus in mouse and hamster cells.

We have provided genetic and biochemical evidence for a suppressor reversion mechanism for an $HGPRT^-$ mutant. The $HGPRT^+$ revertant RJK630 derived from RJK39 expresses shortened mutant and normal molecular weight HGPRT. This is a rare clone since of 20 other revertants studied none had these features. Furthermore, the clone has a chromosomal count which exceeds the normal diploid complement. We are obligated therefore to point out that if these features are the consequence of suppression the suppressor gene expression is rare and may not occur in diploid cells. Some prokaryotic suppressors are also rare since they are recessive lethals (Soll and Berg, 1969) and occur only in diploid cells. Chinese hamster cells are established to be pseudodiploid for a number of autosomal loci. These differing mechanisms for reversion of the suspected amber mutant RJK39 should be amenable to direct examination. We are currently examining RJK630 for suppressor tRNA content.

Using genetic methods of analyzing unstable revertants of $HGPRT^-$, CRM^- mutants we have in the case of revertants of RJK10 demonstrated the requirement of a gene which is not syntenic with G-6PDH for the $HGPRT^+$ of an unstable revertant. Stable revertants preserve this syntenic relationship. We are currently examining the chromosomal character of these revertants. While it is clear that the chromosomal number of unstable revertants is equal to the wild-type, study of banded chromosomes is needed to try to distinguish between X chromosome fragmentation and somatic crossing over of suppresor gene.

Our biochemical and genetic studies provide evidence for both PCT and suppressor mutations in Chinese hamster cells. In both cases the data is consistent with these conclusions but does not exclude other possibilities which are discussed. Our laboratory efforts are now directed toward the confirmation of these interpretations by independent biochemical and genetic methods. The definitive experiments are incomplete at the time of this manuscript preparation but the directions for experimentation are clear.

References

Adetugbo, K., Milstein, C. and Secher, D. (1977). *Nature (London)* **265**, 299.
Astrin, K. and Caskey, C.T. (1976). *Arch. Biochem. Biophys.* **176**, 397.
Beaudet, A.L., Roufa, D.J. and Caskey, C.T. (1973). *Proc. Natl. Acad. Sci.* **70**, 320.

Capecchi, M.R., Hughes, S.H. and Wahl, C.M. (1975). *Cell* **6**, 269.
Capecchi, M.R., Van der Haar, R.A., Capecchi, N.E. and Sveda, M.M. (1977). *Cell* **12**, 371.
Caskey, C.T., Fenwick, R., Chiang, C.S., Astrin, K. and Wasmuth, J. (1975). *Proc. XI Int. Cancer Congress, Florence*, 46.
Chapman, V.M. and Show, T.B. (1976). *Nature (London)* **259**, 665.
Chasin, L. and Urlaub, G. (1975). *Science* **187**, 1091.
Chiang, C.S. (1977). Ph.D. Thesis, Baylor College of Medicine, Houston, Texas.
Ditta, G., Soderberg, K., Landy, F. and Scheffler, I.E. (1976). *Somatic Cell Genet.* **2**, 331.
Fenwick, R., Sawyer, T.H., Kehu, G.D., Astrin, K.H. and Caskey, C.T. (1977) *Cell* **12**, 383.
Fenwick, R.G., Wasmuth, J.J. and Caskey, C.T. (1977). *Somatic cell Genet.* **3**, 207.
Ford, D.K. and Yerganian, G. (1958). *J. Natl. Cancer Institute* **21**, 393.
Fowler, A.V. and Zabin, I. (1968). *J. Mol. Biol.* **33**, 35.
Garen, A. (1968). *Science* **160**, 149.
Gesteland, R.F., Wolfner, M., Gasafi, D., Fink, G., Botstein, D. and Roth, J.R. (1976). *Cell* **7**, 381.
Gesteland, R.F., Willis, N., Lewis, J.B. and Grodzicker, T. (1977). *Proc. Natl. Acad. Sci.* **74**, 4567.
Gillin, F.D., Roufa, D.J., Beaudet, A.L. (1972). *Genetics,* **72**, 239.
Goldstein, J., Beaudet, A.L. and Caskey, C.T. (1970). *Proc. Natl. Acad. Sci.* **67**, 99.
Goodman, H.M., Abelson, J., Landy, S., Brenner, S. and Smith, J.D. (1968). *Nature (London)* **217**, 1019.
Goodman, H.M., Olson, M.V. and Hall, B.D. (1977). *Proc. Natl. Acad. Sci.* **74**, 5453.
Gupta, R.S. and Siminovitch, L. (1976). *Cell* **9**, 213.
Harris, M. (1971). *J. Cell Phys.* **78**, 177.
Henderson, J.F., Kelley, W.N., Rosenbloom, F.M. and Seegmiller, J.E. (1969). *Am. J. Hum. Genetics* **21**, 61.
Hill, C.W. (1975). *Cell* **6**, 419.
Kao, F. and Puck, T. (1967). *Genetics* **55**, 513.
Lowe, J.K., Brox, L. and Henderson, J.F. (1977). *Cancer Res.* **37**, 736.
Milman, G., Krauss, S.W. and Olsen, A.S. (1977). *Proc. Natl. Acad. Sci.* **74**, 926.
Olsen, A.S. and Milman, G. (1974). *J. Biol. Chem.* **249**, 4030.
Roth, J.R. (1974). *Ann. Rev. Genetics* **8**, 319.
Sato, K., Slesinski, R.S. and Littlefield, J.W. (1972). *Proc. Natl. Acad. Sci.* **69**, 1244.
Sherman, F., Stewart, J.W., Parker, J.H., Putterman, G.J., Agrawal, B.B.L. and Margoliash, E. (1970). Symposia of the Society for Experimental Biology XXIV, 85.
Sherman, F., Liebman, S.W., Stewart, J.W. and Jackson, M. (1973). *J. Mol. Biol.* **78**, 157.
Siminovitch, L. (1976). *Cell* **7**, 1.
Smith, J.D. (1973). *British Med. Bull.* **29**, 220.

Soderberg, K.L., Dita, G.S. and Scheffler, I.E. (1977). *Cell* **10**, 697.
Soll, L. and Berg, P. (1969). *Proc. Natl. Acad. Sci.* **63**, 392.
Summers, W.P., Wagner, M. and Summers, W.C. (1975). *Proc. Natl. Acad. Sci.* **72**, 4081.
Suzuki, T. and Garen, A. (1969). *J. Mol. Biol.* **45**, 549.
Szybalski, W. and Smith, M.J. (1959). *Proc. Soc. Exp. Biol. Med.* **101**, 662.
Wahl, G.M., Hughes, S.W. and Capecchi, M.R. (1975). *J. Cell Physiology* **85**, 307.
Wasmuth, J.J. and Caskey, C.T. (1978). *Life Sciences* **22**, 1469.
Yourno, J. and Kohno, T. (1975). *Science* **175**, 650.

MICROINJECTION OF tRNAs INTO SOMATIC CELLS

J.E. CELIS, K. KALTOFT, A. CELIS, R. FENWICK AND †C.T. CASKEY

Division of Biostructural Chemistry, Institute of Chemistry, Aarhus University, Denmark.
†Baylor College of Medicine, Texas Medical Center, Houston, Texas, USA.

Introduction

In prokaryotes and in eukaryotes the nonsense triplets UAA (ochre), UAG (amber) and UGA (opal) signal termination of translation and the release of the complete polypeptide chain from the tRNA on the ribosome (Brenner *et al.*, 1965; Weigert and Garen, 1965; Stretton *et al.*, 1966; Sambrook *et al.*, 1967; Zipser, 1967; Beaudet and Caskey, 1971; Stewart *et al.*, 1972; Stewart and Sherman, 1972; Stewart and Sherman, 1973; Caskey *et al.*, this volume). If, however, as a result of mutation, these nonsense codons appear in phase inside a structural gene, they cause chain termination at the point of mutation and release of an incomplete polypeptide (Sarabhai *et al.*, 1964; Celis *et al.*, 1973). In bacteria and yeast these mutations can be partially reversed by second mutations, in another gene, that codes for an altered tRNA molecule able to decode the chain terminating codon as an amino acid (Hawthorne and Mortimer, 1963; Capecchi and Gussin, 1965; Engelhardt *et al.*, 1965; Andoh and Ozeki, 1968; Goodman *et al.*, 1968; Hawthorne and Mortimer, 1968; Gilmore *et al.*, 1971; Sherman *et al.*, 1973; Hawthorne and Leupold, 1974; Capecchi *et al.*, 1975; Liebman *et al.*, 1975; Gesteland *et al.*, 1976; Piper *et al.*, 1976; Brandriss *et al.*, 1976).

In mammalian cells and their viruses nonsense mutations have so far only been partially characterized in the case of Herpes thymidine

---

Abbreviations: CHO Chinese hamster ovary cells, FITC-BSA Fluoresceine isothiocyanate labelled bovine serum albumin, HAU Hemagglutinating units, HGPRT Hypoxanthine-guanine phosphoribosyl transferase, MEM Eagle's minimal essential medium, PBS Phosphate buffered saline, PEG Polyethylene glycol, U.V. Ultraviolet light.

kinase (Summers *et al.*, 1975); HGPRT (Capecchi *et al.*, 1977; Caskey *et al.*, this volume) and adenoviruses-2-SV40 hybrid virus (Gesteland *et al.*, 1977). In all cases only a handfull of prospective nonsense mutants have been isolated starting with a large number of defective mutants.

The main problem one encounters in searching for nonsense mutants in mammalian cells is the lack of procedures for the rapid screening of potential nonsense mutants among a population of defective mutants (deletions, frameshift, missense and nonsense). One such procedure is to microinject (Kaltoft *et al.*, 1976; Celis, 1977; Capecchi *et al.*, 1977) suppressor tRNAs from yeast (Piper *et al.*, 1976; Piper, this volume) into the defective mutants and determine if the mutation is temporarily rescued after injection. In this paper we will review our work on microinjection of tRNAs into somatic cells by means of the red cell mediated microinjection (Kaltoft *et al.*, 1976) and the direct microinjection with fine micropipettes (Celis, 1977), as well as present preliminary data on the screening of prospective mutants of the CHO HGPRT using the latter technique. Finally we will discuss new lines of experimentation that we believe could be of value when searching for nonsense and suppressor mutations in mammalian cells.

Red Cell Mediated Microinjection of tRNAs

General Features

Furusawa *et al.*, (1974); Loyter *et al.*, (1975); and Schlegel and Rechsteiner (1975). first described a technique to inject macromolecules into somatic cells containing receptors for Sendai virus. The method consists of two steps; first the macromolecules are incorporated into the erythrocytes during hypotonic hemolysis and secondly, the resealed ghosts are fused with the recipient cells using inactivated Sendai virus. Proteins such as ferritin (Loyter *et al.*, 1975), albumin (Schlegel and Rechsteiner, 1975; Wassermann *et al.*, 1976; Wille and Willecke, 1976), thymidine kinase (Schlegel and Rechsteiner, 1975); IgG (Nishimura *et al.*, 1976; Wassermann *et al.*, 1977) and tRNAs (Kaltoft *et al.*, 1976; Capecchi *et al.*, 1977; Schlegel *et al.*, 1978) have been injected into mammalian cells using this method. Recently a similar technique that uses PEG instead of Sendai virus as the fusogen has been described by Kriegler and Livingston (1977) and by Kaltoft and Celis (1978). This technique has the advantage that it can be used for all cell types and that it uses hemoglobin-free erythrocyte ghosts (Kaltoft and Celis, 1978).

In Figure 1 we have summarized the steps involved in the red cell

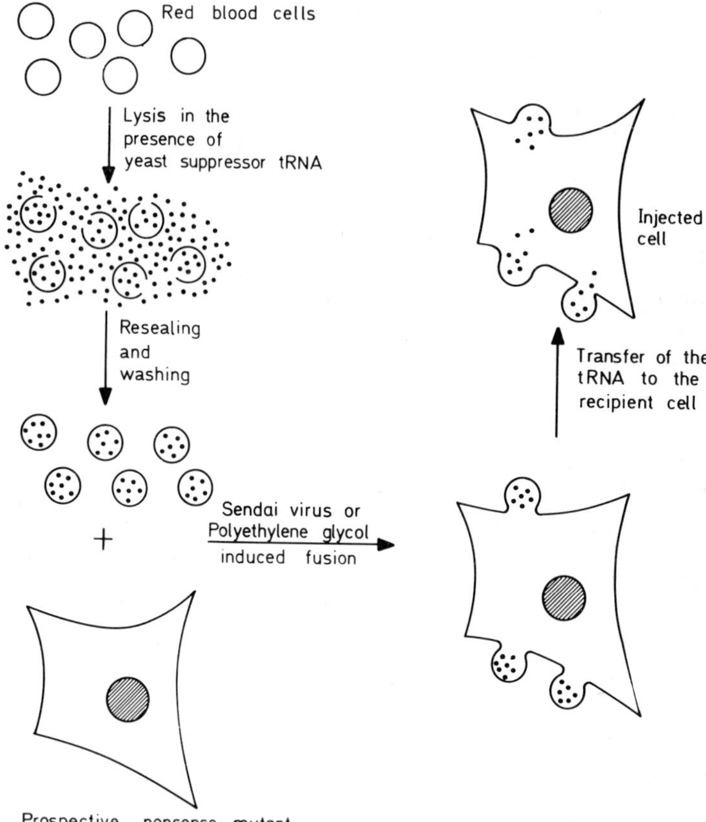

Fig. 1 Scheme for assaying potential nonsense mutations in somatic cells (after Celis, 1977).

mediated microinjection as we proposed that it could be used to assay or screen *in vivo* for the presence of potential nonsense mutations in mammalian cells (Kaltoft *et al.*, 1976; Celis, 1977). This scheme can be used in principle for all cell types although the experimental conditions may not necessarily be the same, especially if Sendai virus is used as fusogen.

Loading of Erythrocytes with tRNA

There are four procedures that can be used to load red blood cells with macromolecules. The 'snap-loading' procedure (Hoffman, 1958; Seeman, 1967; Baker, 1967; Ihler, Glew and Schnure, 1973) consists in lysing the erythrocytes in 10 or more volumes of water containing the macromolecule to be trapped. When hemolysis takes place a fraction of the intracellular content leaks out of the cell and is replaced in part by the entry of the added macromolecules. After a few seconds at 0° the cells reseal and restoration of the saline concentration induces the cells to shrink. The 'dialysis loading' (Loyter, *et al.*, 1975) consists in placing the erythrocytes together with the macromolecule to be trapped inside a dialysis tube and the whole content is then dialysed against hypotonic media. After restoring the saline concentration, the cells are incubated at 37° for 30 min. In the 'preswell loading' (Rechsteiner, 1975) the erythrocytes are first swollen in saline with 0.7-0.9 volumes of water and afterwards lysed in the presence of 1 volume of buffer containing the macromolecule to be loaded. Restoration of the saline concentration is followed by incubation at 37° for 60 min. In all cases the loaded red blood cells are washed several times with isotonic saline prior to fusion. A clear disadvantage of these procedures is that a considerable amount of erythrocyte cytoplasmic proteins, mainly hemoglobin, remains in the ghosts and are therefore transferred to the recipient cells together with the macromolecule to be injected. Recently Kaltoft and Celis (1978) have described a procedure to prepare hemoglobin-free ghosts that consists in resuspending the red cell pellet (1 vol) in 50 volumes of 10m*M* Na-phosphate buffer, 2m*M* ATP pH 7.2, followed by centrifugation at 7,000 × g for 10 min. and 3 washes in the same buffer. Loading of the ghosts is achieved by adding 2 volumes of the solution containing the macromolecules to be trapped to half the volume of pelleted ghosts. These ghosts have so far been used only for PEG induced fusion (Kaltoft and Celis, 1978).

To load tRNAs into rabbit and human red blood cells we have used the 'presswell loading' (Schlegel and Rechsteiner, 1975; Kaltoft *et al.*, 1976), the modification of the 'snap-loading' procedure (Celis, 1977) and the 'hemoglobin-free ghost loading' (Kaltoft and Celis, 1978). In all cases about 10% of the input tRNA is loaded into red cells under conditions in which more than 95% of the red cells lyse (Figure 2). Similar loading efficiencies have been reported by Capecchi *et al.*, (1977), and by Schlegel *et al.*, (1978). The amount of tRNA loaded as well as the number of loaded ghosts is dependent on the concentration of tRNA used as shown in Table I (Celis, 1977). About 9.3×10^6 molecules are loaded per ghost when the final concentration of tRNA is 6 mg/ml. Under these conditions more than 98% of the red cells are lysed as determined by phase contrast microscopy (Figure 2). If the final concentration of tRNA is increased to 12 or 18 mg/ml, then considerably less red cells are lysed, although the ghosts do contain more tRNA (1.13 and 2.75×10^7 molecules respectively).

TABLE I

Incorporation of yeast tRNAs into Rabbit Erythrocyte Ghosts

tRNA final concentration mg/ml	% ghosts	% of tRNA incorporated in red cells after hemolysis, resealing and washing	average number of tRNA molecules loaded per ghost
4	>98	9.8	6.3×10^6
6	>98	9.7	9.3×10^6
12	66	3.9	1.13×10^7
18	21	2.0	2.75×10^7

Rabbit red blood cells (2.5×10^8 cells; 15μl packed volume) were lysed with 6 volumes of distilled $H_2 0$ containing 4S ^{32}P- - tRNA isolated from yeast. The radioactive tRNA was diluted with cold yeast tRNA (methionine and glycine - Boehringer Mannheim GmbH). After the solution had stood 3 min. in ice, enough 10 × PBS was added to it to restore the isotonic saline concentration . The loaded cells were incubated 30 min. at 37° and washed three times before counting. (After Celis, 1977).

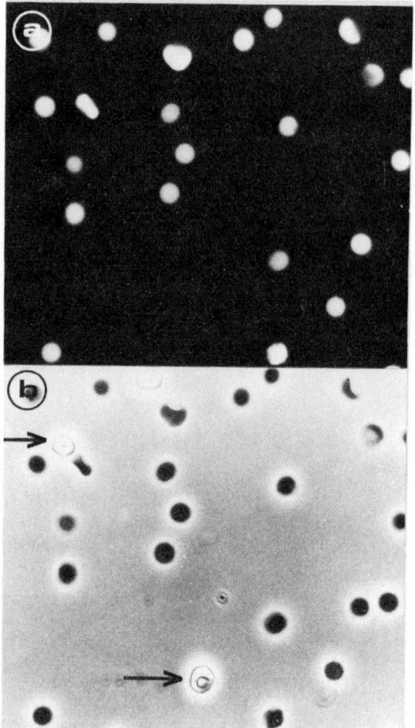

Fig. 2 Staining of yeast tRNA-loaded rabbit erythrocytes with 0.005% acridine orange in PBS. a) Fluorescence b) Phase contrast. Rabbit red blood cells were loaded with yeast methionine and glycine tRNAs at a final concentration of 6 mg/ml using the modification of the 'snap loading' procedure described in the legend of Table I. The arrows show unlysed red blood cells.

Thus, if high concentrations of tRNA are needed to be trapped in the ghosts it will be necessary to purify them after loading in order to eliminate the unlysed red cells (Rechsteiner, 1975).

Cell Fusion Between Loaded Ghosts and Recipient Somatic Cells

a) Sendai virus induced fusion Trypsinized cells are usually fused in suspension at a ratio of about 100 loaded ghosts per recipient cell with 1000 HAU of inactivated Sendai virus (Schlegel and Rechsteiner, 1975; Loyter *et al.*, 1975; Kaltoft *et al.*, 1976) and in the presence of 0.2 - 2m*M* Mn^{++} which inhibits the lysis of the ghosts and stimulates fusion (Peretz *et al.*, 1974; Zakai *et al.*, 1974). It is now clear, however, that the requirement for divalent ions must be optimized for each cell type as there have been reports indicating that Mn^{++} is toxic to some cells (Wassermann *et al.*, 1976) and that La^{++} and Ca^{++} must be used for some cell lines if prolonged survival of the recipient cells is required (Wassermann *et al.*, 1977).

To obtain fusion the cell mixture is placed in the cold for a few min. to induce agglutination followed by a short incubation at 37° to allow fusion. Finally, the cells are washed twice and reseeded in culture flasks for further analysis. The success of the fusion is generally assessed by observing the cells in the inverted microscope. A composite field is shown in Figure 3a where tRNA loaded ghosts have fused with mouse CLID cells (arrows).

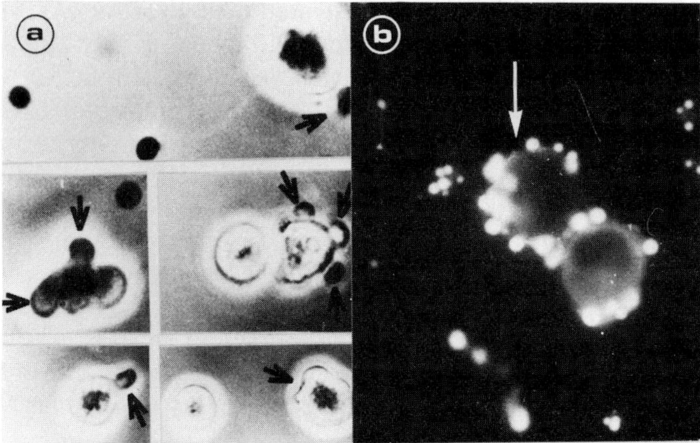

Fig. 3 Fusion of loaded ghosts with recipient somatic cells. a) Phase contrast micrograph showing Sendai virus-mediated fusion of CLID cells with tRNA loaded rabbit ghosts (after Celis, 1977) and b) Fluorescent photograph showing PEG mediated fusion of CHO 16-4 cells with FITC-BSA loaded hemoglobin-free human erythrocyte ghosts.

In average recipient cells fuse with one or two ghosts (Schlegel and Rechsteiner, 1975; Kaltoft *et al.*, 1976; Capecchi *et al.*, 1977), although in rare cases one finds cells fusing with as many as 10 ghosts (Schlegel and Rechsteiner, 1975; Kaltoft *et al.*, 1976).

b) PEG induced fusion In this procedure recipient cells grown in monolayers are fused with loaded ghosts by means of 50% PEG 1000 or 6000 (Kriegler and Livingston, 1977; Kaltoft and Celis, 1978) at a ratio of 100 loaded ghosts per recipient cell. Using this procedure Kaltoft and Celis (1978) have reported an efficient transfer of crude human HGPRT into HGPRT⁻ CHO cells under conditions where nearly 90% of the monoucleated injected cells divided within 30 hrs after fusion.

There are at least three advantages of this procedure over the Sendai virus induced fusion: 1) It is applicable to all somatic cells; 2) it uses hemoglobin-free ghosts (Kaltoft and Celis, 1978); and 3) the average number of ghosts fusing with recipient cells ranges between 5 and 10 (Figure 3b) (Kaltoft and Celis, 1978) as compared to 1 or 2 found in Sendai virus induced fusion (Figure 3a) (Schlegel and Rechsteiner, 1975; Kaltoft *et al.*, 1976; Capecchi *et al.*, 1977). This method will certainly be the procedure of choice for future experiments.

Fate of the Microinjected tRNA

Injection of tRNA into somatic cells by means of the red cell mediated microinjection was first demonstrated by Kaltoft *et al.*, (1976); By fusing rabbit red blood cells loaded with yeast $[^{125}I]$-tRNA^{Phe} with mouse CLID cells they found that about 3% of the tRNA contained in the loaded ghosts was transferred to the mouse CLID cells under conditions where 60% of the cells contained grains above background. In average the cells contained enough grains to have fused with 1 or at least 2 ghosts (Schlegel and Rechsteiner, 1975; Capecchi *et al.*, 1977) and only in a few instances (2% of the cell population) there was indication that the recipient cells have fused with as many as 10 ghosts. In most cells the grains were localized in the cytoplasm (Figure 4) and very often they were observed in clusters. Transfer of tRNA was also demonstrated in the absence of loading but in that case a much smaller fraction of the input tRNA (0.75%) was recovered

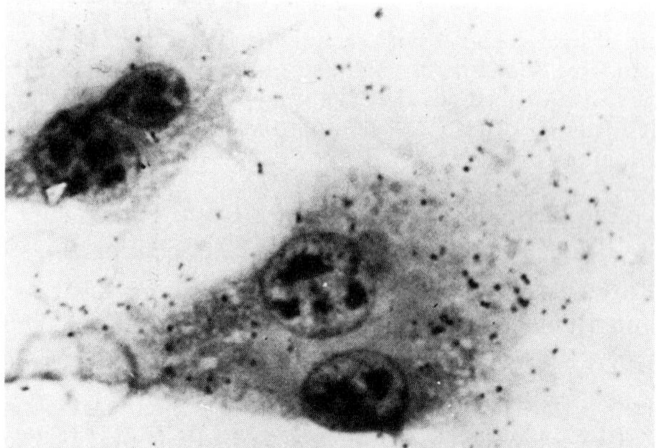

Fig. 4 Autoradiography of CLID cells microinjected with yeast $[^{125}I]$-tRNA^{Phe}

in the mouse cells (Kaltoft *et al.*, 1976). Similar transfer of macromolecules in the absence of loading have been reported by Tanaka *et al.*, (1975), and by Wassermann *et al.*, (1977).

To determine how much of the injected tRNA remained intact after transfer Kaltoft *et al.*, (1976) fused ghosts loaded with ^{32}P-labelled tRNA from yeast (4S RNA) and *E. coli* su$^+$3 tRNA$^{Tyr}_{I}$ with mouse clone 1D cells and analysed the RNAs recovered at different stages of the transfer procedure by means of formamide-polyacrylamide gel electrophoresis. Unfused ghosts were eliminated by washing twice for 10 min. with 0.83% NH_4Cl. Figure 5 shows the gel pattern of samples of yeast tRNA extracted at different stages of the two step transfer procedure.

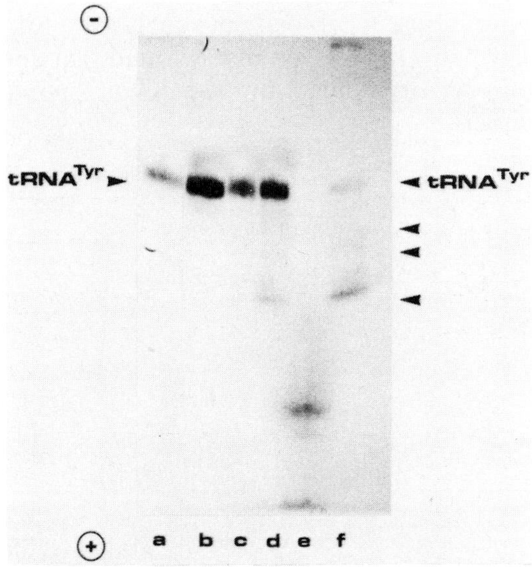

Fig. 5 Gel electrophoresis of [^{32}P]-labelled 4S yeast RNA recovered at different stages of the transfer procedure. a) Control 4S RNA. RNAs recovered from: b) red blood cells, c) supernatant after fusion, d) supernatant after 6 hrs incubation at 37°, e) NH_4Cl wash, f) monolayers. (After Kaltoft *et al.*, 1976).

Figures 5a and b correspond to the control 4S RNA and to the RNA extracted from loaded ghosts respectively. Figure 5c and d shows the RNA extracted from the supernatant after termination of fusion and the supernatant after 6 hrs further incubation. Figure 5e and f shows the RNA extracted from the NH_4Cl wash and the remaining mouse cells in monolayer, respectively. Approximately 50% of the 4S RNA transferred to mouse cells remains intact (Figure 5f) corresponding to about 0.2% of the total input of tRNA used to load the red blood cells. This fraction represents a minimal estimate for the transfer efficiency, as some lysis of mouse cells cannot be avoided due to the NH_4Cl washing step. It is also possible that the small fraction of intact 4S RNA present in the NH_4Cl wash (Figure 5e) is due to lysis of the mouse cells since no intact 4S RNA was detected when loaded ghosts in MEM were incubated for 6 hrs at 37° in the presence of inactivated Sendai virus. Herrera *et al.*, (1970) have demonstrated uptake of tRNA in murine leukemia and human lymphoblastoid cells, but this uptake is most likely a specialized property of such lymphoid cells and is not found in fibroblasts. These authors have also demonstrated that a fraction of the tRNA taken up is active as judged by *in vitro* amino-acylation.

Similar results were obtained when purified *E. coli* su⁺ 3 $tRNA_I^{Tyr}$ was injected into mouse CLID cells and analysed 6 hrs after injection (Kaltoft *et al.*, 1976). In this case it was found that approximately 33% of the radioactivity applied to the gels corresponded to intact tRNA (Figure 6f).

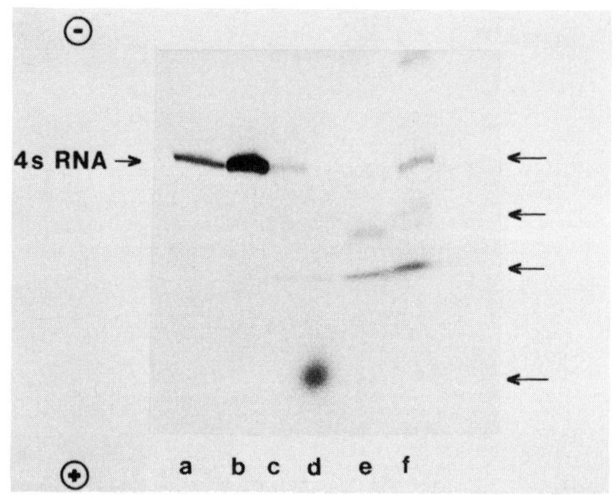

Fig. 6 Gel electrophoresis of ^{32}P-labelled su$^+$ 3 $tRNA_I^{Tyr}$ recovered at different stages of the transfer. a and b) Control tRNA. tRNAs recovered from: c) loaded red blood cells, d) supernatant after fusion, e) supernatant after 6 hrs incubation at 37°, f) monolayers. (After Kaltoft *et al.*, 1976).

This amount corresponds to about 0.1% of the total input of tRNA used to load the red cells. Fingerprinting of T_1 RNase digests of an aliquot of the sample applied to the gel (Figure 6f) demonstrated that this material is intact (Figure 7b). There is, however, one major spot not corresponding to the normal fingerprints of su$^+$ 3 $tRNA_I^{Tyr}$ (Figure 5, arrow) that is derived from the major fragment observed in the formamide gel. Further analysis of the other spots after digestion with pancreatic RNase indicated that they all correspond to the normal sequence (Kaltoft *et al.*, 1976).

Further evidence indicating that tRNAs are stable after injection have been provided by Rechsteiner's group (Schlegel *et al.*, 1976; Schlegel *et al.*, 1978), who have found that microinjected tRNAs turn over with approximately the same rate as endogenous tRNA

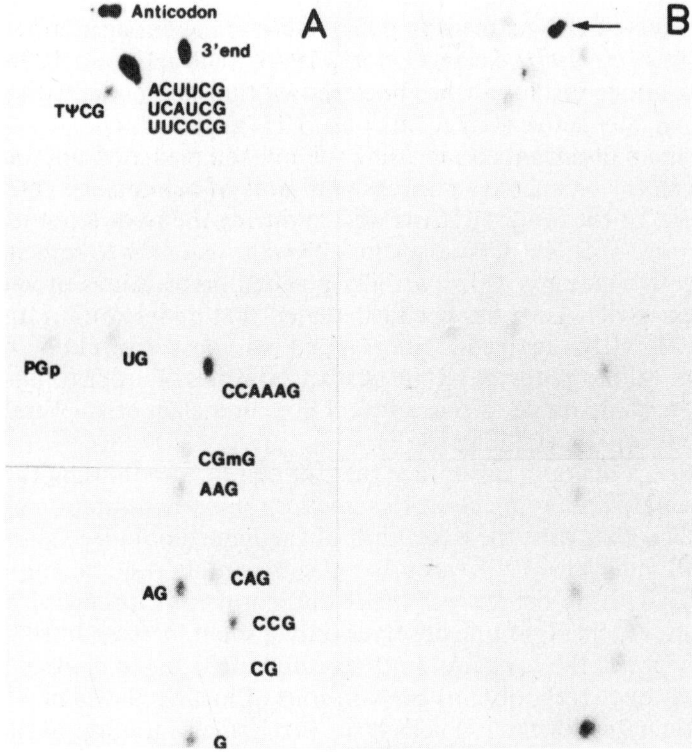

Fig. 7 T_1 RNAse fingerprint of ^{32}P-labelled su$^+$ 3 tRNA$_1^{Tyr}$. a) Control su$^+$3 tRNA$_I^{Tyr}$ and b) tRNA recovered from the monolayers 6 hrs after injection. The arrow shows a new spot which is derived from the faster moving band shown in Figure 6f, (After Kaltoft *et al.*, 1976).

(half life of approximately 1 day). Also there have been reports indicating that yeast tRNAs and *E. coli* su$^+$3 tRNA$_I^{Tyr}$ are stable when injected into *Xenopus oocytes* (Allende *et al.*, 1975) and that the yeast tRNAs are aminoacylated after injection (Gatica *et al.*, 1975). Finally, Capecchi *et al.*, (1977) have shown that microinjection of yeast and *E. coli* ochre suppressor tRNA into a HGPRT$^-$ mutant cell line causes a partial reversion of the phenotype and thus demonstrate that tRNAs are not only stable after injection but are also biologically active.

General Comments on the Red Cell Mediated Microinjection

From the results discussed above it is clear that it is possible to microinject tRNAs into somatic cells under conditions where a large fraction of the cells (up to 60%) are injected in average with the content of one loaded ghost (Kaltoft *et al.*, 1976; Schlegel *et al.*, 1978). The microinjected tRNAs seem to be as stable as the endogenous tRNA (Kaltoft *et al.*, 1976; Schlegel *et al.*, 1976; Schlegel *et al.*, 1978) and at least in one instance it has been shown that microinjected tRNA is biologically active (Capecchi *et al.*, 1977).

The main disadvantages in using the red cell mediated microinjection to screen for prospective nonsense mutants of mammalian cells is that 99% of the input tRNA is wasted during the two-step injection procedure. This is a substantial loss if one is planning to screen many prospective mutants using partially purified preparations of yeast suppressor tRNAs. It has been calculated that under conditions where 9.3×10^6 tRNA molecules are trapped per ghost it would be necessary to purify the suppressor tRNA at least 10 times in order to inject tRNA amounts comparable to those found in mammalian cells (Weinberg and Penman, 1970; Celis, 1977).

Finally, a major disadvantage that we can foresee in using the red cell mediated microinjection to screen for nonsense mutants of the HGPRT is that with the exception of the hemoglobin-free ghosts (Kaltoft and Celis, 1978) all other ghosts contain residual amounts of HGPRT. This we believe will make the scoring of suppression very difficult, especially if one observes only a small increase in HGPRT activity above the background. Even more difficult to control is the fact that one rarely obtains preparations of loaded ghosts in which more than 96% of the red cells lyse. Thus, small variation in the percentage of unlysed red blood cells in each particular loading experiment could also make the experiment difficult to interpret. These problems can be obviated in part by purifying the loaded ghosts from unlysed cells and by determining by means of autoradiography the distribution of grains in injected and control cells.

Despite the problems we have mentioned above we believe that this technique, especially the PEG induced fusion using hemoglobin-free ghosts (Kaltoft and Celis, 1978), could be of value once a potential mutant has been screened by other techniques, as this method can be used for biochemical studies of the suppressed protein.

Microinjection of Suppressor tRNAs into Somatic Cells Using Micropipettes

Effect on Cell Viability and Protein and Nucleic Acid Synthesis

A rapid way to inject suppressor tRNAs into potential nonsense mutations is the direct microinjection of these suppressors with fine micropipettes (Graessmann and Graessmann, 1976; Celis, 1977). If the cells can be grown attached to surfaces and if suppression can be detected by means of autoradiographic techniques one could easily inject and score many mutants in a short time. The setup we have used for microinjection is similar to that described by the Graessmanns (Graessmann and Graessmann, 1976) and it is shown in Figure 8.

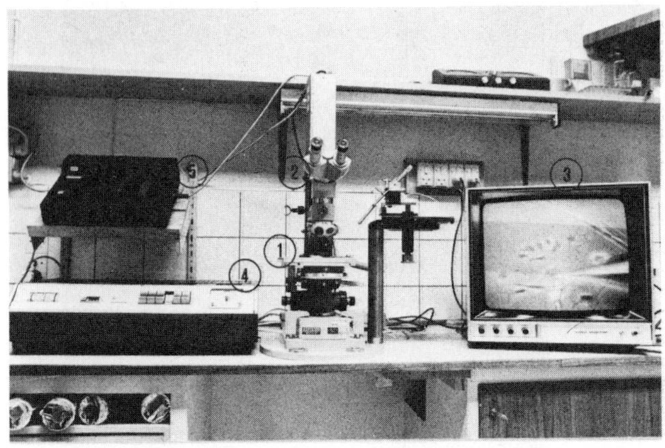

Fig. 8 Microinjection set-up. (1) microscope and micromanipulator, (2) television camera, (3) television monitor, (4) video recorder and (5) sound recorder. The television monitor shows the replay of an injection experiment.

The basic equipment consists of a Carl Zeiss Jena microscope equipped with a micromanipulator. A television camera and a video casette recorder (Philips) have been connected to the microscope in order to record the experiments. The microinjection is shown in Figure 9a and b. In this case 3T3 cells are shown in which the tip of the micropipette (about 0.5μ) has been introduced into the cytoplasm (Figure 9a) and the nucleus (Figure 9b) of two adjacent cells. The volume injected per cell had been determined to be in average 5×10^{-11} ml, corresponding to about 2 - 2.5% of the volume of a CHO cell (Stacey and Alfrey, 1976; Celis, 1977). Volumes of up to 10-30% of the volume of a CHO cell can be injected if necessary, however, under these conditions there is no assurance that a large fraction of the injected cells will survive. Under optimal conditions about 100 cells can be injected in 15 min.

A fluorescent picture of CHO cells microinjected with yeast tRNA is shown in Figure 10. In this case CHO cells have been injected into the cytoplasm (Figure 10a and b) and the nucleus (Figure 10c) with a solution of yeast tRNA reacted with ethidium bromide. As can be seen a large fraction of the tRNA remains in the cytoplasm (Figure 10a and b) although one cannot eliminate the possibility that some tRNA migrates into the nucleus.

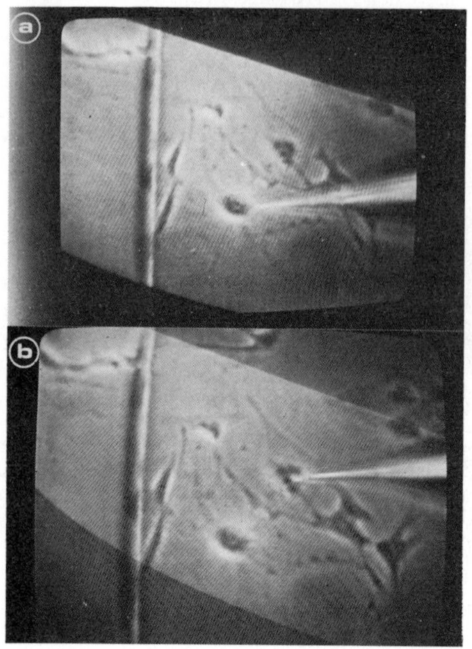

Fig. 9 Micrograph of the television monitor showing cytoplasmic (a) and nuclear (b) injections.

Injection of crude yeast tyrosine suppressor tRNAs (either amber or ochre) at a concentration of 100 mg/ml does not affect viability of CHO cells as essentially all the injected cells divide shortly after injection. Similarly, protein and RNA synthesis are not impaired after injection of suppressor tRNAs (Figure 11a and b) as determined by autoradiographic analysis of injected cells incubated in the presence of [^{3}H]-methionine (Figure 11a) or [^{3}H]-hypoxanthine (Figure 11b) respectively. In both cases we have been unable to find differences between the average number of grains found in injected cells as compared to control ones.

Microinjection of Yeast Tyrosine Amber tRNA into HGPRT$^-$ Mutants

To search for nonsense mutants in mammalian cells we have chosen the X-linked genetic locus coding for the hypoxanthine guanine phosphoribosyl transferase (HGPRT; E.C. 2.4.2.8.) in CHO cells since there is overwhelming evidence indicating that HGPRT structural gene mutants are readily obtained by chemical and u.v. mutagenesis (Beaudet *et al.*, 1973; Sharp *et al.*, 1973; Wahl *et al.*, 1975; Milman

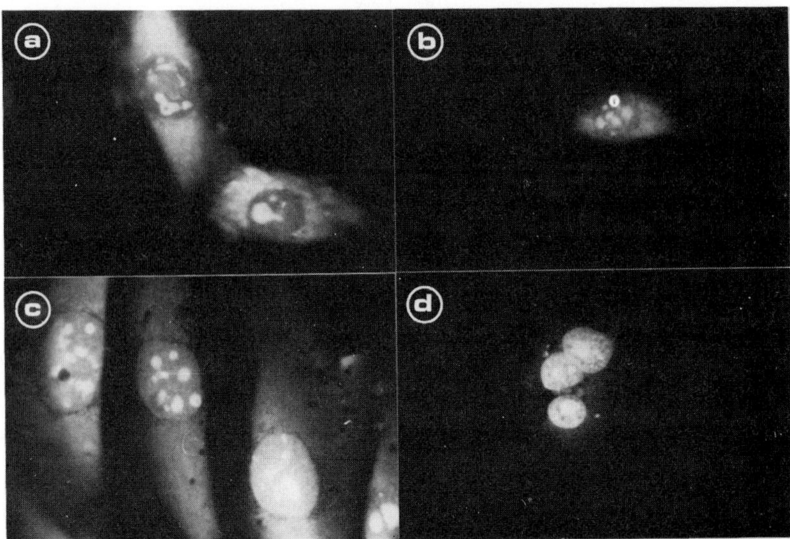

Fig. 10 Microinjection of ethidium bromide yeast tRNA into the cytoplasm and nucleus of CHO cells. a) Cytoplasmic injection, 1 hour after injection, b) Cytoplasmic injection 5 hrs after injection, c) Nuclear injection, 1 hour after injection and d) Cytoplasmic injection of ethidium bromide.

et al., 1976; Siminovitch, 1976; Fenswick *et al.*, 1977; Caskey *et al.*, this volume). The enzyme is non-essential and catalyses IMP or GMP synthesis from hypoxanthine or guanine and 5′-phosphoribosyl 1-pyrophosphate (PRPP). Cells lacking HGPRT activity can be selected by their resistance to purine analogues such as 8-azaguanine and 6-thioguanine (Sybalski *et al.*, 1963) while cells expressing HGPRT can be counterselected by their resistance to chemicals that block the *de novo* purine synthesis pathway (Littlefield, 1963; Thompson and Baker, 1973).

Using ethylmethanesulphonate (EMS) induced mutagenesis (Milman *et al.*, 1976) we have isolated 30 independent HGPRT variants which are resistant to 1m*M* thioguanine. Among these, 28 variants do not incorporate [^{3}H]-hypoxanthine during a 24 hrs incubation period as revealed by autoradiography (Figure 12b) while the other two showed a significant number of grains only in the presence of aminopterin (Figure 12c, d, e and f). The latter class of mutants has been described previously by Fenswick *et al.*, (1977) and most likely correspont to missense mutants. The behaviour of these mutants is most relevant to suppression studies as it is possible that some suppressed HGPRT nonsense mutants will behave in a similar way to these mutants and

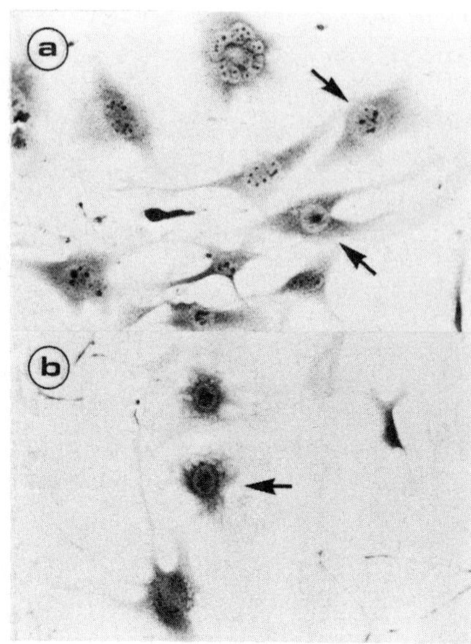

Fig 11 Protein and nucleic acid synthesis in 3T3 cells injected with crude amber tyrosine yeast tRNA. Cells indicated with an arrow have been injected with a crude preparation of yeast amber tyrosine tRNA (100 mg/ml). a) Autoradiography of cells incubated with [^{3}H]-methionine and b) Autoradiography of cells incubated with [^{3}H]-hypoxanthine. Analysis of many injected cells in both cases indicated that there is no significant differences between the number of grains present in injected and uninjected control cells. Similar results are obtained when crude ochre tyrosine yeast tRNA is injected (results not shown).

will require aminopterin for activity.

Among the cell lines that showed no HGPRT activity (Figure 12b) many of them contained a few spontaneous revertants eventhough the mutants have been kept continuously in selective media. In Figure 13 we show revertants of mutant 1-1 that have been detected in the autoradiograms of mutant cells incubated for 24 hrs with [^{3}H]-hypoxanthine. Interestingly enough, if mutant 1-1 cells are incubated in HAT to recover these revertants, one can only recover those having 40% of the enzymatic activity as compared to wild type cells (Figure 13a), but not those having less activity (Figure 13b, 20% activity; Figure 13c, 17% activity; Figure 13d, 5% activity). This is strong evidence indicating that HAT selection may not be suitable to isolate revertants having low levels of HGPRT activity and thus raise doubts as to whether tRNA suppressors can be isolated among the revertants of nonsense mutations of the HGPRT.

So far we have microinjected crude yeast tyrosine amber suppressor tRNA (100 mg/ml) into all the mutants that showed no HGPRT

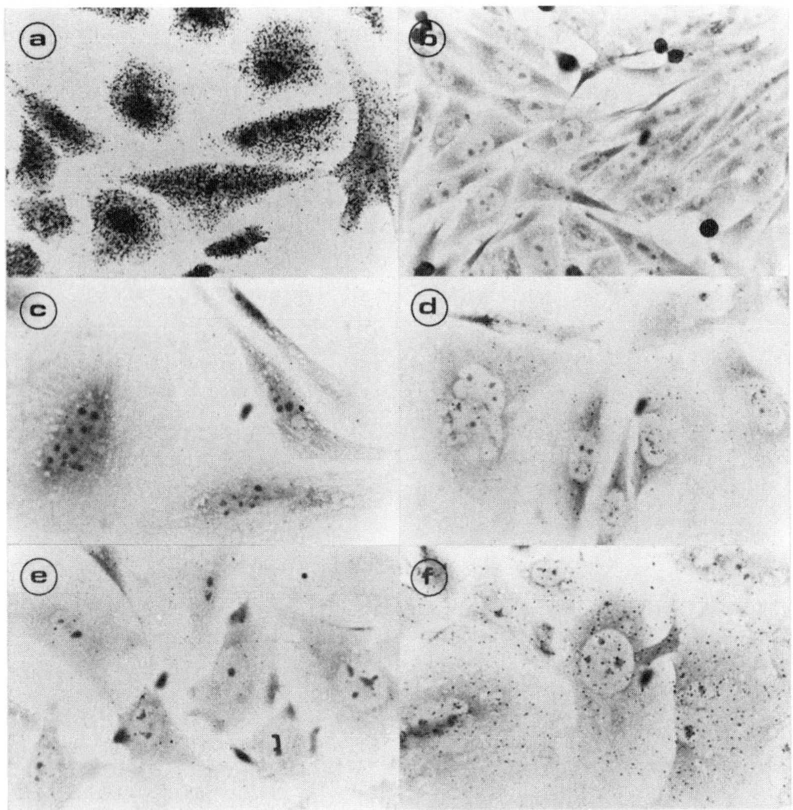

Fig. 12 Autoradiography of CHO cells resistant to 6-thioguanine and incubated in the presence of [^{3}H]-hypoxanthine. a). Wild type CHO cells, b) HGPRT$^-$ variant (28 of the isolated mutants are of this type), c),d) Mutant XIV-1 incubated in the presence of $10^{-5}M$ aminopterin, e) Mutant XIII-1 and f) Mutant XIII-1 incubated in the presence of $10^{-5}M$ aminopterin. Cells were incubated for 24 hrs with 1μCi/ml of [^{3}H]-hypoxanthine (1.8 Ci/mmol, 0.5 μCi/μ1) and the autoradiograms were exposed for 4 days.

activity even in the presence of aminopterin (the class shown in Figure 12b). In all cases microinjected cells were incubated for 24 hrs with 1 μCi/ml of [^{3}H]-hypoxanthine (1.8 Ci/mmol, 0.5 μCi/μ1) in the presence of 10^{-5} M aminopterin. With two exceptions we have found negative results and even in the two cases where there seem to be weak suppression it is not clear whether the increase in the number of grains after injection is statistically significant. Obviously it is possible that our failure to detect suppressible mutants is due to the low concentration of suppressor tRNA we are using and therefore

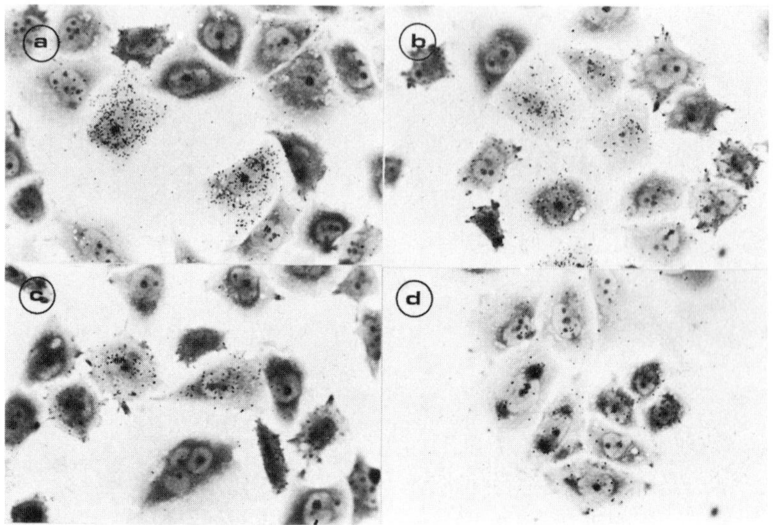

Fig. 13 Autoradiography of spontaneous revertants of CHO HGPRT⁻ mutant 1-1. Mutant 1-1 cells have been kept in $10^{-4}M$ thioguanine since the first passage after cloning. The revertants shown in a) contain about 40% of the enzyme activity present in wild type cells while those of b), c) and d) contain 20, 17 and 5% respectively.

these preliminary experiments need to be repeated using purified preparations of amber as well as ochre suppressor tRNAs.

Somewhat more encouraging preliminary experiments have been obtained with the HGPRT⁻ mutant RJK39 isolated by Fenswick *et al.*, (1977) and discussed in detail in this volume by Caskey *et al.* In this case we have observed partial suppression of the HGPRT activity (Figure 14) only when large volumes of crude yeast tyrosine amber tRNA (about 30% of the cell volume) are injected. Unfortunately, under these conditions many cells die after injection and not all the injected cells present the same number of grains (Figures 14a, b and c) (cells injected first contain more grains than cells injected later on). Since these results are not very reproducible, perhaps because of the large volumes of tRNAs that have to be injected which conflict with cell viability, it would be necessary to microinject purified suppressor tRNA preparations before making a definite statement as to whether in fact one is observing suppression.

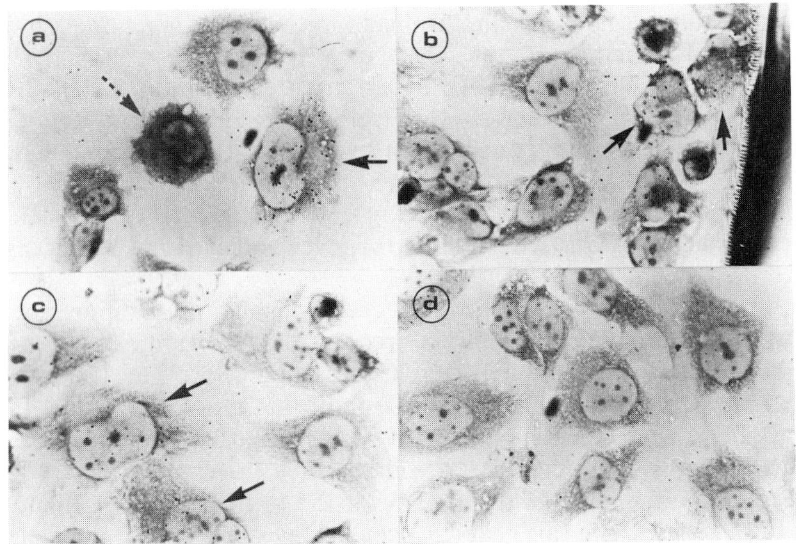

Fig. 14 Autoradiography of Chinese hamster RJK39 cells (Fenswick *et al.*, 1977; Caskey *et al.*, this volume) microinjected with a crude preparation of yeast amber tyrosine tRNA. After injection the cells were incubated for 20 hrs in a MEM containing 1 μCi/ml of [^{3}H]-hypoxanthine (1.8 Ci/mmol, 0.5 μCi/μ1, Amersham) and $10^{-5} M$ aminopterin. Autoradiograms were exposed for 3 days. The cells indicated with an arrow have been microinjected and checked by studying the video tape of the experiment. Volumes of up to 30% of the cell volume were injected in this experiment. Note that there is a clear difference in the number of grains present in the injected cells. Cells injected first contained the higher number of grains (see a, for example). The cell indicated with a broken arrow in a) did not survive the injection. Many other cells did not survive the injection of such a large volume and detached.

Conclusion

Even though the microinjection methods we have described cannot be used yet as routine assays there is encouraging evidence (this paper; Capecchi *et al.*, 1977) indicating that they could be used to attempt a systematic search for nonsense mutations and tRNA suppressors in mammalian cells. It is clear, however, that it will be necessary to have other means of confirming the results obtained by the microinjection techniques especially in those cases where weak suppression is observed. One such way we have started to explore is to fuse by means of PEG, prospective nonsense mutants of the HGPRT with yeast protoplasts harbouring suppressors and which are deficient in HGPRT. Preliminary experiments indicate that this approach is feasible as we have been able to transfer HGPRT from a neurospora slime mutant into HGPRT$^-$ CHO cells (Kaltoft and Celis, unpublished observations). Further experiments should tell whether this is in fact a quicker and easier way of screening for nonsense mutations than the microinjection techniques we have described here.

Even if one succeeds in isolating a nonsense mutant of the HGPRT it is not yet clear whether the HAT selection will pick up revertants harbouring tRNA suppressors as we have found that revertants having about 20% of the enzymatic activity of wild type cells do not grow in HAT. It is important then to develop in the future appropriate selection media that could be used to select for revertants having a wide range of enzymatic activity.

Finally, it is our belief that the successful isolation of suppressor mutations in mammalian cells will require the revertion of two suppressible mutants located in different structural genes. This task is by no means easy and will require the isolation of mutants in two different loci, but in the same cell line. By fusing the two mutant cell lines it will be possible to create a hybrid in which there will be a very low rate of simultaneous spontaneous revertion to the wild type phenotype in both genes. Once this goal has been achieved the stage for isolating a suppressor in mammalian cells will be set.

Acknowledgements

We thank O. Jensen for photography. This work was supported by grants to J.E. Celis from the Danish Natural Science Research Council, NOVO, the Aarhus University Forskningsfond and Euratour.

References

Allende, J.E., Allende, C.C. and Firtel, R.A. (1974). *Cell* **2**, 189.

Andoh, T. and Ozeki, H. (1968). *Proc. Nat. Acad. Sci.* **59**, 792.

Baker, R.F. (1967). *Nature (London)* **215**, 424.

Beaudet, A.L., Roufa, D.J. and Caskey, C.T. (1973). *Proc. Nat. Acad. Sci.* **70**, 320.

Brandriss, M.C., Stewart, J.W., Sherman, F. and Botstein, D. (1976). *J. Mol. Biol.* **102**, 467.

Brenner, S., Stretton, A.O.W. and Kaplan, S. (1965). *Nature (London)* **206**, 994.

Capecchi, M.R. and Gussin, G. (1965). *Science* **149**, 417.

Capecchi, M.R., Hughes, S.H. and Wahl, G.M. (1975). *Cell* **6**, 269.

Capecchi, M.R., Van der Haar, R.A., Capecchi, N.E. and Sveda, M.M. (1977). *Cell* **12**, 371.

Celis, J.E., Smith, J.D. and Brenner, S. (1973). *Nature New Biol.* **109**, 130.

Celis, J.E. (1977). *Brookhaven Symposia in Biology No. 29*, 178.

Engelhardt, D.L., Webster, R.E., Wilhelm, R.C. and Zinder, N.D. (1965). *Proc. Nat. Acad. Sci.* **54**, 1791.

Fenswick, G. Jr., Sawyer, T.H., Kruk, G.D., Astrin, K.H. and Caskey, C.T. (1977). *Cell* **12**, 383.

Furusawa, M., Nishimura, T., Yumaizumi, M. and Okada, Y. (1974). *Nature*

(London). **249**, 449.
Gatica, M., Tarrago, A., Allende, C.C. and Allende, J.E. (1975). *Nature (London).* **256**, 675.
Gilmore, R.A., Stewart, J.W. and Sherman, D. (1971). *J. Mol. Biol.* **61**, 157.
Gesteland, R.F., Willis, N., Lewis, J.B. and Grodzicker, T. (1977). *Proc. Nat. Acad. Sci.* **74**, 4567.
Gesteland, R.F., Wolfner, M., Grisafi, P., Fink, G., Botstein, D. and Roth, J.R. (1976). *Cell* **7**, 381.
Goodman, H.M., Abelson, J., Landy, A., Brenner, S. and Smith, J.D. (1968). *Nature (London).* **217**, 1019.
Graessmann, M. and Graessmann, A. (1976). *Proc. Nat. Acad. Sci.* **73**, 366.
Hawthorne, D.C. and Leupold, U. (1974). *Current Topics Microbiol. Immunol.* **64**, 1.
Hawthorne, D.C. and Mortimer, R.K. (1963). *Genetics* **48**, 617.
Hawthorne, D.C. and Mortimer, R.K. (1968). *Genetics* **60**, 735.
Herrera, F., Adamson, R.H. and Gallo, R.C. (1970). *Proc. Nat. Acad. Sci.* **67**, 1943.
Hoffman, J.F. (1958). *J. Gen. Physiol.* **42**, 9.
Ihler, G., Glew, R. and Schnure, F. (1973). *Proc. Nat. Acad. Sci.* **70**, 2663.
Kaltoft, K. and Celis, J.E. (1978). *Expt. Cell Res.* **115**, 423.
Kaltoft, K., Zeuthen, J., Engbaek, F., Piper, P.W. and Celis, J.E. (1976). *Proc. Nat. Acad. Sci.* **73**, 2793.
Kriegler, M.P. and Livingston, D.M. (1977). *Somatic Cell Genetics.* **3**, 603.
Liebman, S.W., Stewart, J.W. and Sherman, F. (1975). *J. Mol. Biol.* **94**, 595.
Littlefield, J.W. (1963). *Proc. Nat. Acad. Sci.* **50**, 568.
Loyter, A., Zakai, N. and Kulka, R.G. (1975). *J. Cell. Biol.* **66**, 292.
Milman, G., Lee, E., Ghangas, G.S., McLaughlin, J.R. and George, M. (1976). *Proc. Nat. Acad. Sci.* **73**, 4589.
Nishimura, I., Furusawa, M., Yamaizumi, M. and Okuda, Y. (1976). *Cell Structure and Function* **1**, 197.
Peretz, H., Toister, Z., Laster, Y. and Loyter, A. (1974). *J. Cell Biol.* **63**, 1.
Piper, P.W., Wasserstein, M., Engbaek, F., Kaltoft, K., Celis, J.E., Zeuthen, J., Liebman, S. and Sherman, F. (1976). *Nature (London).* **262**, 757.
Rechsteiner, M. (1975). *Exp. Cell. Res.* **93**, 487.
Sambrook, J.F., Fan, D.P. and Brenner, S. (1967). *Nature (London).* **214**, 452.
Sarabhai, A.S., Stretton, A.O.W., Brenner, S. and Bolle, A. (1964). *Nature (London)* **201**, 13.
Schlegel, R., Darrah, L. and Rechsteiner, M. *In:* Molecular Mechanisms in the Control of Gene Expression (eds, Nierlich, D., Rutter, W. and Fox, C.F.), Academic Press, London and New York, 1976, 465.
Schlegel, R.A., Iversen, P. and Rechsteiner, M. (submitted for publication).
Schlegel, R.A. and Rechsteiner, M. (1975). *Cell* **5**, 371.
Seeman, P. (1967). *J. Cell Biol.* **32**, 55.
Sharp, J.D., Capecchi, N.E. and Capecchi, M.R. (1973). *Proc. Nat. Acad. Sci.* **70**, 3145.
Sherman, F., Liebman, S.W., Stewart, J.W. and Jackson, M. (1973). *J. Mol. Biol.* **78**, 157.
Siminovitch, L. (1976). *Cell* **7**, 1.

Stewart, J.W. and Sherman, F. (1972). *J. Mol. Biol.* **68**, 429.
Stewart, J.W. and Sherman, F. (1973). *J. Mol. Biol.* **78**, 169.
Stretton, A.O.W., Kaplan, S. and Brenner, S. (1966). *Cold Spring Harbor Symp. Quant. Biol.* **31**, 173.
Summers, W.P., Wagner, M. and Summers, W.C. (1975). *Proc. Nat. Acad. Sci.* **82**, 4081.
Sybalski, W., Sybalska, E.H. and Ragni, G. (1962). *Nat. Cancer Inst. Monorg.* **7**, 75.
Tanaka, K., Sekiguchi, M. and Okada, Y. (1975). *Proc. Nat. Acad. Sci.* **72**, 4071.
Thompson, L.H. and Baker, R.M. (1973). *In:* Methods in Cell Biology (ed., Prescott, D.M.), Vol 6, 209. Academic Press, New York.
Wahl, G.M., Hughes, S.H. and Capecchi, M.R. (1975). *J. Cell. Physiol.* **85**, 307.
Wassermann, M., Kulka, R. and Loyter, A. (1977). *FEBS Letters* **83**, 48.
Wassermann, M., Loyter, A. and Kulka, R. (1977). *J. Cell Sci.* **27**, 157.
Wassermann, M., Zakai, N., Loyter, A. and Kulka, R. (1976). *Cell* **7**, 551.
Weigart, M.G. and Garen, A. (1965). *J. Mol. Biol.* **12**, 448.
Wille, W. and Willecke, K. (1976). *FEBS Letters* **65**, 59.
Weinberg, R.A. and Penman, S. (1968). *J. Mol. Biol.* **38**, 289.
Zipser, D. (1967). *J. Mol. Biol.* **29**, 441.

USE OF YEAST SUPPRESSORS FOR IDENTIFICATION OF ADENOVIRUS NONSENSE MUTANTS

R. GESTELAND AND N. WILLS

Cold Spring Harbor Laboratory,
Cold Spring Harbor, New York, USA.

Introduction

The practical value of the genetic system in bacteria that utilizes nonsense mutants and their suppressors needs no further testimony than the wealth of detailed genetic information that has accrued as a result. With this in mind we and many others have worked towards developing a suppression system in higher eucaryotes that would hopefully have similar value. (However, recent breakthroughs in cloning and gene transfer have perhaps made the need less urgent). Having neither a nonsense mutant nor a suppressor in mammalian cells we (Gesteland *et al.*, 1976) and Capecchi's group (Capecchi *et al.*, 1975) investigated the yeast nonsense suppressors with the hope that the suppressor tRNA could be used to identify and characterize nonsense mutants in any eukaryote.

In vitro Suppression with Yeast tRNA

Suppression of termination of protein synthesis at a termination codon is a competitive event involving at least the mutationally derived suppressors tRNA specific for the terminator and the termination event involving the release factor. Thus, while suppression can be demonstrated in most, if not all cell-free protein synthesis systems, the efficiency of suppression is quite variable. This depends on the relative efficiency of termination and the concentration of added tRNA. The latter is limited by the purity of the tRNA since crude tRNA is, in general, inhibitory for protein synthesis when moderately large amounts are added. In the case of *E. coli*, suppression can be driven to nearly 100% by inhibiting termination with antibodies to the termination factors (J. Atkins, personal communication). Antibodies against eukaryotic termination factor are not yet available so

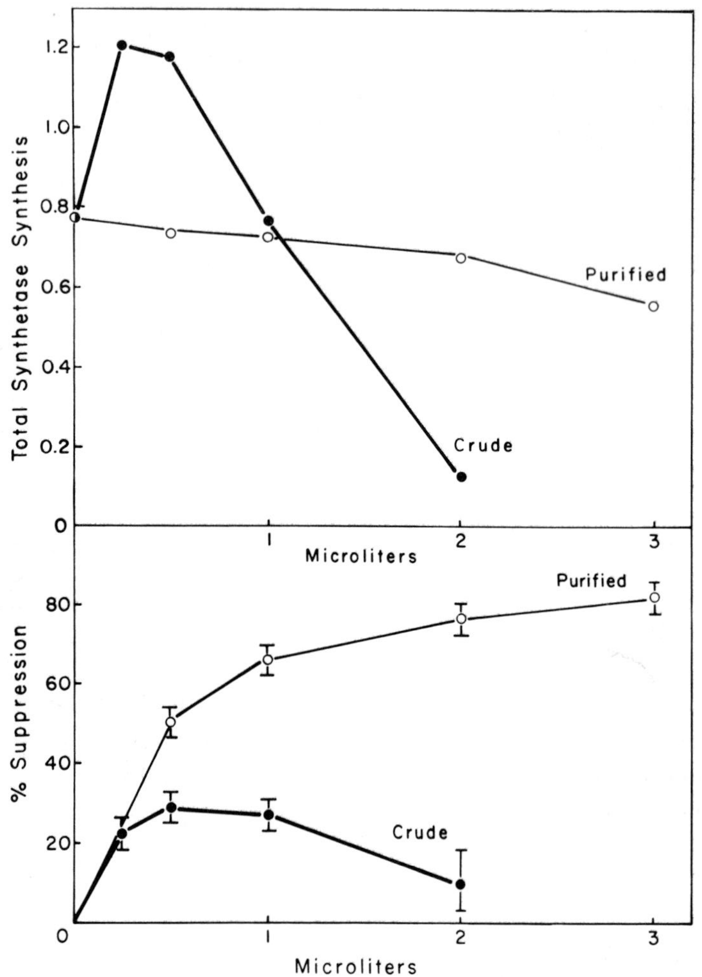

Fig. 1 Effect of amber suppressor tRNA addition on total protein synthesis and on efficiency of suppression of a Qβ synthetase amber mutant. Cell free protein synthesis in the mammalian system of Schreir and Stachelin (1973) was programmed with RNA from the Qβ synthetase amber mutant that makes a 55K polypeptide fragment compared to the 66K wild type product. The relative amounts of both products were measured by densitometer tracing of autoradiograms of SDS polyacrylamide gels. The recessive lethal amber (RL-1) suppressing yeast strain, DBA317, was used to make crude total tRNA and to fractionate by enriching for suppressor activity. The volumes indicated ('crude' 5 mg/m1, 'purified' 0.2 mg/ m1) were added to 25 μl reaction mixtures. (For details of the method see Gesteland *et al.,* 1976).

the only way to achieve efficient suppression is to fractionate tRNA to enrich for suppressor in order to outcompete termination. Figure 1 shows the effect of purification of tRNA on suppression efficiency. Qβ RNA from an amber mutant strain is used for assaying amber suppression and wild type Qβ is used for ochre suppression which results in a read-through product from the UAA terminated synthetase gene. The cell-free system in this case is a mixed system developed by Schreier and Staehelin (1973) consisting of a pH5 fraction and washed ribosomes from mouse ascites cells and rabbit reticulocyte ribosome initiation factors. An efficiency of 50-80% suppression can be obtained both for ochre and amber with fractionated tRNA. Clearly the crude tRNA is inhibitory. Wheat germ extracts respond well to ochre, amber and UGA suppressors (see below). Reticulocyte lystate (Pelham and Jackson, 1976) on the other hand gives efficient amber suppression but only very marginal ochre suppression. Thus the source of the cell-free system influences the efficiency even if purified suppressor tRNA is used.

By analogy with the assay for ochre suppressors using read-through of the Qβ synthetase protein, we can examine the normal termination factors for any messenger that is active in a cell-free system. For instance, we have examined the globin terminators using various suppressors in the wheat germ extracts. Figure 2 shows polyacrylamide gel patterns observed with the crude mixture of α and β rabbit messenger RNA. The UGA suppressing tRNA (from strain 5307, G. Fink) gives one elongated product with slightly faster mobility than the elongated form seen with the ochre suppressor. Rabbit β globin contains 146 amino acids and nucleotide sequencing shows that the terminator is UGA which is followed by 22 translatable codons before a second terminator is encountered (Efstradiadis *et al.*, 1977; Proudfoot, 1977). Thus if the UGA is decoded by a suppressor then an elongated product of 169 residues is predicted. Similarly rabbit α globin has 141 residue and is terminated by UAA, which is followed by 21 translatable codons (Proudfoot *et al.*, 1977). This would result in 163 residues if UAA is read. Thus the sequence predicts that the α read-through protein should be smaller than the β read-through protein. However, the reverse appears to be true (Figure 2) since the α (ochre) product migrates slower than the β (UGA) product as if it is larger. This contradiction is probably only apparent because of the peculiar amino acid composition predicted for the α read-through extension. The nucleotide sequence predicts 9 proline residues out of the 22 amino acids. The β extension is more normal with 2 prolines out of 23. Proline residues are known to cause a relative decrease

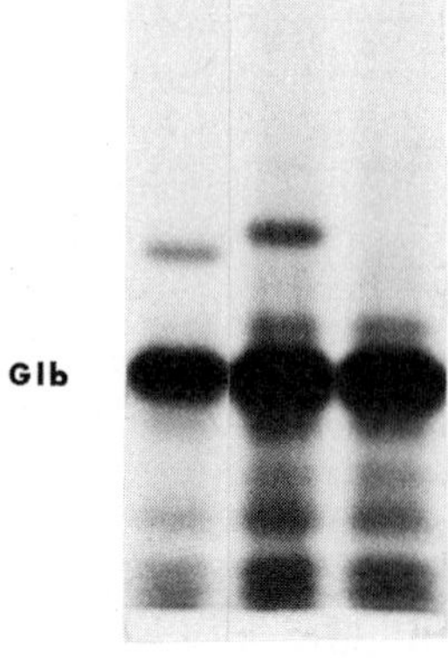

Fig. 2 Suppression of termination at the normal terminators of rabbit globin mRNA. Cell free protein synthesis in wheat germ extracts was programmed with crude rabbit globin mRNA and yeast suppressor was added: 1 μ g of enriched UGA or ochre suppressing tRNA per 25 μl reaction mixture. [^{35}S]-methionine labeled products were fractionated on a 17% SDS polyacrylamide gel and revealed by autoradiography.

in the mobility of proteins during SDS gel electrophoresis. Thus we suspect that the α read-through product migrates anomalously and therefore only appears larger than the β read-through protein.

Thus natural terminators can also be suppressed and if the translatable region is long enough, its length and the identity of the terminator codon can be determined by use of the suppressors *in vitro.* Surprisingly few gene products give detectable elongation with either ochre or amber (UGA has not been so thoroughly tested). One, as yet unidentified adeno 2 protein gives a read-through product with amber suppressing tRNA, but none of the 15 or so others show any effect. One Herpes Virus protein is elongated by a UGA suppressor (Cremer, personal communication). One Brome grass mosaic virus protein gives read-through with amber tRNA. Also as predicted by sequence analysis human α and β globin mRNA give longer products with ochre tRNA (unpublished results with B. Forget). The majority of mRNA's tried however, show no effect. This must mean that very often there is an in-phase terminator not very far downstream from the normal terminator.

Adenovirus Mutants

The growth of Adenovirus in monkey cells is blocked in the synthesis of some late gene products. SV40 virus can supply some function that allows adenovirus to multiply. The SV40 region involved encodes the carboxy-terminal end of the T-antigen from the early region. The rescue function can be supplied either by co-infection, by a resident SV40 genome in a transformed monkey cell line, or by SV40 sequences integrated into the Adenovirus genome in adeno-SV40 hybrid viruses. These hybrids grow equally well on monkey cells and on human cells, the normal host for lytic adeno growth. By contrast, wild type adeno grows about 10^3 fold less well on monkey cells compared to human cells. A number of different derivatives of the SV40 gene product seem to be capable of supplying the rescue function, the only apparent requisite being a small portion of the carboxy-terminal end of the T-antigen. This can even be attached to adeno peptides. How the SV40 information ultimately circumvents the block to adeno multiplication is still very unclear and controversial. Normal synthesis of all of the late adeno proteins is restored but whether the primary act is at the transcriptional, processing or translational level is not sorted out despite much effort. The recent isolation of simple Adenovirus mutants that overcome the block and grow on monkey cells without help from SV40 (Klessig, 1977) may unravel the mystery, since the genomic location of these mutations should tell us what adeno function is inadequate in monkey cells.

Grodzicker realized that the limited host range of Ad2 which is extended in the Ad2 SV40 hybrids provides a classic case for potential isolation of specific host range mutants that should be primarily defective in one gene, the SV40 helper function (Grodzicker *et al.*, 1974). That is, mutants of the hybrid that now fail to grow on monkey cells, but which of course can still be carried and studied by growth on human cells, might be predominantly of the class defective in the SV40 gene of interest. Since this function is totally nonessential on human cells, mutants of any type, missense, nonsense, deletion, frameshift, rearrangement, etc. might be found.

Among a collection of such mutants induced by nitrous acid treatment are three that are nonsense mutants (Gesteland *et al.*, 1977). This identification is based on biochemical experiments that show that *in vitro* synthesis of the normal gene product occurs only in the presence of characterized yeast suppressor tRNAs.

The parental hybrid virus, Ad2nd1 carries an insertion of 17% of the SV40 chromosome (0.11 to 0.28) which includes the distal end of the large T-antigen. This hybrid grows on monkey cells as well as wild type adenovirus grows on human cells. The host range mutants of Ad2nd1 that are now defective for growth on monkey cells show a range of leakage levels on monkey cells, ranging from tight mutants, which include the nonsense mutants with a leakage rate of about 10^{-3}, up to high levels of leakage. The low leakage growth of the tight mutants is not due to reversion since the progeny are still mutant, but rather probably reflects the same low leakage rate seen with wild type adenovirus. Thus, the tight mutants, H71, 140 and 162 grow 10^3 times less well on monkey cells than their parent and all grow normally on human cells. The mutants are defective in the SV40 encoded function since: 1) the block to replication can be overcome by complementation through co-infection with SV40 virus, 2) the mutants are defective in the synthesis of a 30,000 molecular weight protein which is primarily encoded by the SV40 sequences (Grodzicker *et al.*, 1976), 3) revertants of the viral mutants occur with a reasonable frequency suggesting a simple mutation.

The first suggestion that these mutants might include chain terminating nonsense mutants came from investigating the nature of the mutant product. To look at the SV40 encoded material in wild type and mutant infected cells we first purified the SV40 specific mRNA by hybridization using a scheme developed by Lewis *et al.*, (1975). Then this was translated *in vitro* and the protein products were analyzed by electrophoresis on SDS polyacrylamide gels. RNA from the parental Ad2nd1 infected cells produced a 30K protein and the two mutants 140 and 162 gave no 30K but a 19K protein instead and mutant 70 gave a 10K product. Thus the mutants are indeed defective in an SV40 encoded protein. If these mutant products are smaller by virtue of termination at a mutationally acquired nonsense codon within the 30K gene, then suppression in vitro might restore synthesis of some 30K product. Figure 3 shows that 162 is an amber (UAG) mutant since yeast suppressor tRNA gives suppression, and 70 is an ochre mutant since ochre tRNA gives a positive result. These *in vitro* experiments then serve to identify the mutant defect, which we think is the operative defect in vivo. Analysis of the proteins in cells (human or monkey) infected with these mutants show no detectable 30K protein, and with 140 and 162 the 19K fragment can be seen. Moreover, a simple revertant of 140 that has regained growth on monkey cells requires translation of that part of the 30K gene that is distal to the amber site at the 19K position. Since the ochre and amber mutants are tight for growth and 30K cannot be detected we conclude that termination is efficient at these sites. Therefore normal monkey (and human) cells do not harbor efficient suppressors of ochre and amber.

These nonsense mutants could in principle be used for screening of potential suppressor cell lines: either monkey cells that now grow the mutant, or any line permissive for Adenovirus that now makes

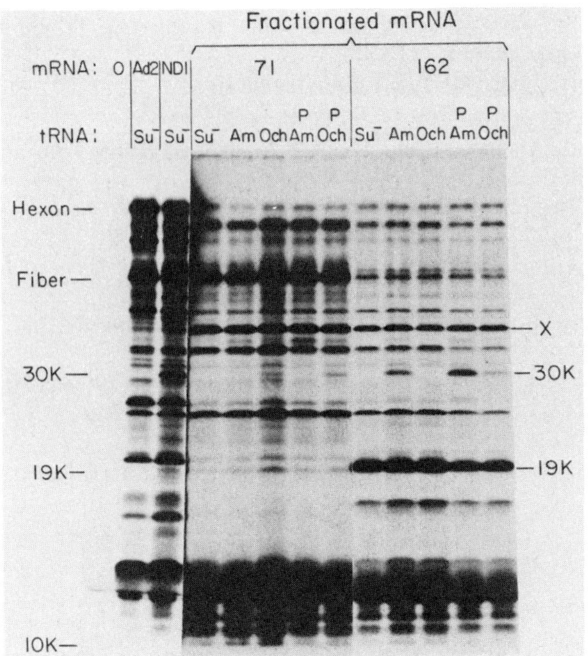

Fig. 3 *In vitro* suppression of mutants 71 and 162 to give 30K protein. For details see Gesteland *et al.*, (1977). *In vitro* protein synthesis was programmed with crude RNA from Ad2 or ND1 infected cells or with RNA from 71 and 162 mutant infected cells fractionated by hybridization to SV40 DNA. Yeast tRNA from nonsuppressing (Su^-), RL1 ('am') or sup 61 ('ochre') were added in crude form or enriched for amber or ochre suppressing activity. The 19K fragment made by mutant 162 is seen and amber tRNA suppressed termination to give some 30K. The 71 fragment of 10K is difficult to see in the low end of the gel, but suppression by ochre can be seen.

30K protein. Obviously neither of these provides a selection system. One possibility is to try to transfer a yeast suppressor gene into monkey cells and use the adenovirus mutants as an assay for success. With this in mind we (with A. Grassmann and T. Grodzicker) are currently asking whether suppression by purified yeast tRNA can work *in vivo* giving active 30K that will provide helper function for adenovirus growth. Thus mutant infected monkey cells are being individually injected with tRNA and pools are being assayed for virus yield. This will hopefully tell us which suppressor to pursue.

References

Capecchi, M.R., Hughes, J.H. and Wahl, G.M. (1975). *Cell* **6**, 269.

Efstradiadis, A., Kafatos, F.C. and Maniatis, T. (1977). *Cell* **10**, 571.

Gesteland, R.F., Wolfner, M., Grisafi, P., Fink, G., Botstein, D. and Roth, J.R. (1976). *Cell* **7**, 381.
Gesteland, R.F., Wills, N., Lewis, J.B. and Grodzicker, T. (1977). *Proc. Natl. Acad. Sci.* **74**, 4567.
Grodzicker, T., Anderson, C.W., Sharp, P.A. and Sambrook, J. (1974). *J. Virol.* **13**, 1237.
Grodzicker, T., Lewis, J.B. and Anderson, C.W. (1976). *J. Virol.* **19**, 559.
Klessig, D.F. (1977). *J. Virol.* **21**, 1234.
Lewis, J.B., Atkins, J.F., Anderson, C.W., Baum, P.R. and Gesteland, R.F. (1975). *Proc. Natl. Acad. Sci.* **72**, 1344.
Pelham, H.R.B. and Jackson, R.J. (1976). *Eur. J. Biochem.* **67**, 247.
Proudfoot, N.J. (1977). *Cell* **10**, 559.
Proudfoot, N.J., Gillam, S., Smith, M. and Longley, J.I. (1977). *Cell* **11**, 807.
Schreier, M.H. and Staehelin T. (1973). *J. Mol. Biol.* **73**, 329.

STRUCTURAL GENE MUTANTS OF IMMUNOGLOBULIN HEAVY CHAINS

D.S. SECHER, R.G.C. COTTON, N.J. COWAN,
R.F. GESTELAND*, AND C. MILSTEIN

*MRC Laboratory of Molecular Biology,
University Medical School, Cambridge, England.*

**Cold Spring Harbor Laboratory,
Cold Spring Harbor, New York, USA.*

Expression of Immunoglobulin Genes

Immunoglobulins, the proteins that carry antibody activity, are a closely related family of serum proteins. Purification of homogeneous immunoglobulins from normal antisera is possible only in exceptional cases and most of our knowledge of immunoglobulin structure is derived from studies on myeloma proteins, the homogeneous products of tumours of immunoglobulin producing cells. A typical immunoglobulin, produced by the mouse myeloma MPOC 21, is illustrated in Figure 1.This tumour has been extensively studied in our laboratory and is the parental line from which the variant lines described below were all derived.

Data from amino acid sequence studies and from classical genetics of polymorphic markers on immunoglobulin chains led to the suggestion that for each polypeptide chain the sequence information from two structural genes is required. One gene (V-gene) codes for the variable region of the polypeptide (V_L or V_H) which defines the antigenic specificity of the immunoglobulin. The other (C-gene) codes for the remainder of the chain (constant region) and determines the biological effector functions of the immunoglobulin. (For recent reviews of the structure of immunoglobulins and their functions see, for example, Secher, 1979; Poljak, 1978; Turner, 1977; Gally, 1973. For reviews of the genetics of immunoglobulins see Rabbitts, 1979; Milstein and Munro, 1970.)

The isolation of a single mRNA for an immunoglobulin light chain (Milstein *et al.,* 1974b) and the codominant expression of immunoglobulin chains in hybrid cells, without the production of chains containing the V-region of one parental chain attached to the C-region of another (Cotton and Milstein, 1973; Köhler and Milstein, 1975), provided strong evidence that the integration of information of V- and C-genes takes place at the level of DNA. Recently it has been directly shown that there is a rearrangement of the DNA coding for immunoglobulin light chains during the differentiation of immunoglobulin

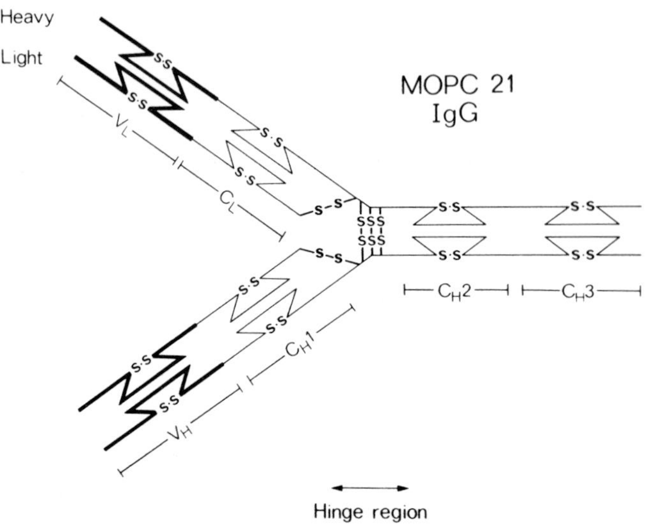

Fig. 1 Diagrammatic representation of an immunoglobulin molecule. The V-regions are indicated by bold lines. The names of the domains and the positions of the disulphide bonds (-S-S-) are indicated (from Secher *et al.*, 1976).

producing cells (Brack *et al.*, 1978; Rabbitts and Forster, 1978). However, even after such rearrangement the DNA coding for immunoglobulin light chains contains interruptions (intervening sequences) that appear to be present in the primary transcription product but are then 'spliced out' during the maturation of the DNA (Rabbitts, 1978).

In the case of the heavy chain genes the somatic rearrangement of DNA and the presence of interruptions in the DNA have not yet been demonstrated owing to the difficulties in purifying heavy chain mRNA, but a similar model seems likely (Rabbitts, 1978).

A possible arrangement of the germ-line genes coding for human and mouse immunoglobulin chains in shown in Figure 2. In both species, there are three linked sets of genes, one for κ light chains, one for λ light chains and one for heavy chains. The three linkage groups are not linked to one another and are probably on different chromosomes (see Secher, 1979). Any one of the V-genes may be associated with a linked C-gene and in the case of the heavy chain linkage group a single V-gene may be associated with more than one C-gene implying either duplication of the V-gene or a controlled switch in the splicing of heavy chain pre-mRNA (Rabbitts, 1978).

The number of V-genes shown in Figure 2 is a minimum number for the germ-line. Each individual can synthesise many more V-regions (>10^3, Secher *et al.*, 1977) and proponents of germ-line theories of

Man.

		V-genes	C-genes
Light Chains	κ	Ia Ib II III	—
	λ	I II III IV	— — —
Heavy Chains		I II III	γ4 γ2 γ3 γ1 α1 α2 μ2 μ1 δ ε

Mouse.

		V-genes	C-genes
Light Chains	κ	I II III IV V VI VII VIII etc.	—
	λ	I II	I II
Heavy Chains		I II III IV	γ_1 γ_{2a} γ_{2b} γ_3 α μ δ ε

Fig. 2 Possible arrangement for a minimum number of genes for human and mouse light and heavy chains, (from Rabbitts, 1979).

antibody diversity predict that the entire repertoire of chains is encoded in the germ-line. The alternative view is that somatic mutation, together with selection, allows for an expansion of a limited V-gene pool (as suggested in Figure 2). This type of model is supported (in the case of mouse λ chains) by evidence from hybridisation kinetics and by the cloning of the $V_{\lambda I}$ and $V_{\lambda II}$ germ-line genes (Brack *et al.*, 1978) and by the amino acid sequences of human κ chains (Milstein, 1967). However, in the case of mouse κ chains the situation is less clear and many more V-genes seem to be involved.

Despite this potentially central role of somatic mutation in the generation of antibody diversity, very little is known about the rate and nature of spontaneous mutation in animal cells. Some years ago we decided to approach this question by studying the clonal diversification of a mouse myeloma cell line in culture (MOPC 21).

Apart from its obvious relevance to immunology the cell line has a number of advantages for studying mutation. Almost all the protein secreted by the cells is the myeloma protein which can be purified in large amounts. Since the protein confers no advantage on the cell that synthesised it a wide range of mutants can be isolated.

A Method for Isolating Spontaneous Mutants of Cells in Culture

The method developed for screening large numbers of subclones of the parental (wild-type) line has been described in detail elsewhere (Secher *et al.*, 1977) and is summarised in Figure 3. Briefly, cells were grown

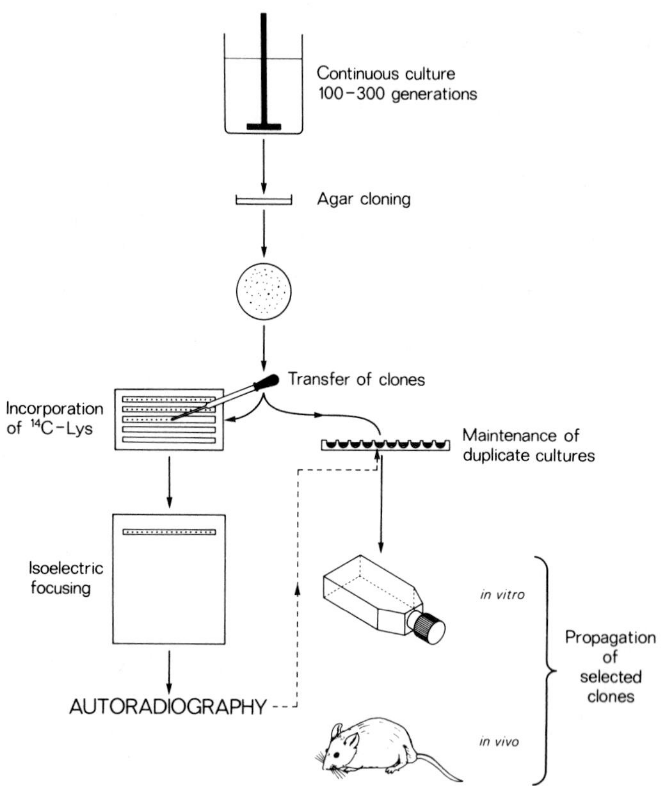

Fig. 3 Procedure for the detection and isolation of mutant clones producing immunoglobulin of altered isoelectric point (from Secher *et al.*, 1976).

in continuous spinner culture and after 100-300 generations aliquots were removed for cloning in soft agar (Cotton *et al.*, 1973). After growth of the subclones they were incubated in the presence of [^{14}C]-lysine and the secreted immunoglobulin analysed by isoelectric focussing and autoradiography of the dried gel. Duplicate cultures were established so that variant clones could be grown up to mass culture or injected into mice, where the cells formed tumours and large amounts of the variant proteins could be recovered from the serum of such tumour-bearing animals. Up to 200 subclones could be tested at a time and a total of over 7,000 were analysed. Examples of autoradiographs of isoelectric focussing gels are shown in Figure 4. The position of application of the clones is visible near the top of the gel and also the characteristic pattern of three bands that defines the pure 'wild-type' MOPC 21 immunoglobulin. In two cases a variant pattern can be seen. Subclones giving rise to variant patterns were recloned as soon as possible to ensure clonal purity.

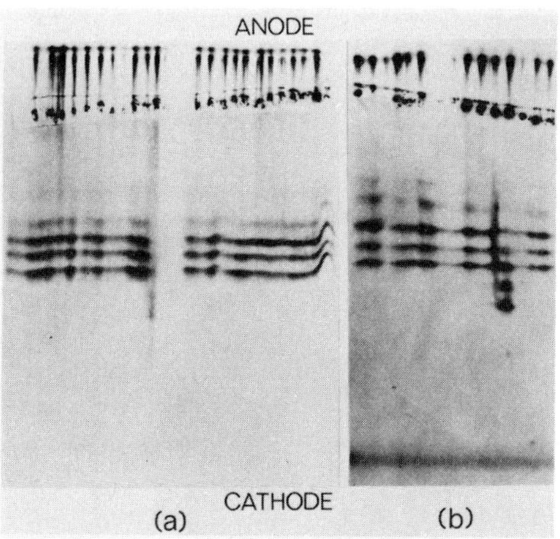

Fig. 4 Autoradiogram of isoelectric focussing gels showing the detection of mutant clones (from Milstein *et al.,* 1974c).

Non-secreting and Non-producing Variants

The first variants to be recovered were a class that secreted no immunoglobulin or immunoglobulin chains. These 'non-secretors' were subdivided (Cowan *et al.,* 1974) according to whether they contain intracellular light chain (NSI), a modified form of the heavy chain (NSII) or no intracellular immunoglobulin chains (NSIII). The molecular defects that give rise to the loss of expression of immunoglobulin chains have not yet been elucidated. Because of the lack of reversion of chain loss it is often assumed to be due to chromosome loss or gene deletion. However, this cannot be a general explanation since two non-secreting clones that contain no intracellular light chains (NSII/1 and NSIII/1), have been shown to contain an RNA that is indistinguishable from the wild-type light chain mRNA by oligonucleotide fingerprinting (Cowan *et al.,* 1974) or by hybridisation to light chain cDNA (T.H. Rabbitts, personal communication). These RNAs, however, are not translated *in vivo* or *in vitro.* The equivalent experiments to look for heavy chain mRNA (or DNA) in cells that have lost the production of heavy chains have not yet been done owing to lack of suitable probes.

TABLE I

Structural Mutants of MOPC 21 Mouse Myeloma

	Wild type MOPC 21	Mutants IFI	IF2	IF3	IF4
Molecular weight	50,800	41,300	40,900	40,700	50,800
H-chain defect		Deletion of C_H3 (residues 358–440)	Deletion of C_H1 (residues 121–214)	Altered sequence of residues 341–354 followed by deletion of C_H3	Asparagine to aspartic acid (residue 415)
Sialic acid	–	+	–	–	–
Binding to F_c -receptor	+	–	+	–	N.D.
H-chain mRNA size	17S	17S	16S	17S	17S
Proposed genetic mechanism		Point mutation ('nonsense')	Intracistronic deletion	Frameshift (–2) Premature 'ochre' termination	A to G transition ('missense')

The molecular weights are calculated from the amino acid sequences, allowing 2,000 daltons for carbohydrate. N.D. = not determined. Data on Fc receptor from Ramasamy *et al.*, (1975). Taken from Secher *et al.*, (1977).

The reason for the non-secretion of free intracellular light chain in NSI is also unknown. Several other light chains are secreted without heavy chains (e.g. Köhler *et al.*, 1976). Cell fusion experiments suggest that apparently any heavy chain may substitute for the MOPC 21 heavy chain as a 'carrier' on which the light chain may be secreted.

The spontaneous rate of mutation to non-secretion of immunoglobulin was about 10^3-10^4/cell/generation (Cotton *et al.*, 1973). A similar rate has been measured for another mouse myeloma cell line, using a different method for identifying variants (Coffino and Scharff, 1971).

Mutant Proteins

We believe the non-secreting variants will prove to be useful in understanding the stages of control of immunoglobulin gene expression, but we have concentrated our efforts on the characterisation of the four mutant cell lines that produce and secrete immunoglobulin of different isoelectric point from the wild-type. The four lines were named IF1, IF2, IF3 and IF4. The preliminary characterisation of the variant proteins was done with radioactive proteins produced by cells in culture (Secher *et al.*, 1973). Later, when large amounts of protein were available from the serum of tumour-bearing mice and the complete sequence of the MOPC21 heavy chain had been elucidated (Adetugbo *et al.*, 1975) a complete definition of the defects in the four mutant proteins was possible. These results are summarised in Figure 5 and Table I. Some further details of the sequencing methods used have been published recently (Adetugbo *et al.*, 1978a, b).

IF1

IF1 has a higher isoelectric point than MOPC 21 immunoglobulin and a smaller heavy chain as judged by SDS-gel electrophoresis (Figure 6). This is due to the deletion of the C_H3 domain from residue 358 to the C-terminus of the heavy chain as shown in Figure 5. The light chain of IF1 is identical to the wild-type and so is the amino acid sequence of the heavy chain up to residue 357. However, the carbohydrate moiety that is attached to residue 291 is modified by the addition of sialic acid (Cowan *et al.*, 1973).

IF2

IF2 also has a smaller heavy chain and an unaltered light chain. The immunoglobulin is more acidic that the wild-type and the alteration in the heavy chain was shown to be a deletion of residues 121-214 (the C_H1 domain). This protein is similar to the heavy chain disease

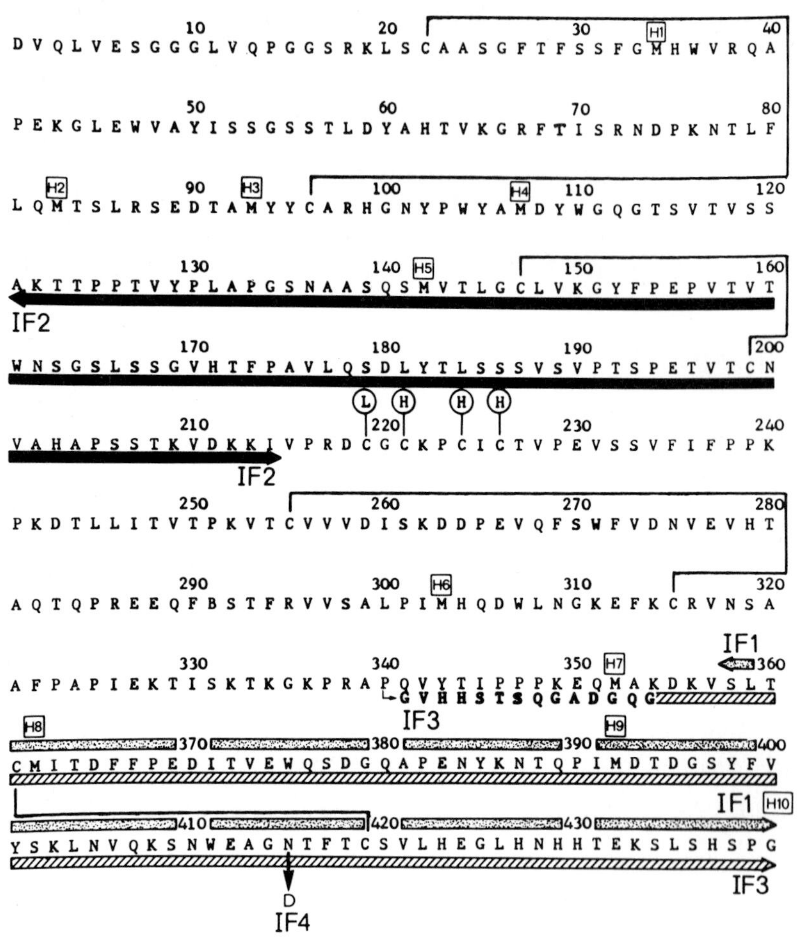

Fig. 5 Sequence of MOPC 21 heavy chain and of the four mutants described in the text. The arrows indicate the extent of deletions. [H1], [H2], etc. indicate the C-terminal residue of the ten CNBr fragments of the wild-type heavy chain (from Milstein *et al.*, 1976).

proteins that have been described in humans (Franklin and Frangione, 1975), especially in the involvement of residue 214 (or in humans the homologous residue 215). We have suggested that this mutant may have arisen from an error in the process of V-C integration (Milstein *et al.*, 1974a).

IF3

The heavy chain of IF3 also lacks the C_H3 domain, but the C-terminus has a stretch of 14 amino acids that does not correspond to any sequence in the wild-type heavy chain (nor to any other known immunoglobulin sequence) (Adetugbo and Milstein, 1978). However,

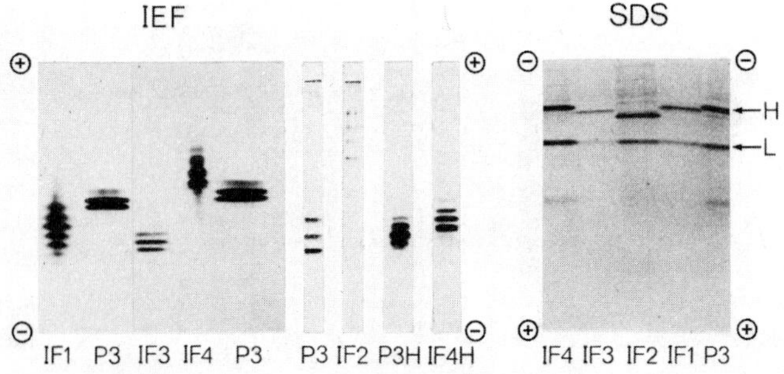

Fig. 6 Comparison of wild-type and mutant immunoglobulins (or immunoglobulin chains) by isoelectric focussing (IEF) or SDS polyacrylamide gel electrophoresis. H, L, indicate the position of heavy and light chains. (Adapted from Adetugbo *et al.*, 1977).

by postulating a frameshift mutation caused by the deletion of two bases it was possible to predict a unique mRNA sequence which could be translated either as the wild-type sequence or as the IF3 sequence depending on the reading frame (Figure 7). This hypothetical mRNA was independently supported by sequence studies on ^{32} P-labelled heavy chain mRNA (Cowan *et al.*, 1976). The oligonucleotide shown in Figure 7D was isolated from both wild-type and IF3 mRNA as predicted whereas that coding for residues 333-342 was found in IF2 mRNA but not in the IF3 mRNA (Cowan *et al.*, 1976; D.S. Secher and N.J. Cowan, unpublished results).

IF4

IF4 heavy chain had the same apparent molecular weight as the wild-type but isoelectric focussing showed that the immunoglobulin and the isolated heavy chain have a lower isoelectric point (Figure 6). The light chain was again unaltered. The only change in amino acid sequence of the heavy chain was the substitution of an aspartic acid for an asparagine at residue 415. In order to show that this substitution was due to a mutation and not to a post-translational deamidation, newly synthesised intracellular immunoglobulin from wild-type and from IF4 cells were compared and seen to have the same difference in isoelectric point as the secreted material (Adetugbo and Milstein, 1977). Furthermore the oligonucleotide previously shown to code for residues 414-416 (h6, Cowan *et al.*, 1976) was absent from fingerprints

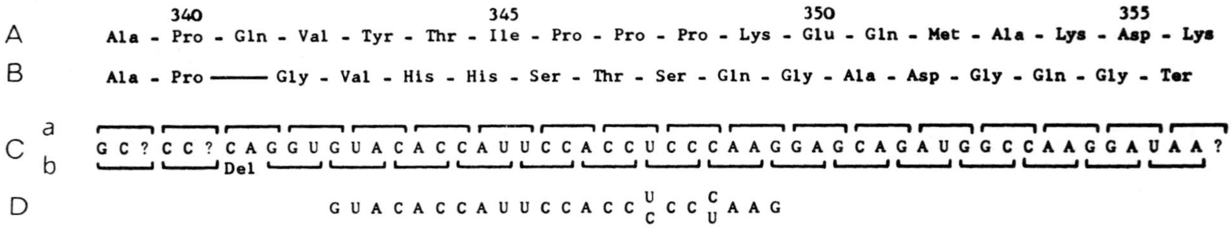

```
                 340                                   345                                    350                                   355
A    Ala - Pro - Gln - Val - Tyr - Thr - Ile - Pro - Pro - Pro - Lys - Glu - Gln - Met - Ala - Lys - Asp - Lys

B    Ala - Pro ——— Gly - Val - His - His - Ser - Thr - Ser - Gln - Gly - Ala - Asp - Gly - Gln - Gly - Ter

C a/b  G C ? C C ? C A G G U G U A C A C C A U U C C A C C U C C C A A G G A G C A G A U G G C C A A G G A U A A ?
                   Del

D                 G U A C A C C A U U C C A C C (U/C) C C (C/U) A A G
```

Fig. 7 IF3 - a frameshift mutant (A) Sequence of a stretch of the MOPC 21 heavy chain (Adetugbo *et al.*, 1975). (B) The equivalent sequence in IF3 (Adetugbo and Milstein, 1978). (C) is a unique, theoretically derived nucleotide sequence that can be translated as MOPC 21 or IF3 depending on the reading frame (C_a or C_b respectively). (D) Sequence of oligonucleotide h6 taken from Cowan *et al.*, (1976). Note that the termination codon for IF3 is UAA (ochre) as predicted by cell-free synthesis (Secher *et al.*, 1975). Taken from Milstein *et al.*, 1976).

of IF4 mRNA (Milstein *et al.*, 1976; Adetugbo *et al.*, 1977) as predicted by an AAU→GAU (Asn→Asp) point mutation.

In Vitro Translation and Suppression of Mutant mRNAs

Immunoglobulin heavy chain mRNA from all four mutant clones was partially purified and translated in cell free systems (Secher *et al.*, 1973; 1975; Cowan *et al.*, 1976). This demonstrated that the deletions

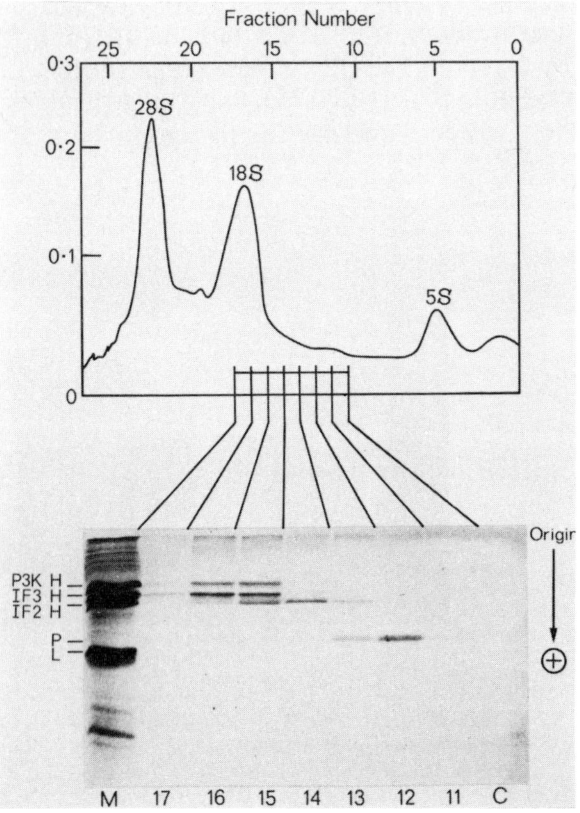

Fig. 8 Size of mutant mRNAs. Membrane derived polysomes were prepared from a mixed cell culture containing equal number of MOPC 21, IF2 and IF3 cells and the RNA resolved by centrifugation on an SDS-sucrose gradient. The RNA from individual fractions was tested for activity in a rabbit reticulocyte cell-free system containing [^{35}S]-methionine. Immunoglobulin chains were recovered from the incubation mixture by immunoprecipitation and resolved in an SDS gel. M = molecular weight marker mixture containing MOPC 21 (P3K), IF2 and IF3 heavy and light chains. C = control (no RNA added to cell-free system). P indicated position of light chain precursor (Milstein *et al.*, 1972). Taken from Cowan *et al.*, (1976).

observed were not due to any post-translational modifications to the proteins. The size of the heavy chain mRNAs was measured in terms of their relative mobilities in sucrose density gradients (Figure 8). IF1 and IF3 heavy chain mRNAs were the same size as the wild-type MOPC 21 mRNA (~17S) but IF2 mRNA was found to be about 16S compatible with a deletion of about 300 bases.

mRNAs from IF1 and IF3 cells were also translated in the presence of added suppressor tRNAs from yeast (see Gesteland *et al.*, this volume). The cell free system was as described by Gesteland *et al.*, (1976) with the concentrations of [^{35}S]-methionine, tRNA, mRNA and Mg^{++} optimised by titration. The tRNAs were prepared from the following yeast strains: Sup^+, RL1 (DBD195); Sup_{am}, RL1 aneuploid (DBA316); Sup_{och}, FM6. Figure 9 shows the translation products of the immuno-

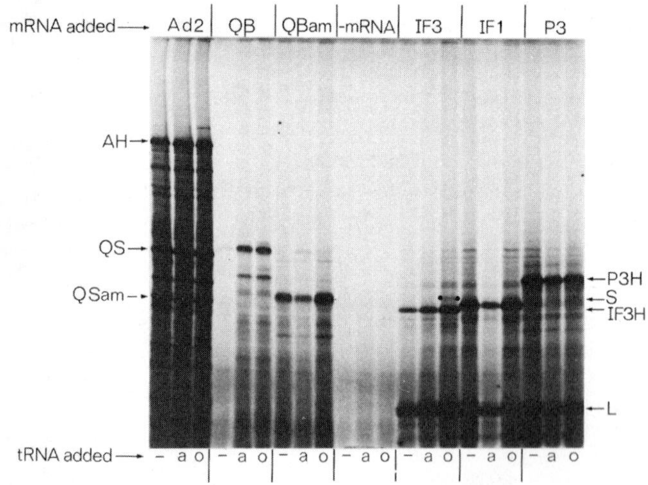

Fig. 9 Suppression of IF3 mutation. mRNAs from different sources were added to a cell-free system containing [^{35}S]-methionine as described in the text. Yeast tRNA from wild-type strain (-), amber suppressing strain (a) or ochre suppressing strain (o) were also added as indicated. Ad2 = late adenovirus 2 mRNA, Qβ is phage RNA from wild-type phage, Qβam is RNA from mutant phage Qβam1. After incubation an aliquot from each reaction was run on a 10% SDS-polyacrylamide gel which was then dried and autoradiographed. IF1H, IF3H and P3H indicate the position of IF1, IF3 and MOPC 21 heavy chains, P is the light chain precursor. AH is the adenovirus 2 hexon, QS is wild-type Qβ synthetase, QSam is the amber mutant Qβ synthetase, S is the presumed read-through product of IF3 heavy chain (indicated on the autoradiograph by two dots).

globulin mRNAs together with the translation products of other mRNAs as controls. The evidence that the suppression systems are working is provided (in the case of ochre suppressing tRNA) by the appearance of read-through products of the Ad2 hexon protein and of the wild-type Qβ synthetase. The amber suppressing tRNA can be seen to suppress the amber mutation in the Qβ synthetase of a mutant strain. The IF3 mRNA clearly gives rise to a new band in the presence of ochre, but not amber or wild-type tRNA. This band (band S, Figure 9) is longer than the IF3 heavy chain but smaller than the wild-type heavy chain and is presumably the read-through product extending the IF3 heavy chain from the UAA termination codon indicated in Figure 7 to the next termination codon in the reading frame C_b of Figure 7. The suppression experiments are thus in complete agreement with those from chemical analysis of protein and nucleic acid of the IF3 cells.

No suppression can be seen in the case of the IF1 translation products with either amber or ochre suppressing tRNA. (Figure 9 and unpublished observations). It is possible that the mutation gives rise to a UGA terminator which would not have been detected in the above experiments. The mRNA sequence for the relevant region has not yet been determined and so the suggestion that the IF1 mutation is a point mutation from a serine codon remains to be confirmed.

The absence of a visible read-through band for the light chain with either amber or ochre suppressor tRNA may now be explained by reference to the sequence of the C-region and 3′ non-coding region of the MOPC 21 light chain mRNA (Hamlyn *et al.*, 1978). The termination codon is UAG (amber) but the amber read-through product would be -(Cys)-Ser-Arg-Gln-Arg-Ser(UGA) where Cys is the C-terminal amino acid of the normal light chain and Ser is the amino acids inserted by the suppressor tRNA. This extension of only five amino acids before coming to the next in phase terminator (UGA) is probably too small a difference to be resolved from the main light chain band on SDS gels.

Significance of the Mutations

The experiments of *in vitro* translation and sequence analysis of the mRNA rule out the possibility that the mutations observed are due to post-translational modifications. The possibility that they result from post-transcriptional modifications of RNA or by the activation of otherwise silent genes have not been formally excluded but seem unlikely for a number of reasons that have been discussed previously (Adetugbo *et al.*, 1977). We conclude that IF1, IF2, IF3 and IF4 represent spontaneous mutations of the mouse $\gamma 1$ structural gene locus. IF1 seems to be a nonsense mutant, IF2 a deletion, IF3 a frameshift and IF4 a missense mutant. These mutational processes have been well described in bacteria, viruses and lower eukaryotes. The above studies show they also take place in mammalian somatic cells in culture. The ability to partially suppress the IF3 mutation *in vitro* provides an opportunity to search for mammalian suppressor tRNAs *in vivo.* From the isolation of the four mutants described above (and the detection of a probable fifth mutation) from a total of 7,000 subclones, we calculated that the rate of mutation for the heavy chain structural genes is about 3×10^{-7}/cell/generation, a considerably higher rate than that reported for spontaneous variations in other mammalian systems which are less than 10^{-8}/cell/generation (see Adetugbo *et al.,* 1977 for original refs.). The isolation of IF4 clearly demonstrates that the screening protocol is sufficiently sensitive to detect a single amino acid substitution giving rise to a change in charge. However, it is likely that more conservative substitutions might be missed and also any mutations that give rise to proteins that are not secreted. Our estimated rate of mutation is, therefore, likely to be an underestimate. For a more detailed discussion of the possible immunological significance of this rate of mutation on the generation of antibody diversity see Secher *et al.,* (1977).

References

Adetugbo, K. (1978a). *J. Biol. Chem.* **253**, 6068.

Adetugbo, K. (1978b). *J. Biol. Chem.* **253**, 6076.

Adetugbo, K. and Milstein, C. (1977). *J. Mol. Biol.* **115**, 75.

Adetugbo, K. and Milstein, C. (1978). *J. Mol. Biol.* **121**, 239.

Adetugbo, K., Milstein, C. and Secher, D.S. (1977). *Nature (London)* **265**, 299.

Adetugbo, K., Poskus, E., Svasti, J. and Milstein, C. (1975). *Eur. J. Biochem.* **56**, 503.

Brack, C., Hirama, M., Lenhard-Schuller, R. and Tonegawa, S. (1978). *Cell* **15**, 1.

Coffino, P. and Scharff, M.D. (1971). *Proc. Nat. Acad. Sci.* **68**, 219.

Cotton, R.G.H. and Milstein, C. (1973). *Nature (London)* **244**, 42.

Cotton, R.G.H., Secher, D.S. and Milstein, C. (1973). *Eur. J. Immunol.* **3**, 135.

Cowan, N.J., Secher, D.S., Cotton, R.G.H. and Milstein, C. (1973). *FEBS Letters* **30**, 343.

Cowan, N.J., Secher, D.S. and Milstein, C. (1974). *J. Mol. Biol.* **90**, 691.

Cowan, N.J., Secher, D.S. and Milstein, C. (1976). *Eur. J. Biochem.* **61**, 355.

Franklin, E.C. and Frangione, B. (1975). *Contemporary Topics in Molecular Immunology* **4**, 89.

Gally, J.A. (1973). *In:* The Antigens (ed. Sela, M.), Vol. 1, 161. Academic Press, London.

Gesteland, R.F., Wolfner, M., Grisafi, P., Fink, G., Botstein, D. and Roth, J.R. (1976). *Cell* **7**, 381.

Hamlyn, P.H., Brownlee, G.C., Cheng, C.C., Gait, M.J. and Milstein, C. (1978). *Cell* **15**, 1067.

Köhler, G., Howe, S.C. and Milstein, C. (1976). *Eur. J. Immunol.* **6**, 292.
Köhler, G. and Milstein, C. (1975). *Nature (London)* **256**, 495.
Milstein, C. (1967). *Nature (London)* **216**, 330.
Milstein, C., Adetugbo, K., Cowan, N.J., Köhler, G., Secher, D. S. and Wilde, C.D. (1976). *Cold Spring Harbor Symp. Quant. Biol.* **41**, (in press).
Milstein, C., Adetugbo, K., Cowan, N.J. and Secher, D.S. (1974a). *Progr. Immunol.* **2** (Vol. I), 157.
Milstein, C., Brownlee, G.G., Cartwright, E.M., Jarvis, J.M. and Proudfoot, N.J. (1974b). *Nature (London)* **252**, 354.
Milstein, C., Brownlee, G.G., Harrison, T.M. and Mathews, M.B. (1972). *Nature New Biol. (London)* **239**, 117.
Milstein, C., Cotton, R.G.H. and Secher, D.S. (1974c). *Ann. Immunol. (Inst. Pasteur)* **125C**, 287.
Milstein, C. and Munro, A.J. (1970). *Ann. Rev. Microbiol.* **24**, 335.
Poljak, R. (1978). *Critical Rev. Biochem.* **5**, 45.
Rabbitts, T.H. (1978). *Nature (London)* **275**, 291.
Rabbitts, T.H. (1979). Defence and Recognition. 2nd edition (ed. Lennox, E.S.) University Park Press, Baltimore, (in press).
Rabbitts, T.H. and Forster, A. (1978). *Cell* **13**, 319.
Ramasamy, R., Secher, D.S. and Adetugbo, K. (1975). *Nature (London)* **253**, 656.
Secher, D.S. (1979). Defence and Recognition, 2nd edition (ed. Lennox, E.S.) University Park Press, Baltimore, (in press).
Secher, D.S., Adetugbo, K., Cowan, N.J., Köhler, G., Milstein, C. and Wilde, C.D. (1976). *In:* Structure-Function Relationships of Proteins (eds. Markham, R and Horne, R.W.), 129. North Holland, Amsterdam.
Secher, D.S., Adetugbo, K. and Milstein, C. (1977). *Immunol. Rev.* **36**, 51.
Secher, D.S., Cotton, R.G.H. and Milstein, C. (1973). *FEBS Letters* **37**, 311.
Secher, D.S., Gesteland, R.F. and Milstein, C. (1975). *Biochem. Soc. Trans.* **3**, 873.
Turner, M.W. (1977). Advanced Immunochemistry (eds. Glynn, L.E. and Steward, M.W.) Wiley, New York.

THE *UNC*-54 MYOSIN HEAVY CHAIN GENE OF *CAENORHABDITIS ELEGANS:* A MODEL SYSTEM FOR THE STUDY OF GENETIC SUPPRESSION IN HIGHER EUKARYOTES

A.R. MACLEOD, J. KARN, R.H. WATERSTON* AND S. BRENNER

MRC Laboratory of Molecular Biology, University Medical School, Cambridge, England.

Introduction

The small free living soil nematode *Caenorhabditis elegans,* has been the subject of genetical and other studies in this laboratory for several years (Brenner, 1974). Mutants have been isolated in many genes affecting its morphology and behaviour and a considerable amount is now known about its cellular anatomy and development (see Sulston and Horvitz, 1977 for references).

A subset of mutants with defective movement have severe alterations in the structure of body wall muscle cells. There are 95 muscle cells disposed in four quadrants which run the length of the animal beneath the cuticle. The musculature is obliquely striated and the sarcomeres are oriented parallel to the long axis of the animal. Disruption of the band structure can easily be detected in the living animal by polarized light microscopy and the defects can be more carefully analysed by electron microscopy. Two of the ten genes with altered muscle phenotypes have been shown to specify major structural proteins of muscle; *unc*-54 codes for a major heavy chain of myosin, while *unc*-15 codes for paramyosin, the core protein of the thick filaments. (Epstein *et al.,* 1974; Waterston *et al.,* 1974, 1977; MacLeod *et al.,* 1977a).

As work with these genes developed, it became evident that it might be possible to use them as model systems for studying gene-protein relationships in a multicellular higher organism. This notion has been reinforced by the isolation of a large number of suppressors in the organism, some of which suppress specific alleles of the myosin and para-myosin genes. Although the molecular mechanism of these

*Present address: Department of Anatomy, Washington University of St. Louis, St. Louis, Mo. 63110.

suppressors are unknown, it will be seen that their detailed study can now be undertaken.

The *unc*-54 Gene

The *unc*-54 gene specifies a major myosin heavy chain present only in body wall muscle cells. All mutants in this gene lead to severe paralysis of the animal by the pharynx is totally unaffected, allowing the animal to feed and survive. The vulval muscle cells are also incompetent and mutants cannot lay eggs; these hatch inside the adult and consume it.

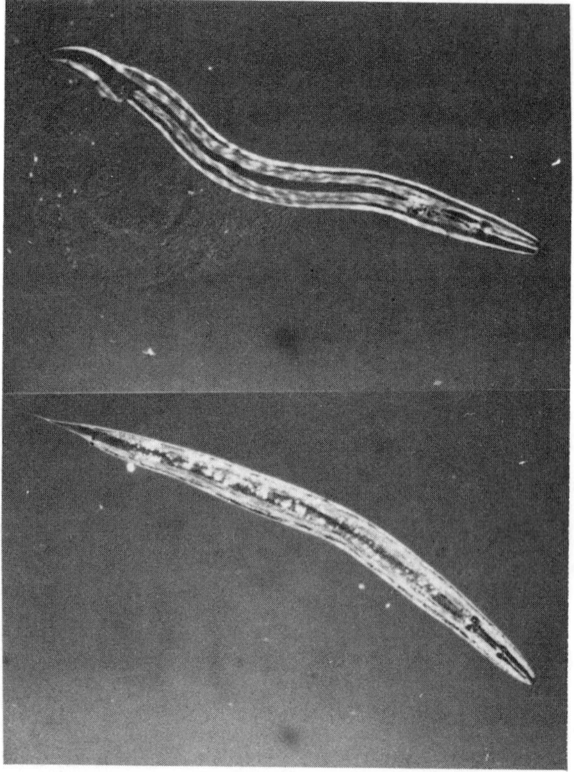

Fig. 1 Polarized light micrographs of wild-type (top) and E190 (bottom) animals. Note the reduction in birefringence associated with the thick filament structures in the E190 animal. Magnification X 30.

Polarized light microscopy reveals that the banded structure seen in normal muscle is disrupted and the birefringence associated with the thick filaments is greatly reduced (Figure 1). Examination of cross sections in the electron microscope shows that the disorganization is accompanied by a drastic reduction in the number of thick filaments. This is found in all mutants of the gene, whether recessive of dominant, and suggests that there are at least two kinds of heavy chain, and possibly two kinds of thick filaments, only one of which is specified by the *unc*-54 gene (Epstein *et al.*, 1974; Waterston *et al.*, 1974). Biochemical studies have confirmed the existence of multiple myosin heavy chains and the assignment of the *unc*-54 gene to a singular major component (MacLeod *et al.*, 1977a).

The heavy chains present in the total myosin fraction from wild type (N2) animals and a large number of different EMS induced mutants of *unc*-54 were characterized by sodium SDS-PAGE before and after chemical cleavage by cyanylation (MacLeod *et al.*, 1977a). Cyanylation cleaves specifically at cysteine residues (Jacobson *et al.*, 1973; Degani and Patchornik, 1974) and, because of the small amount of cysteine in the heavy chains, generates a simple set of large polypeptides which

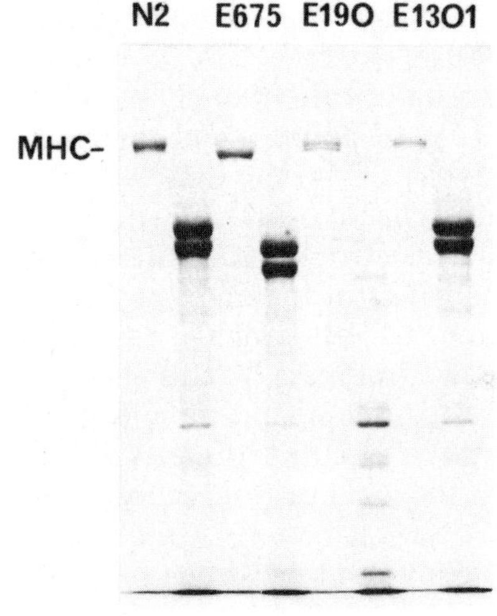

Fig. 2 Electrophoresis of myosin heavy chains and their cyanylation peptides of representative alleles of *unc*-54 in a sodium dodecyl sulphate 6.0% polyacrylamide gel. The position of the wild type *unc*-54 myosin heavy chain is indicated (MHC). The strains from which the myosin heavy chains were isolated are indicated at the top of the figure.

are easily mapped by SDS gel electrophoresis (Figure 2). Wild-type (N2) heavy chains show a major component of 210,000 molecular weight with minor species of slightly lower molecular weight corresponding to pharyngeal and non-muscle myosins. Cyanylation produces two major products of 15,000 and 14,000 moleculare weight with some minor fragments of lower molecule weight.

The mutant E675 gives a normal yield of myosin but analysis by SDS–PAGE shows an altered major heavy chain component which is reduced in molecular weight by 10,000 (Figure 2). Cyanylation of E675 heavy chains produces two major polypeptides both of which are also reduced in molecular weight by 10,000. Peptide mapping proved that the major doublet arises from partial cleavage of the *unc*-54 heavy chain at one of the cysteine residues. The cyanylation fragments contain the C-terminal of the heavy chain molecule and since the difference found in E675 resides in the fragments, a natural suggestion was that these mutations had resulted in a chain terminating nonsense codon. However, comparison of partial digests of N2 and E675 cyanylation fragments labelled specifically at their N-termini showed that the structural defects of E675 was an internal deletion near (but not including) the C-terminus of the heavy chain molecule (MacLeod *et al.*, 1977b). The characterization of the altered *unc*-54 myosin heavy chain found in E675 demonstrates unambiguously that *unc*-54 is the structural gene for the major myosin heavy chain of the body wall musculature.

When myosins isolated from other alleles of *unc*-54 were studied by polyacrylamide gel electrophoresis and chemical cleavage they fell into two classes. Myosins from the first class of mutants appeared identical to the wild-type myosins, showing both the normal yields of myosin and the same electrophoretic patterns as wild type before and after cleavage. An analysis of the myosins isolated from the temperature-sensitive mutant E1301, which is a typical example of this class of mutations, is shown in Figure 2. This class includes all of the dominant, semi-dominant and temperature sensitive mutants. The second class included most of the recessive alleles and were characterized by a large reduction in the yield of total myosin. A striking result was the complete absence of the characteristic *unc*-54 heavy chain doublets after cleavage of the myosin by cyanylation. In Figure 2, this result is illustrated by the analysis of myosin heavy chains isolated from the mutant E190. The absence of normal size *unc*-54 myosin heavy chains and the absence of cyanylation peptides cleaved from the *unc*-54 heavy chains in the myosin fraction isolated from these animals suggests that these mutations lead to a complete failure to synthesize stable *unc*-54 heavy chains.

We have searched unsuccessfully amongst the null mutants of the *unc*-54 gene for mutants which produce prematurely terminated fragments of the myosin molecule. We conclude that if the class of null mutants includes nonsense mutants, then the nonsense fragments have escaped detection either because they are short and thereby fail to coisolate with the myosin fraction, or because they are subjected to rapid *in vivo* degradation. It is useful to point out that the E675 shortened myosin preserves the C-terminus which might be the necessary region of the protein that confers stability. Other explanations for the absence of protein product in these mutants, such as defective messenger RNA production, can be tested, but only with different methods of assay, and we have therefore turned to studies of *in vitro* translation of nematode messengers.

Cell-free Synthesis of *unc*-54 Gene Products

The *in vitro* system chosen was that obtained from wheat germ (Roberts and Paterson, 1973). Total RNA isolated from nematode was found to be an excellent source of nematode messenger RNA and the wheat germ system was programmed with these messenger RNAs under conditions found to be optimal for chain elongation (but not initiation). The [^{35}S] methionine labelled proteins synthesized *in vitro* were analyzed by two dimensional gel electrophoresis (O'Farrell, 1975) and compared to a total protein extract obtained from nematodes labelled *in vivo* with [^{35}S] (Figure 3). A wide range of protein products were synthesized *in vitro* in response to the nematode messenger RNA. The structural proteins of muscle are amongst the abundant proteins seen in both the nematode translation products and in the nematode protein homogenate. Paramyosin (PM), actin(Ac) and tropomyosin (TM) have been tentatively identified on the basis of molecular weight, isoelectric point, and reactivity with antibodies prepared against purified muslce proteins.

Authentic myosin heavy chains are also synthesized when the wheat germ system is programmed with total nematode RNA. To enhance detection, the heavy chain products were enriched by immunoprecipitation with antibcdy prepared against nematode myosin. Figure 4 compares the electrophoretic mobility on SDS polyacrylamide gels of wild type, E675 and E190 immunoprecipitated translation products and ^{35}S-labeled heavy chains purified from N2, E675, and E190 animals. The major *unc*-54 myosin heavy chain of 210,000 molecular weight and two minor heavy chain components of 205,000 and 200, 000 which correspond to the pharangeal and residual myosin heavy chains were synthesized in response to messenger RNA from N2 animals. Each of the heavy chains synthesized *in vitro* (Figure 4, N2, panel 3) comigrates exactly with a corresponding heavy chain synthesized *in vivo* (Figure 4, N2, panel 1) and no distortion of band shape is observed when the *in vivo* and *in vitro* products are mixed (Figure 4, N2, panel 2). When the cell free system is programmed with RNA isolated from E675 animals, the shortened E675 heavy chain is synthe-

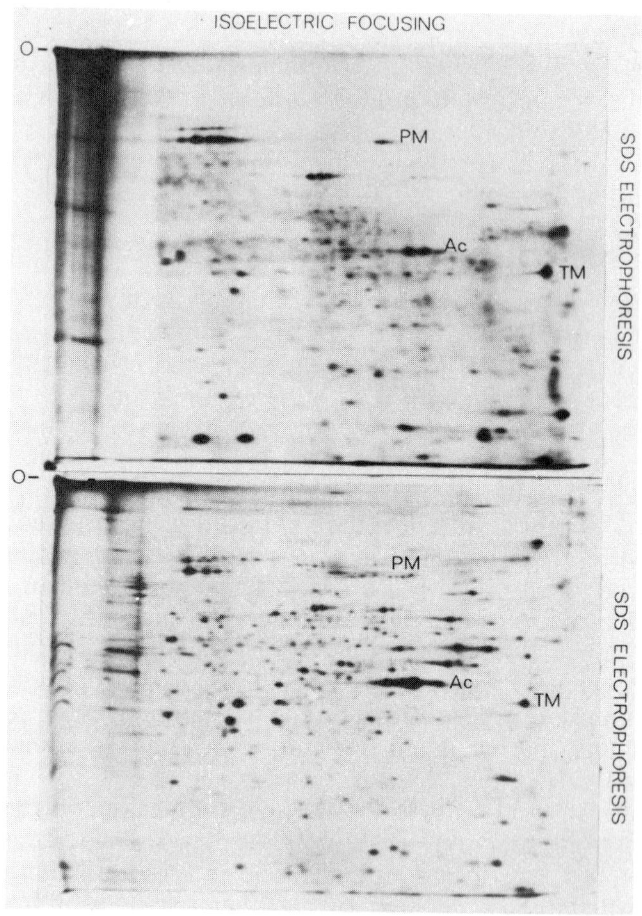

Fig. 3 Two-dimensional gel electrophoresis of [^{35}S] methionine labelled proteins synthesized by a wheat germ cell free system programmed with total nematode RNA (top) and of ^{35}S labelled nematode proteins (bottom). The mobilities of paramyosin (PM), actin (Ac) and tropomyosin (TM) are indicated.

sized, and no full sized *unc*-54 heavy chain is observed (Figure 4, E675, panels 1-3). These data indicate that the deletion observed in the E675 protein results from a deletion of some of the sequence of the E675 messenger RNA and is not a consequence of post translational cleavage. Similarly, the wild-type *unc*-54 heavy chain is not synthesized when the wheat germ system is programmed with E190 RNA (Figure 4, E190, panels 1-3). In addition, the wheat germ system does not synthesize a minor component of 210,000 molecular weight present in the heavy chain fractions isolated from E675 and E190 animals (and also presumably N2). The nature of this component is not known.

The fidelity of translation of myosin heavy messenger RNA by the wheat germ system was also demonstrated by an analysis of the *unc*-54

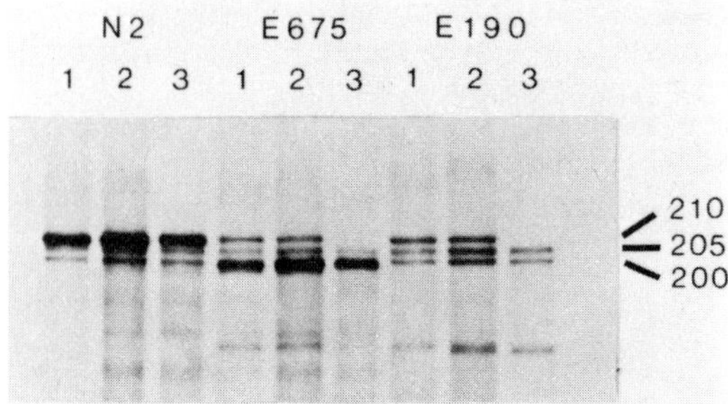

Fig. 4 Electrophoresis of myosin heavy chains synthesized *in vivo* and *in vitro* on a sodium dodecyl sulphate 5% polyacrylamide gel. Myosin heavy chains from N2, E675 and E190 animals were purified from homogenates of ^{35}S labelled nematodes. [^{35}S] methionine labelled myosin heavy chains were synthesized *in vitro* in a wheat germ cell free system programmed with total RNA isolated from N2, E675 and E190 animals and the heavy chains were purified by immuno precipitation with antibody prepared against native nematode myosins. For each of the *unc*-54 alleles, panel 1 shows the electrophoretic monilities of the myosin heavy chains labelled *in vivo*. Panel 3 shows the electrophoretic mobilities of the myosin heavy chains synthesized *in vitro*. Panel 2 is a mixture of the two samples.

myosin heavy chains after chemical cleavage (data not shown). Cleavage of N2 and E675 myosin heavy chains synthesized *in vitro* by cyanylation generated the characteristic *unc*-54 cyanylation peptides with molecular weights identical to the peptides derived from active myosins and demonstrates the correct placement of the internal cysteine residues.

The proceeding experiments demonstrate that cell free translation of nematode messenger RNAs in a heterologous protein synthesizing system provides a reliable assay for *unc*-54 messenger RNA activity. We applied this assay to an analysis of several of the recessive alleles of the *unc*-54 gene. In many cases examined (Figure 5), including E190, novel allele-specific peptides, smaller than the wild-type *unc*-54 myosin heavy chains are detected amongst the translation products precipitated by antimyosin antibody. These products are synthesized in low amount *in vivo* suggesting that the mutations which lead to a failure in synthesis of the complete *unc*-54 myosin heavy chain also result in a defective messenger synthesis or survival. Thus if these are chain terminating nonsense mutants, this would suggest that the messenger is also subjected to degradation if it is not completely translated. However, the fragments could be produced by frameshift or deletion mutations and the structure of each fragment therefore needs

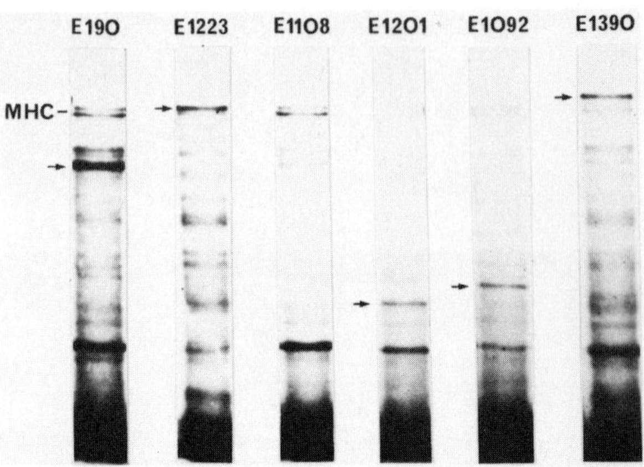

Fig. 5 Identification of myosin heavy chain fragments synthesized *in vitro* by a wheat germ cell free system programmed with RNA isolated from various recessive alleles of the *unc*-54 gene. The cell-free translation products were immunoprecipitated with antibody prepared against native nematode myosin and fractionated on sodium dodecyl sulphate 5.0% polyacrylamide gels. The arrows indicate the position of novel cross reacting peptides synthesized in response to RNA isolated from the different nematode strains. No fragments of the *unc*-54 myosin heavy chain were detected in E1108 animals. E1390 is a revertant in which full length *unc*-54 heavy chain is synthesized.

to be defined. The best way to do this is by *in vitro* suppression (Gesteland *et al.*, 1976; Capecchi *et al.*, 1975) which would also allow us to identify the nonsense codon present by using the appropriate suppressor tRNA. This work is in progress.

There are also other explanations for the effects of the null mutations. The primary effect may be defective messenger production and many mechanisms may be imagined for this, particularly if the gene contains intervening or other sequences which require processing. We are, therefore, attempting to clone the *unc*-54 gene, using a variety of approaches. One approach is to purify the myosin heavy chain mRNA and use this either as a probe or as a source of DNA for clon-

Fig. 6 (see opposite) Fractionation of nematode messenger RNAs by zonal centrifugation. Forty milligrams of total nematode RNA was applied to a 600 ml 10-40% isokinetic sucrose gradient generated in a MSE B–XIV zonal rotor. After centrifugation for 16 hours at 40,000 rpm, 10 ml fractions were collected. The absorbance at 260 nm was recorded and aliquots of every second fraction of the heavy end of the gradient were assayed in the wheat germ system. The translation products were analysed by electrophoresis in a sodium dodecyl sulphate, 7.5% acrylamide gel. The

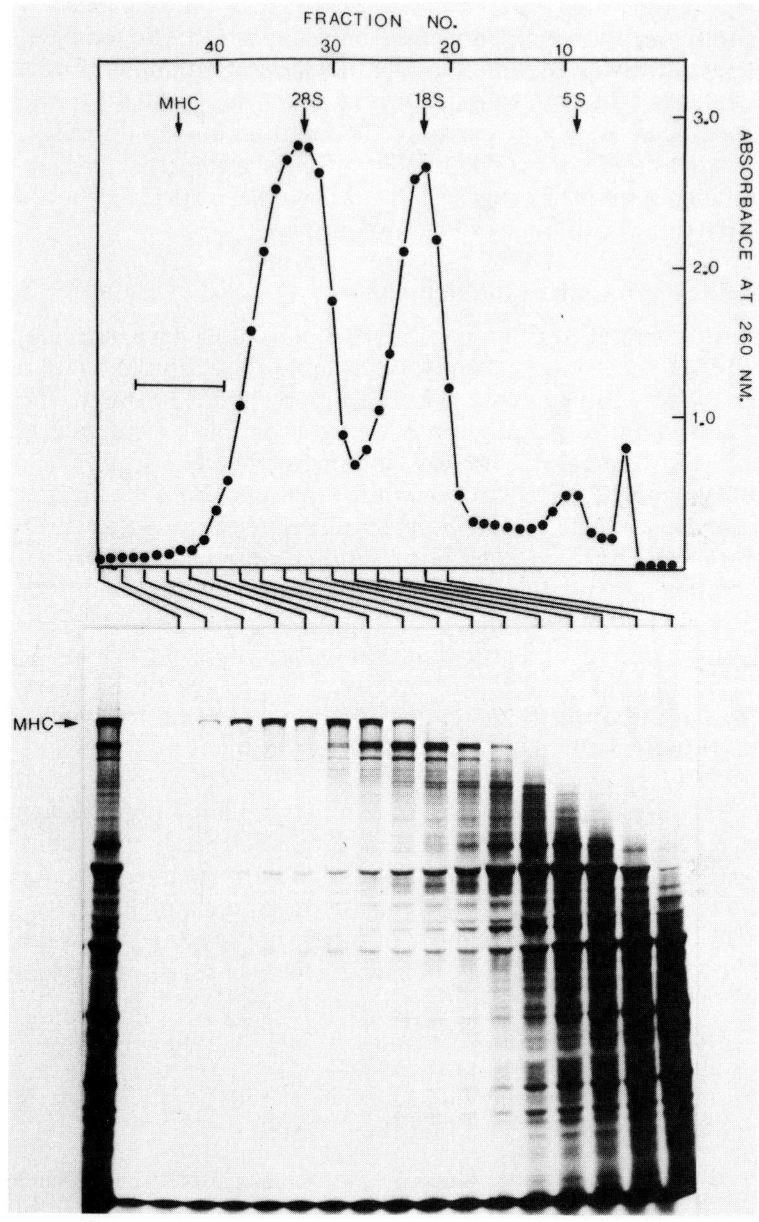

column on the far left shows the cell free translation products synthesized in response to unfractionated nematode RNA. The mobility of the myosin heavy chains (MHC) is indicated by the arrow. The bar in the top panel indicates the region of the sucrose gradient where the myosin messenger RNA sediments.

ing. Using the *in vitro* protein synthesis assay, we have partially purified the messenger by zonal sedimentation velocity centrifugation (Figure 6 The fraction is enriched in *unc*-54 mRNA but contaminated with lower molecular weight aggregates. Repeated centrifugation fails to increase the specific activity, due partly to the fact that *unc*-54 degrades at the same rate as the more stable smaller mRNAs aggregate. Our inability to obtain a pure probe suggests that we will need special procedures to identify clones containing *unc*-54 sequences.

Genetic Suppression in the Nematode

Recently a subset of genetic suppressor mutations have been identified which have the capacity to restore protein production in a null mutant of the paramyosin gene, E1214. The suppression is not very strong; and the amount of paramyosin obtained is only 5-10% of wild type levels. This suppressor, *e*1464, is dominant and allele specific and by phenotypic testing has been shown to suppress a subset of mutants in a number of different genes, many unrelated to muscle. This is the property expected of an informational suppressor. In particular, two mutants of the *unc*-54 gene, *e*1108 and *e*576, are phenotypically suppressed, and in the case of *e*1108 it has been shown that there is restoration of *unc*-54 heavy chain synthesis, but again at a low level about 5% of wild type. Unfortunately, when messenger RNA from *e*1108 was tested for fragment production *in vitro* no fragment was found (Figure 5). In addition, two of the fragment producers, *e*190 and *e*1201, are not suppressed by *e*1464. The experiments are still incomplete, and it is not useful to consider explanations too deeply now. We think it would be best first to try to identify the changes responsible for the null phenotypes and *in vitro* fragment production of the *unc*-54 mutants, and then to use these to characterize not only *e*1464 but also the large number of other suppressor genes in the nematode. Genetic suppressors, of nonsense codons proved enormously valuable in molecular genetic analysis in prokaryotes and are likely to be equally valuable in eukaryotes. However, the processes of information transfer from gene to proteins may have added complexity in eukaryotes, and there may be new types of genetic suppressors and if these exist we should try to find them as well.

Acknowledgement

A.R. MacLeod was supported by post-doctoral fellowships from the Medical Research Council of Canada and from the Muscular Dystrophy Association of America. J. Karn was a recipient of a Helen Hay Whitney Foundation post-doctoral fellowship. During the course of this work R.H. Waterson was supported by a post-doctoral fellowship from the Muscular Dystrophy Association of America. We wish to thank Mrs. R.M. Fishpool for her expert technical assistance.

References

Brenner, S. (1974). *Genetics* **77**, 71.

Capecchi, M.R., Hughes, S.M., and Wahl, G.M. (1975). *cell,* **6**, 269.

Degani, Y. and Patchornik, A. (1974). *Biochemistry* **13**, 1.

Epstein, H.F., Waterston, R.H. and Brenner, S. (1974). *J. Mol. Biol.* **90**, 291.

Gesteland, R.F., Wolfner, M., Grisafi, P., Fink, G., Botstein, D. and Roth, J.R. (1976). *Cell* **7**, 381.

Jacobson, G.R., Schafter, M.H., Stark, G.R. and Vanaman, J.C. (1973). *J. Biol. Chem.* **298,** 6583.

MacLeod, A.R., Waterston, R.H., Fishpool, R.M. and Brenner, S. (1977a). *J. Mol. Biol.* **114**,133.

MacLeod, A.R., Waterston, R.H. and Brenner, S. (1977b). *Proc. Nat. Acad. Sci.* **74**, 5336.

O'Farrell, P.H. (1975). *J. Biol. Chem.* **250**, 4007.

Roberts, B.E. and Paterson, B.M. (1973). *Proc. Nat. Acad. Sci.* **70**, 2330.

Sulston, H. and Horvitz, H.R. (1977). *Devel. Biol.* **56**, 110.

Waterston, R.H., Epstein, H.F. and Brenner, S. (1974). *J. Mol. Biol.* **90**, 285.

Waterston, R.H., Fishpool, R.M. and Brenner, S. (1977). *J. Mol. Biol.* **117**, 679.

CAN SUPPRESSOR tRNA CONTROL TRANSLATION IN MAMMALIAN CELLS?

L. PHILIPSON

Department of Microbiology, University of Uppsala, Sweden.

Introduction

Studies on animal virus genomes have indicated that polycistronic messenger RNA (mRNA) is not operating in eukaryotic cells. Translation initiated usually only at the proximal initiator codon close to the 5′-terminus of a mRNA irrespective of its size (Jacobson and Baltimore, 1968). Additional mRNA species are usually synthesized in different amounts to allow for independent expression of viral genes according to the relative need for the gene products. When equimolar amounts of the products are required, a post-translational proteolytic cleavage mechanism is available which generated the proteins from a polyprotein (Jacobson and Baltimore, 1968). Similar strategies have been described for both plant (Shih and Kaesberg, 1973; Knowland *et al.*, 1975; van Vloten-Doting *et al.*, 1975) and animal viruses (Cancedda *et al.*, 1975; Glanville *et al.*, 1976; Smith *et al.*, 1976).

The synthesis of functional RNA-tumour virus (retrovirus) proteins involves both independent generation of mRNAs and post-translational cleavage mechanisms (Figure 1). There are four known genetic regions present in the genomes of non-defective retroviruses located in the 5′ -3′ direction: *gag*, the region encoding the internal structural proteins of the virion; *pol*, the region encoding the reverse transcriptase; *env*, the region encoding the glycoprotein of the outer surface of the virion envelope; and *src*, the region probably responsible for neoplastic transformation (Baltimore, 1974).

There appears to be at least two mRNAs used for translation for both murine and avian retrovirus proteins (Weiss *et al.*, 1977; van Zaane *et al.*, 1977). One is indistinguishable from the virion RNA and it obviously codes for the *gag* and the *pol* products. The other is smaller and it appears to express the *env*-region. In one avian system the *src* region appears also to be expressed from a separate mRNA (Weiss *et al.*, 1977). Each of the four genetic regions encodes pre-

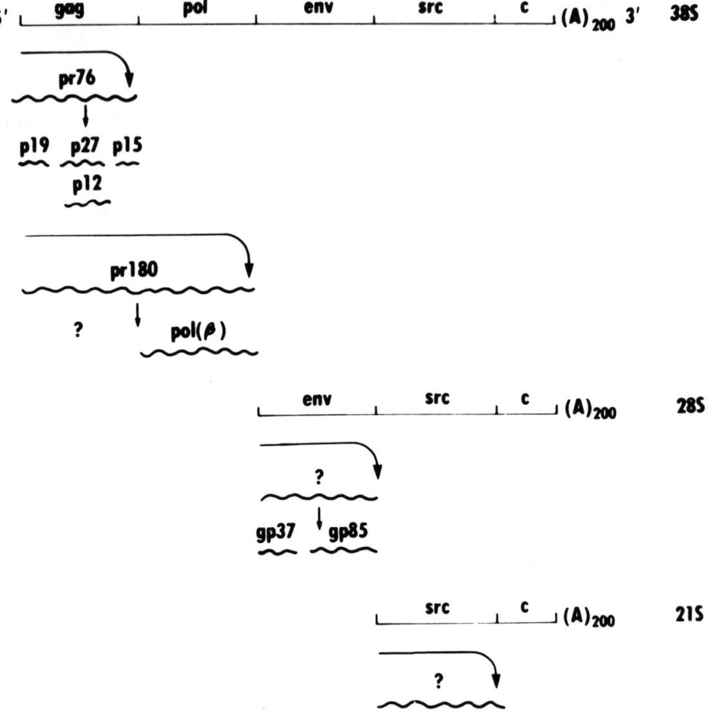

Fig. 1 A scheme for the expression of avian sarcoma virus genes (Weiss *et al.*, 1977). The figure illustrates the identified cytoplasmic RNAs for avian sarcoma viruses (ASV), their genetic composition and their putative protein products, assuming that these RNAs serve as mRNAs. The core proteins of the virus (p19, p15, and p12) are generated by specific proteolytic cleavages of pr76, the primary product of translation from *gag*. The primary product of *pol* is the β subunit of reverse transcriptase, generated by processing of pr180. The glycoproteins of the viral envelope (gp37 and gp85) are probably derived from a single precursor encoded in *env*. The primary translation product of *src* is not yet established.

cursors to the functional proteins and these precursors are proteolytically cleaved to generate the mature products.

The most interesting control mechanism in retrovirus translation concerns the expression of the *gag* and the *pol* genes. It appears that no discrete mRNA is available for the synthesis of reverse transcriptase, rather the 35S mRNA appears to encode both, the *gag* protein ($Pr78^{gag}$ - the precursor of the structural proteins of the virus with a major partial product, $Pr65^{gag}$) and the polypeptide ($Pr180^{gag\text{-}pol}$)

containing antigenic determinants for both the *gag* proteins and the reverse transcriptase (*pol*). The latter is obviously the precursor protein for the enzyme (Naso *et al.*, 1975; Mueller-Lantzsch and Fan, 1976; Kerr *et al.*, 1976; Opperman *et al.*, 1977). The molarity of reverse transcriptase made in infected cells is 0.01–0.1 of the molarity of virion structural proteins suggesting a translational control at expression of these two polypeptides (Panet *et al.*, 1975; Jamjoom *et al.*, 1977). These facts raise the question: what controls whether a ribosome having initiated translation of 35S mRNA will generate the *gag* precursor or - by read-through of the end of the *gag* coding region - will generate the fused *gag-pol* polypeptide (Pr180)? To account for the difference in molarity between the two products there probably is a control involved at the read-through sequence. One possibility is that the codon specifying termination of protein synthesis follows the coding region for the *gag* precursor but that a mechanism is available that allows for occasional read-through of this termination codon ('suppression'). A similar model has been shown to operate for the synthesis of the read-through protein of Qβ phage (Weber and Konigsberg, 1975). We have analyzed this possibility in a cell-free translation system programmed by virion RNA of murine leukemia virus (MuLV) (Philipson *et al.*, 1978).

Effect of Suppressor tRNAs on Translation of MuLV-RNA

One well established mechanism for controlling translational read-through utilizes the ability of suppressor tRNAs to read termination codons as an amino acid (Gorini, 1970; Smith, this volume). The only eukaryotic suppressor tRNAs characterized today are those from yeast (Capecchi *et al.*, 1975; Gesteland *et al.*, 1976; Piper *et al.*, 1976; Piper, this volume). We therefore tested the ability of yeast suppressor tRNAs to increase the yield of the read-through product $Pr180^{gag\text{-}pol}$ as an indication of the existence of a suppressible termination codon between *gag* and *pol*. When a cell-free system was programmed by MuLV-RNA the usual faint band of the read-through product was observed (Figure 2). After supplementation with a purified yeast amber suppressor tRNA, the Pr180 showed a marked increase. A concomitant loss of radioactivity in the *gag* precursor polypeptide band ($Pr78^{gag}$) was also produced by the amber-tRNA. Yeast ochre suppressor tRNA had a less dramatic effect on the generation of Pr180 but depressed the intensity of the *gag* precursor protein. Immunoprecipitation studies of the suppressed products revealed that the Pr180 was in fact a read-through product containing both *gag* and *pol* antigenic determinants. Translation in the nuclease-treated reticulocyte system of Qβ-RNA with an amber mutation in the synthetase gene gave no detectable complete synthetase. Addition of yeast ochre suppressor tRNA stimulated some synthesis of the

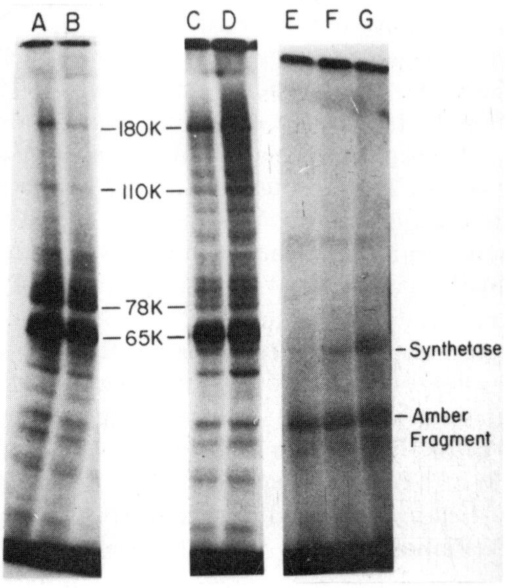

Fig. 2 Yeast suppressor tRNA causes read-through of the *gag-pol* genes of the MuLV genome. Translation mixtures with MuLV RNA were fortified with purified yeast suppressor tRNA at 40 μg/ml. (A) No addition; (B) yeast non-suppressor tRNA; (C) yeast ochre suppressor tRNA; (D) yeast amber suppressor tRNA. The suppressor activity of the yeast tRNA was controlled by concurrent translation of a Qβ (Qβ aml) mutant in the same cell-free system. (E) No addition; (F) ochre suppressor tRNA; (G) amber suppressor tRNA.

complete product while the amber suppressor tRNA produced a three-fold higher yield of this protein (Gesteland *et al.*, 1976). The MuLV *gag* termination signal therefore behaved identically to an established UAG codon in its response to the yeast suppressor tRNAs. Gesteland *et al.*, (1977) had previously shown that a mutant codon in an adeno-SV40 hybrid virus behaved in this fashion and concluded that it was a UAG. We therefore believe it very probable that the *gag-pol* junction of MuLV RNA contains a UAG-codon. Preliminary results with the avian Rous sarcoma virus (RSV) RNA indicate however that although amber suppressor tRNA allows the synthesis *in vitro* of a slightly larger polypeptide than the precursor $Pr76^{gag}$ gene product, it does not increase the yield of the Pr180 which also is the putative *gag-pol* precursor in this system (Pelham, 1978; M. Bishop, personal communication).

A read-through system of functional significance has on the other hand recently been described (Pelham, 1978). When Tobacco Mosaic Virus RNA is translated in a cell-free reticulocyte system, two large polypeptides of molecular weights 110,000 (110K) and 160,000 (160K) are generated (Pelham, 1978), where the smaller one is more abundant. Similar products are found in infected plants (Bruening *et al.*, 1976; Sakai and Takebe, 1972) and protoplasts (Paterson and Knight, 1975) and they may be components of the viral replicase (Zaitlin *et al.*, 1973; Hunter *et al.*, 1976). The synthesis of these two proteins appears to be initiated at the same site and the larger product may be generated by partial read-through of an amber termination codon at the end of the 110K polypeptide since the presence of yeast amber suppressor tRNA increased the synthesis of the 160K protein at the expense of the 110K protein (Pelham, 1978). A similar but less conspicuous read-through may also be achieved by increase of the magnesium concentration in the *in vitro* translation system (Pelham, 1978). Taken together these results suggest that at least in two systems from plant and animal cells a suppression has been demonstrated *in vitro* which probably also operates *in vivo* to create a functional product. These results may therefore suggest that suppressor tRNA could have an important function in mammalian cells to control the expression of two genes in juxtaposition on the same mRNA.

Suppression or 'Two out of Three' Codon Reading

An alternative to faithful suppression with suppressor tRNA which controls a functional read-through in mammalian cells has recently been suggested (Lagerkvist, 1978). In some tRNA classes only the first two nucleotides in the anticodon are recognized by the codon of MS2-RNA during protein translation *in vitro* (Lagerkvist, 1978); a method referred to as 'two out of three' reading. If the third codon position has an absolutely discriminatory influence *in vivo* or *in vitro* one would not expect amber (UAG) suppressors to be able to suppress ochre (UAA) mutations because this would involve the recognition of A in the third codon position by C in the anticodon. Nevertheless, there is evidence both in the results reported here and in other systems (Strigini and Brickman, 1973; Feinstein and Altman, 1977) that this suppression takes place although the efficiency is much lower than that of amber suppression. The 'two out of three' reading may not involve a threat to translational fidelity *in vivo* if it only occurs when there is no competition with other tRNAs for the codon sequence. However, for nonsense suppressors this would not be the case because there would be no competing tRNA. The eukaryotic cells may therefore possibly have access to tyrosine tRNAs which could recognize amber or ochre codons when functional read-through is required. The primary or secondary structures of the mRNA and also the tRNA would then be of importance for execution of the read-through process. A tRNA which fulfill these requirements and which can recognize an opal (UGA) codon has been described from a bacterial system (Hirsch, 1971; Hirsch and Gold, 1971). It is a tryptophan nonsense suppressor tRNA which recognizes the terminator codon UGA without any corresponding change in the anticodon sequence thus certainly read by the 'two out of three' method. The mutation lies outside the anticodon region with a change of G25 into A25. Although the 'two out of three' method for reading codons may pose a significant threat to the fidelity of protein synthesis if it occurs indiscriminatory *in vivo*, it might as suggested here also provide a control mechanism for read-through of genes in juxtaposition without the involvement of suppressor tRNA. We are at present analyzing purified tyrosine tRNA from eukaryotic cells to establish whether the 'two out of three' reading method has general significance, or whether suppressor-like tRNA can exist in normal cells. A simple assay method is to use *in vitro* translation of murine leukemia virus RNA.

References

Baltimore, D. (1974). *Cold Spring Harbor Symp. Quant. Biol.* **39**, 1187.

Bruening, G., Beachy, R.N., Scalla, R. and Zaitlin, M. (1976). *Virology* **71**, 498.

Cancedda, R., Villa-Komoroff, L., Lodish, H.F. and Schlesinger, M. (1975). *Cell* **6**, 215.

Capecchi, M.R., Hughes, S. H. and Wahl, G.M. (1975). *Cell* **6**, 269.

Feinstein, S.I. and Altman, S. (1977). *J. Miol. Biol.* **112**, 453.

Gesteland, R.F., Wolfner, M., Grisafi, P., Fink, G., Botstein, D. and Roth, J.R. (1976). *Cell* **7**, 381.

Gesteland, R.F., Wills, N., Lewis, J.B. and Grodzicker, T. (1977). *Proc. Natl. Acad. Sci.* **74**, 4567.

Glanville, N., Ranki, M., Morser, J., Kääriänen, L. and Smith, A. (1976). *Proc. Natl. Acad. Sci.* **73**, 3059.

Gorini, L. (1970). *Ann. Rev. Gen.* **4**, 107.

Hirsch, D. (1971). *J. Mol. Biol.* **58**, 439.

Hirsch, D. and Gold, L. (1971) . *J. Mol. Biol.* **58**, 459.

Hunter, A.R., Hunt, T., Knowland, J. and Zimmern, D. (1976). *Nature (London)* **260**, 759

Jacobson, M.F. and Baltimore, D. (1968). *Proc. Natl. Acad. Sci.* **61**, 77.

Jamjoom, G.A., Naso, R.B. and Arlinghaus, R.B. (1977). *Virology* **78**, 11.

Kerr, I.M., Olshevsky, U., Lodish, H. and Baltimore, D. (1976). *J. Virol.* **18**, 627

Knowland, J., Hunter, T., Hunt, T. and Zimmern, D. (1975). *INSERM* **47**, 211.

Lagerkvist, U. *Proc. Natl. Acad. Sci.* **75**, 1759.
Mueller-Lantzsch, N. and Fan, H. (1976). *Cell* **9**, 579.
Naso, R.B., Arcement, L.J., Wood, T.G., and Arlinghaus, R. (1975). *Biochim. Biophys. Acta* **383**, 195.
Opperman, H., Bishop, J.M., Varmus, H.E. and Levintow, L. (1977). *Cell* **12**, 993.
Panet, A., Baltimore, D. and Hanafusa, H. (1975). *J. Virol.* **16**, 146.
Paterson, R. and Knight, C.A. (1975). *Virology* **64**, 10.
Pelham, H.R.B. (1978). *Nature (London)* **272**, 469.
Philipson, L., Andersson, P., Olshevsky, U., Weinberg, R. and Baltimore, D. (1978). *Cell* **13**, 189.
Piper, P.W., Wasserstein, M., Engbaek, F., Kaltoft, K., Celis, J.E., Zeuthen, Liebman, S.W. and Sherman, F. (1976). *Nature (London)* **262**, 757.
Sakai, F. and Takebe, I. (1972). *Molec. gen. Genet.* **118**, 93.
Shih, D.S. and Kaesberg, P. (1973). *Proc. Natl. Acad. Sci.* **70**, 1799.
Smith, A.E., Kamen, R., Mangel, W.F., Shure, H. and Wheeler, T. (1976). *Cell* **9**, 481.
Strigini, P. and Brickman, E. (1973). *J. Miol. Biol.* **75**, 659
van Vloten-Doting, L., Rutgers, T., Neeleman, L. and Bosch, C. (1975). *INSERM* **47**, 233.
Van Zaane, E., Gielkens, A.L.J., Hesselink, W.G. and Bloemers, H.P.J. (1977). *Proc. Natl. Acad. Sci.* **74**, 1855.
Weber, K. and Konigsberg, W. (1975). Cold Spring Harbor, New York: Cold Spring Harbor Laboratory Press, pp. 51.
Weiss, S.R., Varmus, H.E. and Bishop, J.M. (1977). *Cell* **12**, 983.
Zaitlin, M., Duda, C.T. and Petti, M.A. (1973). *Virology* **53**, 300.

COMPILATION OF tRNA SEQUENCES*

M. SPRINZL, F. GRÜTER AND D.H. GAUSS

Max-Planck-Institut für Experimentelle Medizin,
Göttingen, W. Germany.

The tRNA sequences are given in a linear form using the numbering of $tRNA^{Phe}$ from yeast in order to underline the relationship to the three dimensional structure of $tRNA^{Phe}$, the only structure known from X-ray analysis. Since there is no information about the structure of tRNA species with a large extra arm the numbering in this part of the tRNA sequences is arbitrary. The secondary structure of tRNAs is indicated by specific underlining of the corresponding base paired sequences. Regarding the cloverleaf type of presentation the reader is referred to either the original literature cited or to other tRNA sequence compilations (1-5).

In this compilation, unknown nucleosides are always given as N; odd nucleosides with lengthy names the common abbreviation of which requires more than three letters, are given with the prefix x and specified in detail in the corresponding footnote. A nucleoside followed by a nucleoside in brackets, or an index indicating a modification in brackets, denotes the fact that both types of nucleosides can occupy this position in varying amounts. Part of a sequence in brackets designates a piece of sequence not unambigously analyzed. The abbreviations of the modified nucleosides are given according to the IUPAC-IUB rules.

---

*Reprinted from Nucleic Acids Research with permission from IRL. This collection of sequences appears in *Nucleic Acids Res.* 5, r15 (1978). An up-to-date collection of tRNA sequences will be published every year by Nucleic Acids Research.

References

Barciszewski, J., Rafalski, A.J. (1978). Atlas of Transfer Ribonucleic Acids and Modified Nucleosides, Poznan, (in press).

Barrell, B.G., Clark, B.F.C. (1974). Handbook of Nucleic Acid Sequences, Joynson-Bruvvers Ltd. Oxford.

Dirheimer, G., Ebel, J.P., Bonnet, J., Gangloff, J., Keith, G., Krebs, B., Kuntzel, B., Roy, A., Weissenbach, J., Werner, C. (1972). *Biochimie* **54**, 127.

Sodd, M.A., Doctor, B.P. (1974). *Methods Enzymol.* **29**, 741.

Sodd, M.A. (1976). *In:* CRC Handbook of Biochemistry and Molecular Biology, 3rd Edition, (ed. G.D. Farman), Nucleic Acids Vol. II, 423. The Chemical Rubber Company, Cleveland.

			Aminoacyl Stem									D Stem				D Loop														D Stem					Anticodon Stem					Anticodon Loop							Anticodon Stem				
			1	2	3	4	5	6	7	8	9	10	11	12	13	14	15	16	17	a	b	18	19	a	b	c	d	20	21	22	23	24	25	26	27	28	29	30	31	32	33	34	35	36	37	38	39	40	41	42	43
ALANINE																																																			
001	1A	E.coli	G	G	G	G	G	C	A	s^4U	A	G	C	U	C	A	G	C	D			G	G					G	A	G	A	G	C	G	C	C	U	G	C	U	U	o^5U[a]	G	C	A	C	G	C	A	G	G
002	1	T.utilis	G	G	G	C	G	U	G	U	m^1G	G	C	G	U	A	G	D	D			G	G					D	A	G	C	G	C	m^2_2G	Ψ	U	C	G	C	U	U	I	G	C	m^1I	Ψ	G	C	G	A	A
003[b]	1	Yeast	G	G	G	C	G	U	G	U	m^1G	G	C	G	U	A	G	D	C			G	G					D	A	G	C	G	C	m^2_2G	C	U	C	C	C	U	U	I	G	C	m^1I	Ψ	G	G	G	A	G
004	1/2	Bombyx mori	G	G	G	G	G	C	G	U	A	m^2G	C	U	C	A	G	A	D			G	G					U	A	G	A	G	C	m^2_2G	C	U	C	G	C	Um	U	I	G	C	m^1I	Ψ	Gm	C[c]	G	A	G
ARGININE																																																			
011	1	E.coli	G	C	A	U	C	C	G	s^4U	A	G	C	U	C	A	G	C	D			G	G					(U)D	A	G	A	G	U	A	C	U	C	G	G	s^2C	U	I	C	G	m^2A	A	C	C	G	A	G
012	2	E.coli B	G	C	A	U	C	C	G	s^4U	A	G	C	U	C	A	G	C	D			G	G	A				(U)D	A	G	A	G	U	A	C	U	C	G	G	C	U	I	C	G	m^2A	A	C	C	G	A	G
013	2	Yeast	Ψ	U	C	C	U	C	G	U	m^1G	m^2G	C	C	C	A	A	D				G	G	D				C	A	C	G	G	C	m^2_2G	Ψ	C	U	G	G	C	U	I	C	G	A	A	C	C	A	G	A
014	3	Yeast	G	C	G	C	U	C[e]	G	U	m^1G	m^2G	C	G	U	A	A	D				G	G					C	A	A	C	G	C	m^2_2G	Ψ	C	U	G	A	C	U	xU[f]	C	U	t^6A	A	Ψ	C	A	G	A
ASPARAGINE																																																			
021		E.coli	U	C	C	U	C	U	G	s^4U	A	G	U	U	C	A	G	D	C			G	G					D	A	G	A	A	C	G	G	C	G	G	A	C	U	Q	U	U	t^6A	A	Ψ	C	C	G	U
ASPARTIC ACID																																																			
031	1	E.coli	G	G	A	G	C	G	G	s^4U	A	G	U	U	C	A	G	D	C			G	G	D				D	A	G	A	A	U	A	C	C	U	G	C	C	U	Q	U	C	m^2A	C	G	C	A	G	G
032		Yeast	U	C	C	G	U	G	A	U	A	G	U	U	Ψ	A	A	D				G	G	D				C	A	G	A	A	U	G	G	G	C	G	C	Ψ	U	G	U	C	m^1G	C	G	U	G	C	C
CYSTEINE																																																			
041		Yeast	G	C	U	C	G	U	A	U	G	G	C	G	C	A	G	D				G	G					D	A	G	C	G	C	A	G	C	A	G	A	Ψ	U	G	C	A	i^6A	A	Ψ	C	U	G	U
GLUTAMINE																																																			
051	1	E.coli K12	U	G	G	G	G	U	A	s^4U	C	G	C	C	A	A	G	C				Gm	G					D	A	A	G	G	C	A	C	C	G	G	U	Um	U	N[g]	U	G	m^2A[l]	Ψ	A	C	C	G	G
052	2	E.coli K12	U	G	G	G	G	U	A	s^4U	C	G	C	C	A	A	G	C				Gm	G					D	A	A	G	G	C	A	C	C	G	G	A	Um	U	C	U	G	m^2A	Ψ	Ψ	C	C	G	G
053		Phage T4	U	G	G	G	A	A	U	s^4U	A	G	C[h]	C	A	A	G	D	D			G	G					D	A	A	G	G	C	A	U	A	G	C	A	C	U	N[k]	U	G[l]	m^2A	C	Ψ	G[n]	C	U	A
054		Phage T4[m]	U	G	G	G	A	A	U	s^4U	A	G	C	C	C	A	A	G	D	D		G	G	D				A	A	G	G	G	C	A	U	A	G	C	A	C	U	N[k]	U	G	m^2A	C	Ψ	G	C	U	A

001 R.J.Williams,W.Nagel,B.Roe,B.Dudock (1974) Biochem.Biophys. Res.Commun. 60, 1215-1221.
002 S.Takemura,K.Ogawa (1973) J.Biochem. 74, 322-333.
003 J.R.Penswick,R.Martin,G.Dirheimer (1975) FEBS-Lett. 50, 28-31.
004 K.U.Sprague,O.Hagenbüchle,M.C.Zuniga (1977) Cell 11, 561-570.
011 K.Murao,T.Tanabe,F.Ishii,M.Namiki,S.Nishimura (1972) Biochem.Biophys.Res.Commun. 47, 1332-1337.
012 K.Chakraburtty (1975) Nucleic Acids Res. 2, 1787-1792.
013 J.Weissenbach,R.Martin,G.Dirheimer (1975) Eur.J.Biochem. 56, 527-532.
014 B.Kuntzel,J.Weissenbach,G.Dirheimer (1974) Biochimie 56, 1069-1087.

021 K.Ohashi,F.Harada,Z.Ohashi,S.Nishimura,T.S.Stewart,G.Vögeli, T.McCutchan,D.Söll (1976) Nucleic Acids Res. 3, 3369-3376.
031 F.Harada,K.Yamaizumi,S.Nishimura (1972) Biochem.Biophys.Res.Commun. 49, 1605-1609.
032 J.Gangloff,G.Keith,J.P.Ebel,G.Dirheimer (1972) Biochim.Biophys.Acta 259, 210-222.
041 N.J.Holness,G.Atfield (1976) Biochem. J. 153, 447-454.
051 +052 M.Yaniv,W.R.Folk (1975) J.Biol.Chem. 250, 3243-3253.
053 J.G.Seidman ,M.M.Comer,W.H.McClain (1974) J.Mol.Biol. 90, 677-689.
054 C.Guthrie (1975) J.Mol.Biol. 95, 529-548.

	Extra Arm																				TΨ Stem					TΨ Loop							TΨ Stem					Aminoacyl Stem										
	44	45	a	b	c	d	e	f	g	46	47	a	b	c	d	e	f	g	h	i	j	48	49	50	51	52	53	54	55	56	57	58	59	60	61	62	63	64	65	66	67	68	69	70	71	72	73	74	75	76
001	A	G								m^7G	U											C	U	G	C	G	G	T	Ψ	C	G	A	U	C	C	C	G	C	G	C	G	C	U	C	C	C	A	C	C	A
002	A	G								G	D											C	U	C	C	G	G	T	Ψ	C	G	m^1A	C	U	C	C	G	G	A	C	U	C	G	U	C	C	A	C	C	A
003	A	G								G	D(U)											C	U	C	C	G	G	T	Ψ	C	G	A	U	U	C	C	G	G	A	C	U	C	G	U	C	C	A	C	C	A
004	A	G								m^7G	U											A	m^5C	C	G	G	G	A	Ψ	C	G	m^1A	U	A	C	C	C	G	G	C	G	C	C	U	C	C	A	C	C	A
011	C	G								m^7G	X[d)]											C	G	G	A	G	G	T	Ψ	C	G	A	A	U	C	C	U	C	C	C	G	G	A	U	G	C	A	C	C	A
012	C	G								m^7G	X[d)]											C	G	G	A	G	G	T	Ψ	C	G	A	A	U	C	C	U	C	C	C	G	G	A	U	G	C	A	C	C	A
013	A	G								A	D											U	m^5C	C	A	G	G	T	Ψ	C	A	m^1A	G	U	C	C	U	G	G	C	G	G	G	G	A	A	G	C	C	A
014	A	G								A	D											U	A	U	G	G	G	T	Ψ	C	G	m^1A	C	C	C	C	C	A	U	C	G	A	G	U	G	C	G[e)]	C	C	A
021	A	U								m^7G	U											C	A	C	U	G	G	T	Ψ	C	G	A	G	U	C	C	A	G	U	C	A	G	A	G	G	A	G	C	C	A
031	G	G								m^7G	U											C	G	C	G	G	G	T	Ψ	C	G	A	G	U	C	C	C	G	Ψ	C	C	G	U	U	C	C	G	C	C	A
032	A	G								A												U	m^5C	G	G	G	G	T	Ψ	C	A	A	U	U	C	C	C	C	G	U	C	G	G	G	G	A	G	C	C	A
041	U	G								m^7G	D											m^5C	C	U	U	A	G	T	Ψ	C	G	m^1A	U	C	C	U	G	A	G	U	G	C	G	A	G	C	U	C	C	A
051	C	A								U	U											C	C	C	U	G	G	T	Ψ	C	G	A	A	U	C	C	A	G	G	U	A	C	C	C	C	A	G	C	C	A
052	C	A								U	U											C	C	G	A	G	G	T	Ψ	C	G	A	A	U	C	C	U	C	G	U	A	C	C	C	C	A	G	C	C	A
053	G	A								U	G											C	A	A	A	G	G	T	Ψ	C	G	A	G	U	C	C[h)]	U	U	U	A	U	U	C	C	C	A	G	C	C	A
054	G	A								U	G											C	A	A	A	G	G	T	Ψ	C	G	A	G	U	C	C	U	U	U	A	U	U	C	C	C	A	G	C	C	A

a) o^5U is uridine-5-oxyacetic acid.

b) Compare R.W. Holley et al. (1965) Science 147, 1462-1465.

c) Isoacceptor 2 has C_{40}, isoacceptor 1 has U_{40} or Ψ_{40}.

d) X is 3N-(3-amino-carboxypropyl)uridine, Z. Ohashi, M. Maeda, J.A. McCloskey, S. Nishimura (1974) Biochem. 13, 2620-2625.

e) Second $tRNA^{Arg}$ has U_6 and U_{73}.

f) Identified as 5-methoxycarbonylmethyluridine, mcm^5U, B. Kuntzel, J. Weissenbach, R.E. Wolff, T.D. Tumaitis-Kennedy, B.G. Lane, G. Dirheimer (1975)

g) N is likely a derivative of 2-thiouridine.

h) Mutations: $C_{11} \rightarrow U_{11}$, $G_{40} \rightarrow A_{40}$, $C_{62} \rightarrow U_{62}$.

i) Double mutation: $G_{36} \rightarrow A_{36}$, $N_{34} \rightarrow C_{34}$, M.M. Comer, K. Foss, W.H. McClain (1975) J.Mol.Biol. 99, 283-293.

k) N is an unkown derivative of uridine.

l) Mutation: $G_{36} \rightarrow A_{36}$.

m) From precursor.

			Aminoacyl Stem									D Stem				D Loop														D Stem					Anticodon Stem					Anticodon Loop							Anticodon Stem				
			1	2	3	4	5	6	7	8	9	10	11	12	13	14	15	16	17	a	b	18	19	a	b	c	d	20	21	22	23	24	25	26	27	28	29	30	31	32	33	34	35	36	37	38	39	40	41	42	43
	GLUTAMIC ACID																																																		
061	1	E.coli B	G	U	C	C	C	C	U	U	C	G	U	C	Ψ	A	G	A				G	G	C	C			C	A	G	G	A	C	A	C	C	G	C	C	C	U	xU[a]	U	C	m^2A	C	G	G	C	G	G
062	2	E.coli	G	U	C	C	C	C	U	U	C	G	U	C	Ψ	A	G	A				G	G	C	C			C	A	G	G	A	C	A	C	C	G	C	C	C	U	xU[b]	U	C	m^2A	C	G	G	C	G	G
063	3	**Yeast**	U	C	C	G	A	U	A	U	A	G	U	G	Ψ	A	A	C				G	G	C				D	A	U	C	A	C	A	Ψ	C	A	C	G	C	U	xU	U	C	A	C	C	G	U	G	G
	GLYCINE																																																		
071	1	E.coli[c]	G	C	G	G	G	C	G	s^4U[c]	A	G	U	U	C	A	A	U				G[c]	G					D	A	G	A	A	C	G	A	G	A	G	C	U	U	C	C[d]	C	A	A	G	C	U	C	U
072	2	E.coli	G	C	G	G	G	C	A	U	C	G	U	A	U	A	A	U				G	G	C				U	A	U	U	A	C	C	U	C	A	G	C	C	U	U(N)[e]	C	C[f]	A[f]	A	G	C	U	G	A
073	3	E.coli	G	C	G	G	G	A	A	U	A	G	C	U	C	A	G	D	D			G	G					D	A	G	A	G	C	A	C	G	A	C	C	U	U	G[g]	C	C[gh]	A[h]	A	G	G	U	C	G
074	1A	S.epidermidis[i]	G	C	G	G	G	A	G	s^4U	A	U	U	U	C	A	A	C	U			U	U					D	A	G	A	A	U	A	C	G	U	U	C	C	U	U	C	C	C	G	G	A	A	C	G
075	1B	S.epidermidis[i]	G	C	G	G	G	A	G	s^4U	A	G	U	U	C	A	A	U				U	U					D	A	G	A	A	C	A	C	A	U	U	C	C	U	U	C	C	C	G	G	A	A	U	G
076		Phage T4	G	C	G	G	A	U	A	U	C	G	U	A	U	A	A	U				Gm	G					D	A	U	U	A	C	C	U	C	A	G	A	C	U	xU[k]	C	C	A	A	Ψ	C	U	G	A
077		Yeast	G	C	G	Cm	A	A	G	U	m^1G	G	U	U	Ψ	A	G	D				G	G					D	A	A	A	A	U	C	C	A	A	C	G	Ψ	U	G	C	C	A	Ψ	C	G	U	U	G
078	1	Wheat germ	G	C	A	Cm	C	A	G	U	m^1G	G	U	C	Ψ	A	G	D				G	G					U	A	G	A	A	U	A	G	U	A	C	C	C	U	G	C	C	A	m^5C	G	G	U	A	C
079	1	Bombyx mori	G	C	A	Um	C	G	G	U	m^1G	G	U	U	C	A	G	U				G	G					D	A	G	A	A	U	G	C	U	C	G	C	C	U	G	C	C	A	m^5C	G	C	G	G	G
	HISTIDINE																																																		
081	1	E.coli[l]	G G	G	U	G	G	C	U	A	s^4U	A	G	C	U	C	A	G	D	D			G	G					D	A	G	A	G	C	C	C	U	G	G	A	U	U	Q	U	G	m^2A	Ψ[m]	Ψ[m]	C	C	A	G
	ISOLEUCINE																																																		
091	1	E.coli	A	G	G	C	U	U	G	U	A	G	C	U	C	A	G	G	D(U)			G	G	D				D	A	G	A	G	C	G	C	A	C	C	C	C	U	G	A	U	t^6A	A	G	G	G	U	G
092		T.utilis	G	G	U	C	C	C	U	U	G	m^2G	C	C	C	A	G	D	D			G	G	D				D	A	A	G	G	C	m^2_2G	Ψ	G	G	U	G	C	U	I	A	U	t^6A	A	C	G	C	C	A

061 **M.Uziel,A.J.Weinberg (1975) Nucleic Acids Res. 2, 469-476.**
062 Z.Ohashi,F.Harada,S.Nishimura (1972) FEBS-Lett. 20, 239-241;
K.O.Munninger, S.H.Chang (1972) Biochem.Biophys.Res.Commun. 46, 1837-1842.
063 T.Kobayashi,T.Irie,M.Yoshida,K.Takeishi,T.Ukita (1974) Biochim.Biophys. Acta 366, 168-181.
071 C.W.Hill,G.Combriato,W.Steinhart,D.L.Riddle,J.Carbon (1973) J.Biol.Chem. 248, 4252-4262.
072 **J.W.Roberts**,J.Carbon (1975) J.Biol.Chem. 250, 5530-5541.
073 **C.Squires**,J.Carbon (1971) Nature New Biology 233, 274-277.
074 +075 R.J.Roberts (1974) J.Biol.Chem. 249, 4787-4796.
076 S.Stahl, G.V.Paddock,J.Abelson (1974) Nucleic Acids **Res.** 1, **1287-1304**;
B.G.Barell,A.R.Coulson,W.H.McClain (1973) FEBS-L**ett. 37, 64-69.**
077 M.Yoshida (1973) Biochem.Biophys.Res.Commun. 50, 779-784.
078 K.B.Marcu, R.E.Mignery,B.S.Dudock (1977) Biochemistry 16, 797-806.
079 J.P.Garel,G.Keith (1977) Nature 269, 350-352;
M.C.Zuniga,J.A.Steitz (1977) Nucleic Acids Res. 4, 4175-4196.
081 C.E.Singer, G.R.Smith (1972) J.Biol.Chem. 247, 2989-3000.
091 M.Yarus, B.G.Barrell (1971) Biochem.Biophys.Res.Commun. 43, **729-734.**
092 S.Takemura,M.Murakami,M.Miyazaki (1969) J.**Biochem. 65, 553-566.**

	Extra Arm																						TΨ Stem					TΨ Loop							TΨ Stem					Aminoacyl Stem										
	44	45	a	b	c	d	e	f	g	46	47	a	b	c	d	e	f	g	h	i	j	48	49	50	51	52	53	54	55	56	57	58	59	60	61	62	63	64	65	66	67	68	69	70	71	72	73	74	75	76
061	U	A								A												C	A	G	G	G	G	T	Ψ	C	G	A	A	U	C	C	C	C	U	G	G	G	G	G	A	C	G	C	C	A
062	U	A								A												C	A	G	G	G	G	T	Ψ	C	G	A	A	U	C	C	C	C	U	A	G	G	G	G	A	C	G	C	C	A
063	A	G								A												C	m^5C	G	G	G	G	T	Ψ	C	G	A	C	U	C	C	C	C	G	U	A	U	C	G	G	A	G	C	C	A
071	A									U	A											C	G	A	G	G	G	T	Ψ	C	G	A	U	U	C	C	C	U	U	C	G	C	C	C	G	C	U	C	C	A
072	U									G	A											U	G	C	G	G	G	T	Ψ	C	G	A	U	U	C	C	C	G	C	U	G	C	C	C	G	C	U	C	C	A
073	G	G								m^7G	U											C	G	C	G	A	G	T	Ψ	C	G	A	G	U	C	U	C	G	U	U	U	C	C	C	G	C	U	C	C	A
074	A									G	A											U	A	U	A	G	G	U	G	C	A	A	A	U	C	C	U	A	U	C	U	U	C	C	G	C	U	C	C	A
075	A									G	G											U	A	U	A	G	G	U	G	C	A	A	G	U	C	C	U	A	U	C	U	U	C	C	G	C	U	C	C	A
076	U									G	A											U	G	U	G	A	G	T	Ψ	C	G	A	U	U	C	U	C	A	U	U	A	U	C	C	G	C	U	C	C	A
077	-									G	G											C	m^5C	C	C	G	G	T	Ψ	C	G	A	U	U	C	C	G	G	G	C	U	U	G	C	G	C	A	C	C	A
078	A									G	A											m^5C	m^5C	m^5C	G	G	G	U	Ψ	C	G	m^1A	U	U	C	C	C	G	G	C	U	G	G	U	G	C	A	C	C	A
079	C									G	G											m^5C	m^5C	m^5C	G	G	G	T	Ψ	C	G	m^1A	U	U	C	C	C	G	G	C	C	G	A	U	G	C	A	C	C	A
081	U	U								m^7G	U											C	G	U	G	G	G	T	Ψ	C	G	A	A	U	C	C	C	A	U	U	A	G	C	C	A	C	C	C	C	A
091	A	G								m^7G	X[n)]											C	G	G	U	G	G	T	Ψ	C	A	A	G	U	C	C	A	C	Ψ	C	A	G	G	C	C	U	A	C	C	A
092	A	G								A	D											m^5C	A	G	C	A	G	T	Ψ	C	G	m^1A	U	C	C	U	G	C	U	A	G	G	G	A	C	C	A	C	C	A

a) xU is 5-N-methylaminomethyl-2-thiouridine, mam^5s^2U.

b) xU is 5-methoxycarbonylmethyl-2-thiouridine, mcm^5s^2U.

c) And Salmonella typhimurium; in position 8 is probably s^4U in E.coli but U in Salmonella typhimurium, G_{18} is partly Gm in Salmonella typhimurium.

d) Salmonella typhimurium sufD frameshift suppressor mutation has CCCC at the anticodon, D.L. Riddle, J. Carbon (1973) Nature New Biology 242, 230-234. Mutation $C_{35} \rightarrow U_{35}$, C.W. Hill, G. Combriato, W. Dolph (1974) J.Bacteriol. 117, 351-359.

e) N_{34} is an unidentified derivative of uridine.

f) Mutation: $C_{36} \rightarrow U_{36}$, $A_{37} \rightarrow A^{*}_{37}$, J. Roberts, J. Carbon (1974) Nature 250, 412-414.

g) Mutation: $G_{34} \rightarrow N_{34}$, $C_{36} \rightarrow N_{36}$, N are unknown derivatives of uridine and cytidine, respectively.

h) Mutation: $C_{36} \rightarrow A_{36}$, $A_{37} \rightarrow ms^2i^6A_{37}$, N^6-(Δ^2-isopentenyl)-2-methylthioadenosine, J. Carbon, E.W. Fleck (1974) J.Mol.Biol. 85, 371-391; J. Roberts, J. Carbon (1974) Nature 250, 412-414.

i) Staphylococcus epidermidis Texas 26.

k) xU is probably related to 5-methylaminomethyl-2-thiouridine, mam^5s^2U.

l) And Salmonella typhimurium.

m) HisT mutation $\Psi_{38} \rightarrow U_{38}$, $\Psi_{39} \rightarrow U_{39}$, C.E. Singer, G.R. Smith, R. Cortese, B.N. Ames (1972) Nature New Biology 238, 72-74.

n) Probably X, 3N-(3-amino-3-carboxypropyl)uridine, S. Friedman, H.J. Li, K. Nakanishi, G. van Lear (1974) Biochem. 13, 2932-2937.

No.		tRNA	1	2	3	4	5	6	7	8	9	10	11	12	13	14	15	16	17	a	b	18	19	a	b	c	d	20	21	22	23	24	25	26	27	28	29	30	31	32	33	34	35	36	37	38	39	40	41	42	43
		LEUCINE																																																	
101	1	E.coli B[a]/K 12[b]	G	C	G	A	A	G	G	U	G	G	C	G	G	A	A	D	D			Gm	G	D				A	G	A	C	G	C	G	C	U	A	G	C	U	U	C	A	G	N[c]	Ψ	G	Ψ	U	A	G
102	2	E.coli K12[b]	G	C	C	G	A	G	G	U	G	G	U	G	G	A	A	D	D			Gm	G	D				A	G	A	C	A	C	G	C	U	A	C	C	U	U	G	A	G	N[c]	Ψ	G	G	U	A	G
103		Phage T4	G	C	G	A	G	A	A	U	G	G	U	C	A	A	A	D	D			Gm	G	D				A	A	A	G	G	C	A	C	A	G	C	A	C	U	N[d]	A	A	xA[e]	A	Ψ	G	C	U	G
104	3	Yeast	G	G	U	U	G	U	U	U	G	m^2G	C	ac^4C	G	A	G	C				Gm	G	D	C			D	A	A	G	G	C	m^2_2G	C	C	U	G	A	Ψ	U	m^5C	A	A	m^1G	C	Ψ	C	A	G	G
105		Yeast	G	G	G	A	G	U	U	U	G	m^2G	C	ac^4C	G	A	G	D				Gm	G	D	D			D	A	A	G	G	C	m^2_2G	Ψ	C	A	G	A	Ψ	U	U	A	G	m^1G	C	Ψ	C	U	G	A
		LYSINE																																																	
111		E.coli B	G	G	G	U	C	G	U	U	A	G	C	U	C	A	G	D	D			G	G					D	A	G	A	G	C	A	G	U	U	G	A	C	U	xU[f]	U[h]	U	t^6A	A	Ψ	C	A	A	U
112		Bacillus subtilis	G	A	G	C	C	A	U	U	A	G	C	U	C	A	G	U	D			G	G					D	A	G	A	G	C	A	U	C	U	G	A	C	U	U(N)	U	U	N[c]	A	Ψ	C	A	G	A
113	1	Yeast (haploid)	G	C	C	U	U	G	U	U	m^2G	m^2G	C	G	C	A	A	D	C			G	G					D	A	G	C	G	C	m^2_2G	Ψ	A	U	G	A	C	U	C[i]	U	U	t^6A	A	Ψ	C	A	U	A
114	2	Yeast[j]	Ψ	C	C	U	U	G	U	U	A	m^2G	C	U	C	A	G	D	D			G	G					D	A	G	A	G	C	m^2_2G	Ψ	Ψ	C	G	G	C	U	xU	U	U	t^6A	A	C	C	G	A	A
		METHIONINE-INITIATOR																																																	
121		E.coli CA 265	C	G	C	G	G	G	G	s^4U	G	G	A	G	C	A	G	C	C	U		G	G					D	A	G	C	U	C	G	U	C	G	G	G	Cm	U	C	A	U	A	A	C	C	C	G	A
122		Thermus thermophilus	C	G	C	G	G	G	G	s^4U	G	G	A	G	C	A	G	C	C	U		Gm	G					D	A	G	C	U	C	G	U	C	G	G	G	C	U	C	A	U	A	A	C	C	C	G	A
123		Bacillus subtilis	C	G	C	G	G	G	G	U	G	G	A	G	C	A	G	U	U	C		G	G					D	A	G	C	U	C	G	U	C	G	G	G	C	U	C	A	U	A	A	C	C	C	G	A
124		Mycoplasma	C	G	C	G	G	G	G	s^4U	A	G	A	G	C	A	G	U	D(U)			G	G					D	A	G	C	U	C	G	C	C	G	G	G	C	U	C	A	U	A	A	C	C	C	G	G
125		Neurospora crassa (mitochondrial)	U	G	C	G	G	A	U	U	A	U	U	G	U	A	A	D				A	G					D	A	A	C	A	U	A	U	U	U	G	G	C	U	C	A	U	m^1G	N[l]	C	C	G	A	A
126		Neurospora crassa	A	G	C	U	G	C	A	U	m^1G	G	C	G	C	A	G	C				G	G					A	A	G	C	G	C	m^2_2G	C	N[m]	G	G	G	C	U	C	A	U	t^6A	A	C	C	C	G	G
127		Anacystis nidulans	C	G	C	G	G	G	G	U	A	G	A	G	C	A	G	C	C	U		G	G					D	A	G	C	U	C	G	U	C	G	G	G	Cm	U	C	A	U	A	A	C	C	C	G	A
128		Yeast	A	G	C	(C	G)	C	G	U	m^1G	m^2G	C	G	C	A	G	D				G	G					A	A	G	C	G	C	m^2_2G	C	A	G	G	G	C	U	C	A	U	t^6A	A	C	C	C	U	G
129		Mammalian[r]	A	G	C	A	G	A	G	U	m^1G	m^2G	C	G	C	A	G	C				G	G					A	A	G	C	G	U	m^2G[s]	C	U	G	G	G	C	C	C	A	U	t^6A	A	C	C	C	A	G
		METHIONINE																																																	
131	1	E.coli CA 265	G	G	C	U	A	C	G	s^4U	A	G	C	U	C	A	G	D(u)	D			G(m)	G	D				D	A	G	A	G	C	A	C	A	U	C	A	C	U	ac^4C	A	U	t^6A	A	Ψ	G	A	U	G
132	3	Yeast	G	C	U	U	C	A	G	U	A	m^2G	C	U	C	A	G	D	A			G	G					A	A	G	A	G	C	$(m^2_2)G$	Ψ(u)	C	A	G	Ψ	C	U	C	A	U	t^6A	A	Ψ	C	U	G	A
133	4	Mammalian[t]	G	C	C	U	C	m^2G	U	U	A	m^2G	C	G	C	A	G	D	A			G	G					D	A	G	C	G	C	m^2_2G	Ψ	C	A	G	Ψ	C	U	Cm	A	U	t^6A	A	Ψ	C	U	G	A

101 H.U.Blank,D.Söll (1971) Biochem.Biophys.Res.Commun. 43, 1192-1197; S.K.Dube,K.A.Marcker,A.Yudelevich (1970) FEBS-Lett. 9, 168-170.

102 H.U.Blank,D.Söll (1971) Biochem.Biophys.Res.Commun. 43, 1192-1197.

103 T.C.Pinkerton,G.Paddock,J.Abelson (1973) J.Biol.Chem. 248, 6348-6365.

104 S.H.Chang,S.Kuo,E.Hawkins,N.R.Miller (1973) Biochem.Biophys.Res.Commun. 51, 951-955.

105 K.Randarath,L.S.Y.Chia,R.C.Gupta,E.Randerath,E.R.Hawkins,C.K.Brum,S.H.Chang (1975) Biochem.Biophys.Res.Commun. 63, 157-163.

111 K.Chakraburtty,A.Steinschneider,R.V.Case,A.H.Mehler (1975) Nucleic Acids Res. 2, 2069-2075.

112 Y.Yamada,H.Ishikura (1977) Nucleic Acids Res. 4, 4291-4303.

113 S.J.Smith,H.S.Teh,A.N.Ley,P.D'Obrenan (1973) J. Biol. Chem. 248, 4475-4485.

114 J.T.Madison,S.J.Boguslawski (1974) Biochemistry 13, 524-527.

121 S.K.Dube,K.A.Marcker (1969) Eur.J.Biochem. 8, 256-262.

122 K.Watanabe,T.Oshima,S.Nishimura (1976) Nucleic Acids Res. 3, 1703-1713.

123 Y.Yamda,H.Ishikura (1975) FEBS-Lett. 54, 155-158.

124 R.T.Walker,U.L.RajBhandary (1978) Nucleic Acids Res. 5, 57-70.

125 J.E.Heckman, L.I.Hecker, S.D.Schwartzbach,W.E.Barnett, B.Baumstark, U.L.RajBhandary (1978) Cell 13, 83-95.

126 A.M.Gillum,L.I.Hecker,M.Silberklang,S.D.Schwartzbach,U.L.RajBhandary, W.E.Barnett (1977) Nucleic Acids Res. 4, 4109-4131.

127 B.Ecarot-Charrier,R.J.Cedergren (1976) FEBS-Lett. 63, 287-290.

128 M.Simsek,U.L.RajBhandary (1972) Biochem.Biophys.Res.Commun. 49, 508-515.

129 M.Simsek,U.L.RajBhandary,M.Boisnard,G.Petrissant (1974) Nature 247, 518-520; A.M.Gillum,N.Urquhart,M.Smith,U.L.RajBhandary (1975) Cell 6, 395-405; A.M.Gillum,B.A.Roe,M.P.J.S. Anandaraj,U.L.RajBhandary (1975) Cell 6, 407-413; P.W.Piper,B.F.C.Clark (1974) Eur.J.Biochem. 45, 589-600; M.Wegnez, A.Mazabraud,H.Denis,G.Petrissant,M.Boisnard (1975) Eur.J.Biochem. 60, 295-302.

131 S.Cory,K.A.Marcker (1970) Eur.J.Biochem. 12, 177-194.

132 H.Gruhl,H.Feldmann (1976) Eur.J.Biochem. 68, 209-217; O.Koiwai,M.Miyazaki (1976) J.Biochem. 80, 951-959.

133 P.W.Piper (1975) Eur.J.Biochem. 51, 283-293; G.Petrissant,M.Boisnard (1974) Biochimie 56, 787-789.

	Extra Arm																				
	44	45	a	b	c	d	e	f	g	46	47	a	b	c	d	e	f	g	h	i	j	48
101	U	G	U	C	C					U	U	A	C	G	G	A	C	G				U
102	U	G	C	C	C					A	A	U	A	G	G	G	C	U				U
103	C	G	G	A	A					U	G	A	U	U	U	C	C	U				U
104	U	A	U	C						G	U	A	A	G	A	U	G					m5C
105	U	A	U	C						U	U	C	G	G	A	U	G					m5C
111	U	G								m7G	X[g)]											C
112	G	G								m7G	U											C
113	A	G								m7G	U											U
114	A	U								m7G	D(U)											m5C
121	A	G								m7G[k)]	U											C
122	A	G								m7G	U											C
123	A	G								G	U											C
124	A	G								G	C											C
125	U	G									A											C
126	A	G								m7G	U(D)											C
127	A	G								m7G	U											C
128	A	U								m7G	D											m5C
129	A	G								m7G	D											m5C
131	G	G								m7G	X[g)]											C
132	A	G								m7G	D(U)											m5C
133	A	G								m7G	D											m5C

	TΨ Stem					TΨ Loop							TΨ Stem					Aminoacyl Stem										
	49	50	51	52	53	54	55	56	57	58	59	60	61	62	63	64	65	66	67	68	69	70	71	72	73	74	75	76
101	G	G	G	G	G	T	Ψ	C	A	A	G	U	C	C	C	C	C	C	C	C	U	C	G	C	A	C	C	A
102	A	C	G	G	G	T	Ψ	C	A	A	G	U	C	C	C	G	U	C	C	U	C	G	G	U	A	C	C	A
103	G	U	G	G	G	T	Ψ	C	G	A	G	U	C	C	C	A	C	U	U	C	U	C	G	C	A	C	C	A
104	A	A	G	A	G	T	Ψ	C	G	A	A	U	C	U	C	U	U	A	G	C	A	A	C	C	A	C	C	A
105	A	A	G	G	G	T	Ψ	C	G	m1A	A	U	C	C	C	U	U	A	G	C	U	C	U	C	A	C	C	A
111	G	C	A	G	G	T	Ψ	C	G	A	A	U	C	C	U	G	C	A	C	G	A	C	C	C	A	C	C	A
112	G	A	A	G	G	T	Ψ	C	G	A	G	U	C	C	U	U	C	A	U	G	G	C	U	C	A	C	C	A
113	A	G	G	G	G	T	Ψ	C	G	m1A	G	C	C	C	C	C	U	A	C	A	G	G	G	C	U	C	C	A
114	A	G	G	G	G	T	Ψ	C	G	m1A	G	C	C	C	C	C	U	A	Ψ	G	A	G	G	A	G	C	C	A
121	G	U	C	G	G	T	Ψ	C	A	A	A	U	C	C	G	G	C	C	C	C	C	G	C	A	A	C	C	A
122	G	C	G	G	G	s2T	Ψ	C	A	m1A	A	U	C	C	C	G	C	C	C	C	C	G	C	A	A	C	C	A
123	G	C	A	G	G	T	Ψ	C	A	A	A	U	C	C	U	G	C	C	C	C	C	G	C	A	A	C	C	A
124	G	C	A	G	G	U	Ψ	C	G	A	G	U	C	C	U	G	C	C	C	C	C	G	C	A	A	C	C	A
125	A	U	A	G	G	U	G	C	A	A	A	U	C	C	U	G	U	A	U	C	C	G	C	A	U	C	C	A
126	A	C	U	C	G	A	U	C	G	m1A	A	A	C	G	A	N[n)]	U	U	G	C	A	G	C	U	A	C	C	A
127	A	G	A	G	G	T	Ψ	C	A	A	A	U	C	C	U	C	U	C	C	C	C	G	C	C	A	C	C	A
128	m5C	U	C	G	G	A	U	C	G	m1A	A	A	C	C	G	N[o)]	N[p)]	C	G	(C	G)	G	C	U	A	C	C	A
129	G	A	U	G	G	A	U	C	G	m1A	A	A	C	C	A	U	C	C	U	C	U	G	C	U	A	C	C	A
131	A	C	A	G	G	T	Ψ	C	G	A	A	U	C	C	C	G	U	C	G	U	A	G	C	C	A	C	C	A
132	G	A	G	A	G	T	Ψ	C	G	m1A	A	C	C	U	C	U	C	C	U	G	G	A	G	C	A	C	C	A
133	G	U	G	A	G	T	Ψ	C	G	m1A	U	C	C	U	C	A	C	A	C.G		G	G	G	C	A	C	C	A

a) Identical with Salmonella typhimurium LT2 $tRNA_1^{Leu}$; His T mutant of S.t. $tRNA_1^{Leu}$ has $\Psi_{38} \rightarrow U_{38}, \Psi_{40} \rightarrow U_{40}$, H.S. Allaudeen, S.K. Yang, D. Söll (1972) FEBS-Lett. 28, 205-208.

b) For numbering of E.coli leucine tRNAs see R.E. Hurd, G.T. Robillard, B.A. Reid (1977) Biochem. 16, 2095-2100.

c) N is an unknown derivative of guanosine.

d) N is an unknown derivative of uridine.

e) xA is N^6-(Δ^2-isopentenyl)2-methylthioadenosine,ms^2i^6A, Y. Yamada, S. Nishimura, H. Ishikura (1971) Biochim.Biophys.Acta 247, 170-174.

f) xU is 5-N-methylaminomethyl-2-thiouridine,mam^5s^2U.

g) X is 3N-(3-amino-3-carboxypropyl)uridine.

h) U_{34} is partially replaced by N, which is probably a derivative of 2-thiouridine.

j) Is identical with Saccharomyces cerevisiae, haploid 2 , C.J. Smith, H.-S. Teh, A.N. Ley, P. D'Obrenan (1973) J.Biol.Chem. 248, 4475-4485.

k) Is A_{46} in the minor species of $tRNA^{fMet}$ from E.coli, S.K. Dube, K.A. Marcker, B.F.C. Clark, S. Cory (1968) Nature 218, 231-233; B.Z. Egan, J.F. Weiss, A.D. Kelmers (1973) Biochem.Biophys.Res.Commun. 55, 320-327.

l) N is most probably pseudouridine.

m) N_{28} is an unidentified derivative of pyrimidine.

n) N_{64} is an unidentified derivative of guanosine.

o) N_{64} is an unidentified derivative of adenosine.

p) N_{65} is an unidentified derivative of guanosine.

r) Rabbit liver, sheep mammary glands, salmon testes, salmon liver, human placenta, mouse myeloma cells, oocytes and somatic cells of Xenopus laevis.

		Aminoacyl Stem									D Stem				D Loop														D Stem					Anticodon Stem					Anticodon Loop							Anticodon Stem				
		1	2	3	4	5	6	7	8	9	10	11	12	13	14	15	16	17	a	b	18	19	a	b	c	d	20	21	22	23	24	25	26	27	28	29	30	31	32	33	34	35	36	37	38	39	40	41	42	43
	PHENYLALANINE																																																	
141	**E.coli**	G	C	C	C	G	G	A	s^4U	A	G	C	U	C	A	G	D	C			G	G					D	A	G	A	G	C	A	G	G	G	G	A	Ψ	U	G	A	A	xA[a)]	A	Ψ	C	C	C	C
142	B.stearothermophilus	G	C	U	C	G	U	G	U	A	G	C	U	C	A	G	U	C			G	G					D	A	G	A	G	C	A	A	A	G	G	A	C	U	Gm	A	A	xA[a)]	A	Ψ	C	C	U	U
143	Bacillus subtilis	G	G	C	U	C	G	G	U	A	G	C	U	C	A	G	U	D			G	G					D	A	G	A	G	C	A	A	C	G	G	A	C	U	Gm	A	A	xA[a)]	A	Ψ	C	C	G	U
144	Mycoplasma	G	G	U	C	G	U	G	U	A	G	C	U	C	A	G	U	C			G	G					D	A	G	A	G	C	A	G	C	A	G	A	C	U	G	A	A	m^1G	C	Ψ	C	U	G	C
145	Euglena grac. chloro.	G	C	U	G	G	G	A	U	A	G	C	U	C	A	G	D	U			G	Gm					U(D)	A	G	A	G	C	G	G	A	G	G	A	C	U	G	A	A	xA[a)]	A	Ψ	C	C	U	U
146	**Bean chloroplast**	G	U	C	G	G	G	A	U	A	G	C	U	C	A	G	U	D			Gm	G					D	A	G	A	G	C	A	G	A	G	G	A	C	U	G	A	A	xA[a)]	A	Ψ	C	C	U	C
147	Yeast	G	C	G	G	A	U	U	U	A	m^2G	C	U	C	A	G	D	D			G	G					G	A	G	A	G	C	G	C	C	A	G	A	Cm	U	Gm	A	A	Y[d)]	A	Ψ	m^5C	U	G	G
148	Wheat germ	G	C	G	G	G	G	A	U	A	m^2G	C	U	C	A	G	D	D			G	G					G	A	G	A	G	C	m^2_2G	Ψ	C	A	G	A	Cm	U	Gm	A	A	Y[e)]	A	Ψ	C	U	G	A
149	Mammalian[f)]	G	C	C	G	A	A	A	U	A	m^2G	C	U	C	m^1A	G	D	D			G	G					G	A	G	A	G	C	m^2_2G	Ψ	Ψ	A	G	A	Cm	U	Gm	A	A	Y[e)]	A	Ψ	C	U	A	A
	PROLINE																																																	
151	**Phage T4**	C	U	C	C	G	U	G	$(s^4)U$	A	G	C	U	C	A	G	U	U	U		G	G					D	A	G	A	G	C	G	C	C	U	G	A	Um	U	N[h)]	G	G	m^1G	A	Ψ	C	A	G	G
	SERINE																																																	
161	E.coli 1	G	G	A	A	G	U	G	s^4U	G	G	C	C	G	A	G	C				Gm	G	D	D			G	A	A	G	G	C	A	C	C	G	G	U	Cm	U	o^5U[i)]	G	A	xA[a)]	A	A	C	C	G	G
162	E.coli 3	G	G	U	G	A	G	G	s^4U	G	G	C	C	G	A	G	A				G	G	C	D			G	A	A	G	G	C	G	C	U	C	C	C	s^2C[k)]	U	G	C	U	t^6A	A	G	G	G	A	G
163	Phage T4	G	G	A	G	G	C	G	s^4U	G	G	C	A	G	A	G	U				Gm	G	D	D			U	A	A	U	G	C	A	C	C	G	G	U	Cm	U	N	G	A	xA[a)]	A	A	C	C	G	G
164	Yeast II (I)[l)]	G	G	C	A	A	C	U	U	G	G	C	ac^4C	G	A	G	D				Gm	G	D				D	A	A	G	G	C	m^2_2G	A	A	A	G	A	Ψ	U	I	G	A	i^6A	A	Ψ	C	U	U	U
165	Rat liver 1	G	U	A	G	U	C	G	U	G	G	C	ac^4C	G	A	G	D				Gm	G	D				D	A	A	G	G	C	m^2_2G	A	Ψ	G	G	A	m^3C	U	I	G	A	i^6A	A	Ψm	C	C	A	U
166	Rat liver 3	G	A	C	G	A	G	G	U	G	G	C	ac^4C	G	A	G	D				Gm	G	D				D	A	A	G	G	C	m^2_2G	A	Ψ	G	G	A	m^3C	U	G	C	U	mt^6A	A	Ψ	C	C	A	U
	THREONINE																																																	
171	E.coli	G	C	U	G	A	U	A	U	A	G	C	U	C	A	G	D	D			G	G					D	A	G	A	G	C	G	C	A	C	C	C	U	U	G	G	U	mt^6A	A	G	G	G	U	G
172	Bacillus subtilis	G	C	C	G	(G	U)	G	U	A	G	C	U	C	A	A	U	D			G	G					(u)D	A	G	A	G	C	A	A	C	U	G	A	C	U	mo^5U	G[m)]	U	t^6A	A	Ψ	C	A	G	U
173	Yeast **1 a, 1b**	G	C	U	U	C	U	A	U	G	m^2G	C	C	A	A	G	D	D			G	G					D	A	A	G	G	C	m^2_2G	C	C	A	C	A	m^3C	U	I	G	U	t^6A	A	Ψ	G	U	G	G

141 **B.G.**Barrell,F.Sanger (1969) FEBS-Lett. 3, 275-278.
142 G.Keith,C.Guerrier-Takada,H.Grosjean,G.Dirheimer (1977) FEBS-Lett. 84, 241-244.
143 H.Arnold,G.Keith (1977) Nucleic Acids Res. 4 , 2821-2829.
144 M.E.Kimball,K.S.Szeto,D.Söll (1974) Nucleic Acids Res. 1 ,1721-1732.
145 S.H.Chang,L.Hecker,M.Silberklang,C.K.Brum,W.E.Barnett,U.L.RajBhandary (1976) Cell 9, 717-724.
146 P.Guillemaut,G.Keith (1977) FEBS-Lett. 84, 351-356.
147 U.L.RajBhandary,S.H.Chang (1968) J.Biol.Chem. 243, 598-608.
148 G.A.Everett,J.T.Madison (1976) Biochemistry 15, 1016-1021; B.S.Dudock,G.Katz (1969) J.Biol.Chem. 244, 3069-3074.
149 G.Keith, F.Picaud,J.Weissenbach, J.P.Ebel, G.Petrissant,G.Dirheimer (1973) FEBS-Lett. 31, 345-347; G.Keith,J.P.Ebel,G.Dirheimer (1974) FEBS-Lett. 48, 50-52; B.A.Roe,M.P.J.S.Anandaraj,L.S.Y.Chia,E.Randerath,R.C.Gupta,K.Randerath (1975) Biochem.**Biophys**.Res.Commun. 66, 1097-1105.
151 J.G.Seidman,**B.G.Barrell**,W.H.McClain (1975) J.Mol.Biol. **99, 733-760**.
161 H. Ishikura, Y. Yamada, S. Nishimura (1971) FEBS-Lett. 16, 68-70; Y. Yamada, H. Ishikura (1975) Biochim.Biophys.Acta 402, 285-287.
162 Y. Yamada, H. Ishikura (1973) FEBS-Lett. 29, 231-234; D. Ish-Horowicz, B.F.C. Clark (1973) J.Biol.Chem. 248, 6663-6673.
163 W.H. McClain, B.G. Barrell, J.G. Seidman (1975) J.Mol.Biol. 99, 717-732.
164 H.G. Zachau, D. Dütting, H. Feldmann (1966) Hoppe-Selyer's Z.Physiol.Chem. 347, 212-235.
165 T. Ginsberg, H. Rogg, M. Staehelin (1971) Eur.J.Biochem. 21, 249-257.
166 H. Rogg, P. Müller, M. Staehelin (1975) Eur.J.Biochem. 53, 115-127.
171 L. Clarke, J. Carbon (1974) J.Biol.Chem. 249, 6874-6885.
172 T. Hasegawa, H. Ishikura (1978) Nucleic Acids Res. 5, 537-548.
173 J. Weissenbach, I. **Kiraly, G. Dirheimer (1977) Biochimie 59, 381-391**.

	Extra Arm																				
	44	45	a	b	c	d	e	f	g	46	47	a	b	c	d	e	f	g	h	i	j	48
141	G	U								m^7G	X[b)]											C
142	G	U								m^7G	U(N)[c)]											C
143	G	U								m^7G	U											C
144	G	U								m^7G	U											C
145	G	U								m^7G	X[b)]											C
146	C	U								m^7G	X[b)]											C
147	A	G								m^7G	U											C
148	A	G								m^7G	D											C
149	A	G								m^7G	(D)											C
151	A	G								m^7G	U											C
161	C	G	A	C	C	C				G	A	A	A	G	G	G	U	U				C
162	U	A	U	G	C	G	G	U		C	A	A	A	A	G	C	U	G	C	A	U	C
163	C	A	G	U	C	G	C			U	C	C	G	G	C	G	A	C	U			C
164	Um	G	G	G	C					U	C[1)]	U	G	C	C	C	G					m^5C
165	Um	G	G	G	G					U	m^3C	U	C	C	C	C	G					m^5C
166	Um	G	Ψ	G	C					U	m^3C	U	G	C	A	C	G					m^5C
171	A	G								m^7G	U											C
172	A	G								m^7G	U											U
173	A	G								A	D											$(m^5)C$

	TΨ Stem					TΨ Loop							TΨ Stem					Aminoacyl Stem										
	49	50	51	52	53	54	55	56	57	58	59	60	61	62	63	64	65	66	67	68	69	70	71	72	73	74	75	76
141	C	U	U	G	G	T	Ψ	C	G	A	U	U	C	C	G	A	G	U	C	C	G	G	G	C	A	C	C	A
142	G	G	C	G	G	T	Ψ	C	G	A	U	U	C	C	G	U	C	C	G	C	G	A	G	C	A	C	C	A
143	G	G	C	G	G	T	Ψ	C	G	A	U	U	C	C	G	U	C	C	C	G	A	G	C	C	A	C	C	A
144	G	G	C	G	G	U	Ψ	C	A	A	U	U	C	C	G	U	C	C	A	C	G	A	C	C	A	C	C	A
145	A	C	C	A	G	T	Ψ	C	A	A	A	U	C	U	G	G	U	U	C	C	U	A	G	C	A	C	C	A
146	A	C	C	A	G	T	Ψ	C	A	A	A	U	C	U	G	G	U	U	C	C	U	G	G	C	A	C	C	A
147	m^5C	U	G	U	G	T	Ψ	C	G	m^1A	U	C	C	A	C	A	G	A	A	U	U	C	G	C	A	C	C	A
148	G	C	G	U	G	T	Ψ	C	G	m^1A	U	C	C	A	C	G	C	U	C	A	C	C	G	C	A	C	C	A
149	m^5C	C	U	G	G	T(U)[g)]	Ψ	C	G	m^1A	U	C	C	C	G	G	G	U	U	U	C	G	G	C	A	C	C	A
151	C	A	A	G	G	T	Ψ	C	A	A	A	U	C	C	U	U	G	U	A	U	G	G	A	G	A	C	C	A
161	C	A	G	A	G	T	Ψ	C	G	A	A	U	C	U	C	U	G	C	G	C	U	U	C	C	G	C	C	A
162	C	G	G	G	G	T	Ψ	C	G	A	A	U	C	C	C	C	G	C	C	U	C	A	C	C	G	C	C	A
163	A	U	A	G	G	T	Ψ	C	A	A	A	U	C	C	U	A	U	C	G	C	C	U	C	C	G	C	C	A
164	G	C	A	G	G	T	Ψ	C	A[1)]	A	A[1)]	U	C	C	U	G	C	A	G	U	U	G	U	C	G	C	C	A
165	G	C	A	G	G	T	Ψ	C	G	m^1A	A	U	C	C	U	G	C	C	G	A	C	U	A	C	G	C	C	A
166	G	U	G	G	G	T	Ψ	C	G	m^1A	A	U	C	C	C	A	U	C	C	U	C	G	U	C	G	C	C	A
171	G	G	C	A	G	T	Ψ	C	G	A	A	U	C	U	G	C	C	U	A	U	C	A	G	C	A	C	C	A
172	G	G	G	G	G	T	Ψ	C	A	A	G	U	C	C	U	C	U	U	G	C	C	G	G	C	A	C	C	A
173	A(G)	U	C	G	G	T	Ψ	C	A	m^1A	A	U	C	C	G	A	U(c)	U	G	G	A	A	G	C	A	C	C	A

a) xA_{37} is N^6-(Δ^2-isopentenyl)-2-methylthioadenosine, ms^2i^6A.

b) X is 3N-(3-amino-3-carboxypropyl)uridine, S. Friedman, H.I. Li, K. Nakanishi, G. van Lear (1974) Biochem. 13, 2932-2937; S. Nishimura, Y. Taya, Y. Kuchino, Z. Ohashi (1974) Biochem.Biophys.Res. Commun. 57, 702-708.

c) U_{47} may be modified.

d) Y_{37} is wybutosine, yW, K. Nakanishi, N. Furutachi, M. Funamizu, D. Grunberger, I.B. Weinstein (1970) J.Amer.Chem.Soc. 92, 7617-7619.

e) oY_{37} is wybutoxosine, o_2yW, S.H. Blobstein, D. Grunberger, I.B. Weinstein, K. Nakanishi (1973) Biochem. 12, 188-193; A.M. Feinberg, K. Nakanishi, J. Barciszewski, A.J. Rafalski, H. Augustyniak, M. Wiewiorowski (1974) J.Amer.Chem.Soc. 96, 7797-7800.

f) Rabbit liver, calf liver and human placenta.

g) Content of T is different for different species.

h) N_{34} is an unidentified derivative of uridine.

i) o^5U is uridine-5-oxyacetic acid.

k) In the position 32 is most probably 2-thiocytidine.

l) Isoacceptor has U_{47}, G_{57} and G_{59}.

m) mo^5U is 5-methoxyuridine.

No.	tRNA	Aminoacyl Stem 1	2	3	4	5	6	7	8	9	D Stem 10	11	12	13	D Loop 14	15	16	17	a	b	18	19	a	b	c	d	20	21
	TRYPTOPHAN																											
181	E.coli CA244 (su⁻)[a]	A	G	G	G	G	C	G	s^4U[b]	A	G	U	U	C	A	A	D	D			G	G					D	A
182	Yeast	G	A	A	G	C	G	G	U	m^1G	m^2G	C	U	C	A	A	D				Gm	G					D	A
183	Chicken Cells[d]	G	A	C	C	U	C	G	U	m^1G	m^2G	C	G	C	A	A	C				Gm	G					D	A
	TYROSINE																											
191	E.coli[e]	G	G	U	G	G	G	G	s^4U	s^4U	C	C	C	G	A	G	C				Gm	G	C	C			A	A
192	B.stearothermophilus	G	G	A	G	G	G	G	s^4U	A	G	C	G	A	A	G	U				Gm	G	C	U			A	A
193	Yeast[h]	C	U	C	U	C	G	G	U	A	m^2G	C	C	A	A	G	D	D			Gm	G	D	D			D	A
194	T.utilis	C	U	C	U	C	G	G	U	m^1G	m^2G	C	C	A	A	G	D	D			Gm	G	D	D			D	A
	VALINE																											
201	E.coli 1 K12,B	G	G	G	U	G	A	U	s^4U	A	G	C	U	C	A	G	C	D			G	G					G	A
202	E.coli 2a	G	C	G	U	C	C	G	s^4U	A	G	C	U	C	A	G	D	D			G	G	D				D	A
203	E.coli 2b	G	C	G	U	U	C	A	s^4U	A	G	C	U	C	A	G	D	D			G	G	D				D	A
204	B.stearothermophilus	G	A	U	U	C	C	G	U	A	G	C	U	C	A	G	C	D			G	G					G	A
205	Yeast 1	G	G	U	U	U	C	G	U	m^1G	G	U	C	ψ	A	G	D	C			G	G	D				D	A
206	Yeast 2a	G	G	U	C	C	A	A	U	m^2G	G	U	C	C	A	G	D				G	G	D	D			C	A
207	Yeast 2b	G	U	U	C	C	A	A	U	A	$(m^2)G$	U	G	ψ	A	G	C				G	G	C				D	A
208	T.utilis 1	G	G	U	U	U	C	G	U	m^1G	G	U	C	ψ	A	G	D	D			G	G	D				C	A
209	Mammalian[1]	G	U	U	U	C	C	G	U	A	G	U	G	ψ	A	G	D				G	G	D				D	A

No.	tRNA	D Stem 22	23	24	25	26	Anticodon Stem 27	28	29	30	31	Anticodon Loop 32	33	34	35	36	37	38	Anticodon Stem 39	40	41	42	43
	TRYPTOPHAN																						
181	E.coli CA244 (su⁻)[a]	G	A	G	C	A	C	C	G	G	U	Cm	U	C	C	A	xA[c]	A	A	C	C	G	G
182	Yeast	G	A	G	C	ψ	ψ	ψ	C	G	A	Cm	U	Cm	C	A	A	A	ψ	C	G	A	A
183	Chicken Cells[d]	G	C	G	C	m^2_2G	ψ	C	U	G	A	Cm	U	Cm	C	A	m^1G	A	ψm	C	A	G	A
	TYROSINE																						
191	E.coli[e]	A	G	G	G	A	G	C	A	G	A	C	U	Q[f]	U	A	xA[g]	A	ψ	C	U	G	C
192	B.stearothermophilus	m^1A	C	G	C	G	G	C	G	G	A	C	U	Q[f]	U	A	xA[c]	A	ψ	C	C	G	C
193	Yeast[h]	A	G	G	C	m^2_2G	C	A	A	G	A	C	U	G	ψ	A	i^6A	A	ψ	C	U	U	G
194	T.utilis	A	G	G	C	m^2_2G	ψ	C	A	G	A	C	U	G	ψ	A	i^6A	A	ψ	C	U	G	A
	VALINE																						
201	E.coli 1 K12,B	G	A	G	C	A	C	C	U	C	C	C	U	o^5U[i]	A	C	m^6A	A	G	G	A	G	G
202	E.coli 2a	G	A	G	C	A	C	C	A	C	C	U	U	G	A	C	A	U	G	G	U	G	G
203	E.coli 2b	G	A	G	C	A	C	C	A	C	C	U	U	G	A	C	A	U	G	G	U	G	G
204	B.stearothermophilus	G	A	G	C	G	C	(C	A)	C	C	U	U	G	A	C	m^6A	G	G	G	U	G	G
205	Yeast 1	U	G	G	C	A	ψ	C	U	G	C	ψ	U	I	A	C	A	C	G	C	A	G	A
206	Yeast 2a	A	G	A	C	m^2_2G	ψ	C	G	C	C	ψ	U	N[k]	A	C	A	C	G	G	C	G	A
207	Yeast 2b	U	C	A	C	m^2G	ψ	ψ	G	C	C	ψ	U	C	A	C	A	C	G	G	C	A	A
208	T.utilis 1	U	G	G	C	A	ψ	C	U	G	C	ψ	U	I	A	C	A	C	G	C	A	G	A
209	Mammalian[1]	U	C	A	C	m^2G	ψ	U	C	G	C	Cm	U	I	A	C	A	m^5C	G	C	G	A	A

181 D. Hirsch (1971) J.Mol.Biol. 58, 439-458.

182 G. Keith, A. Roy, J.P. Ebel, G. Dirheimer (1972) Biochimie 54, 1405-1426.

183 F. Harada, R.C. Sawyer, J.E. Dahlberg (1975) J.Biol.Chem. 250, 3487-3497.

191 B.P. Doctor, J.E. Loebel, M.A. Sodd, D.B. Winter (1969) Science 163, 693-695.

192 R.S. Brown, J.R. Rubin, D. Rhodes, H. Guilley, A. Simoncsits, G.G. Brownlee (1978) Nucleic Acids Res. 5, 23-36.

193 J.T. Madison, H.-K. Kung (1967) J.Biol.Chem. 242, 1324-1330.

194 S. Hashimoto, S. Takemura, M. Miyazaki (1972) J.Biochem. 72, 123-134.

201 M. Yaniv, B.G. Barrell (1969) Nature 222, 278-279;
F. Kimura, F. Harada, S. Nishimura (1971) Biochemistry 10, 3277-3283.

202+203 M. Yaniv, B.G. Barrell (1971) Nature New Biol. 233, 113-114.

204 C. Takada-Guerrier, H. Grosjean, G. Dirheimer, G. Keith (1976) FEBS-Lett. 62, 1-3.

205 J. Bonnet, J.P. Ebel, G. Dirheimer, L.P. Shershneva, A.I. Krutilina, T.V. Venkstern, A.A. Bayev (1974) Biochimie 56, 1211-1213.

206 V.D. Axel'rod, V.M. Kryukov, S.N. Isaenko, A.A. Bayev (1974) FEBS-Lett. 45, 333-336.

207 V.G. Gorbulev, V.D. Axel'rod, A.A. Bayev (1977) Nucleic Acids Res. 4, 3239-3258.

208 T. Mizutani, M. Miyazaki, S. Takemura (1968) J.Biochem. 64, 839-848.

209 P.W. Piper (1975) Eur.J.Biochem. 51, 295-304;
P. Jank, N. Shinda-Okada, S. Nishimura, H.J. Gross (1977) Nucleic Acids Res. 4, 1999-2008.

	Extra Arm																					TΨ Stem					Tψ Loop							TΨ Stem					Aminoacyl Stem										
	44	45	a	b	c	d	e	f	g	46	47	a	b	c	d	e	f	g	h	i	j	48	49	50	51	52	53	54	55	56	57	58	59	60	61	62	63	64	65	66	67	68	69	70	71	72	73	74	75	76
181	G	U								m^7G	U											U	G	G	G	A	G	T	Ψ	C	G	A	G	U	C	U	C	U	C	C	G	C	C	C	C	U	G	C	C	A
182	G	G								m^7G	D											U	G	C	A	G	G	T	Ψ	C	A	m^1A	U	U	C	C	U	G	Ψ(U)	C	C	G	U	U	U	C	A	C	C	A
183	A	G								m^7G	C											U	G	C	G	U	G	Ψ	Ψ	C	G	m^1A	A	U	C	A	C	G	U	C	G	G	G	G	U	C	A	C	C	A
191	C	G	U	C						A	U	C	G	A	C	U	U					C	G	A	A	G	G	T	Ψ	C	G	A	A	U	C	C	U	U	C	C	C	C	C	A	C	C	A	C	C	A
192	U	C	C	C						U	U	U	G	G	G	U	U					C	G	G	C	G	G	T	Ψ	C	G	A	A	U	C	C	G	U	C	C	C	C	C	U	C	C	A	C	C	A
193	A	G								A	D											m^5C	G	G	G	C	G	T	Ψ	C	G	m^1A	C	U	C	G	C	C	C	C	C	G	G	G	A	G	A	C	C	A
194	A	C								A	D											m^5C	G	G	G	C	G	T	Ψ	C	G	m^1A	A	U	C	G	C	C	C	C	C	G	A	G	A	G	A	C	C	A
201	G	G								m^7G	U											C	G	G	C	G	G	T	Ψ	C	G	A	U	C	C	C	G	U	C	A	U	C	A	C	C	C	A	C	C	A
202	G	G								m^7G	X[j)]											C	G	G	U	G	G	T	Ψ	C	G	A	G	U	C	C	A	C	U	C	G	G	A	C	G	C	A	C	C	A
203	G	G								m^7G	X[j)]											C	G	U	U	G	G	T	Ψ	C	G	A	G	U	C	C	A	A	U	U	G	A	A	C	G	C	A	C	C	A
204	A	G								m^7G	U											C	G	C	U	G	G	T	Ψ	C	G	A	G	C	C	C	A	G	U	C	G	G	A	A	U	C	A	C	C	A
205	A	C								m^7G	D											C	m^5C	C	C	A	G	T	Ψ	C	G	m^1A	U	C	C	U	G	G	G	C	G	A	A	A	U	C	A	C	C	A
206	A	G								A	D											C	m^5C	C	G	A	G	T	Ψ	C	G	m^1A	A	C	C	U	C	G	G	U	U	G	G	A	U	C	A	C	C	A
207	A	G								m^7G	D											C	m^5C	C	G	A	G	T	Ψ	C	G	m^1A	U	C	C	U	C	G	G	U	U	G	G	A	U	C	A	C	C	A
208	A	C																				m^5C	C	C	C	A	G	T	Ψ	C	G	m^1A	U	C	C	U	G	G	G	C	G	A	A	A	U	C	A	C	C	A
209	A	G								m^7G	D											m^5C	m^5C	C	C	G	G	U[m)]	Ψ	C	G	m^1A	A	A[m)]	C	C	G	G	G	C	G	G	A	A	A	C	A	C	C	A

a) In the CA I6 44 UGA suppressors strain $tRNA^{Trp}$ has $G_{24} \rightarrow A_{24}$; $tRNA^{Trp}$ from E.coli su^+(UAG) has $C_{35} \rightarrow U_{35}$, $G_{24} \rightarrow A_{24}$ and $tRNA^{Trp}$ from E.coli su^+(UAA/G) has $C_{34} \longrightarrow U_{34}$, $C_{35} \longrightarrow U_{35}$, M. Yaniv, W.R. Folk, P. Berg, L. Soll (1974) J.Mol.Biol. 86, 245-260.

b) Identification os s^4U is not unambiguous, only the $s^4U_8 \rightarrow C_{13}$ cross link was identified.

c) xA_{37} is N^6-(Δ^2-isopentenyl)2-methylthioadenosine, ms^2i^6A, Y. Yamada, S. Nishimura, H. Ishikura (1971) Biohim.Biophys.Acta 247, 170-174.

d) The sequence was determined on primer RNA for initiation of in vitro Rous Sarcoma virus DNA synthesis; $tRNA^{Trp}$ from chicken cells has an identical composition, L.C. Waters, W.-K. Yang (1975) J.Biol.Chem. 250, 6627-6629.

e) For mutations see: B.G. Barrell, B.F.C. Clark (1974) Handbook of Nucleic Acid Sequences (Joynson Bruvvers Ltd., Oxford); J.D. Smith (1976) Prog.Nucl.Ac.Res.Mol.Biol. 16, 25-73; E.J. Celis, M. Squire, K. Kaltoft, E. Riisom (1977) Nucleic Acids Res. 4, 2799-2809.

f) Q_{34} is 7-(4,5-cisdihydroxy-1-cyclopenten-3-ylaminomethyl)-7-deazaguanosine, H. Casai, Z. Ohashi, F. Harada, S. Nishimura, N.J. Oppenheimer, P.F. Crain, J.G. Liehr, D.L. von Minden, J.A. McCloskey (1975) Biochem. 14, 4198-4208.

g) xA_{37} was identified as ms^2i^6A, F. Harada, H.J. Gross, F. Kimura, S.H. Chang, S. Nishimura, U.L. RajBhandary (1968) Biochem.Biophys.Res.Commun. 33, 299-306.

h) Yeast sup5a, amber suppressor has $G_{33} \longrightarrow C_{33}$, P.W. Piper, M. Wasserstein, F. Engback, K. Kattoft, J.E. Celis, Y. Zeuthen, S. Liebman, F. Sherman (1976) Nature 261, 757-761.

i) o^5U is uridine 5-oxyacetic acid

j) X is 3-N(3-amino-3-carboxypropyl)uridine, S. Friedman, H.J. Li, K. Nakanishi, G. van Lear (1974) Biochem. 13, 2932-2937.

k) N_{34} is an unknown derivative of uridine

l) Mouse myeloma and rabbit liver

m) The $U_{54} \cdot A_{60}$ base pair was detected by P. Jank, D. Riesner, H.J. Gross (1977)

COMPILATION OF MUTANT tRNA SEQUENCES

J.E. CELIS

Division of Biostructural Chemistry,
Department of Chemistry,
University of Aarhus, Denmark.

Compilation of Mutant tRNA Sequences

As a supplement to the compilation of tRNA sequences published by M. Sprinzl, F. Grüter and D.H. Gauss, this section presents the nucleotide sequence of published mutant tRNAs together with a brief description of properties of these mutants that could be of interest to the reader.

Mutant tRNAs have been derived by mutation of suppressor tRNAs genes present in *E. coli*, ϕ80 or bacteriophage T4. In all cases mutants having different levels of suppressor activity have been isolated. These mutants present single base substitutions and in many cases (those mutants having no suppressor activities) it has been possible to isolate double mutants among the revertants having suppressor activity. Double mutants have also been constructed by genetic recombination.

To designate a particular nucleotide substitution one refers to the identity of the new nucleotide followed by a number indicating its position in the tRNA sequence starting from the 5′ end. For example, the mutant A31 in tyrosine suppressor tRNA refers to a tRNA that contains an A at position 31 instead of the usual base (G).

In the case of the mutant T4 tRNAs, some of these mutants have been sequenced in the precursor as they do no produce any detectable amount of mature tRNA. In this case I have listed these mutants in a separate table and have indicated which part of the tRNA precursor is the one carrying the mutation. The position of the nucleotide substitution in this case has been numbered from the 5′-terminus of the altered tRNA in the precursor sequence.

Mutant Suppressor Tyrosine tRNAs (su^{+}3, amber)

Mutant[a)]	Base change from	to	Interesting properties[b)]	Reference number
A1	G	A	Mischarger; inserts glutamine *in vivo*. Strong suppressor.	1.2
A1U81	G C	A U	Mischarger; inserts tyrosine and glutamine *in vivo*. Strong suppressor.	2,3
A1G82	G A	A G	Mischarger; inserts glutamine *in vivo*. Strong suppressor.	4
A2	G	A	ts; mischarger; inserts tyrosine and glutamine *in vivo*. Weak suppressor.	5,2
A2U80	G C	A U	tr; indistinguishable from original suppressor. Strong suppressor.	5,2
A15	G	A	Defective in a step subsequent to its binding to the ribosome; accumulates tRNA precursor. Weak suppressor.	6
A15D19	G C	A D	Strong suppressor.	5
A15D20	G C	A D	Strong suppressor.	5
A17	Gm	A	Accumulates tRNA precursor. Weak suppressor	6

A25	G	A	Accumulates tRNA precursor. No detectable suppressor activity.	5
A25U11	G C	A U	Indistinguishable from original suppressor. Strong suppressor.	5
A25U19	G C	A U	Strong suppressor.	5
A31	G	A	Altered Km of aminoacylation: accumulates precursor. Weak suppressor.	6,7
A31U16	G G	A U	Strong suppressor.	7
A31U41	G C	A U	Strong suppressor.	7
A31U45	G C	A U	Strong suppressor.	7
U31	G	U	Accumulates tRNA precursor. Weak suppressor.	7
U31U16	G C	U U	Strong suppressor.	7
U31A41	G C	U A	Strong suppressor.	7
U31U45	G C	U U	Strong suppressor.	7
A46	G	A	Altered Km of aminoacylation; accumulates tRNA precursor. Weak suppressor.	8
A46U54	G C	A U	Strong suppressor.	8

Mutant Suppressor Tyrosine tRNAs (su^+3, amber) cont'd.

Mutant[a)]	Base change from	to	Interesting properties[b)]	Reference number
A62	G	A	Accumulates tRNA precursor. No detectable suppressor activity.	9
+C(73-78)	C insertion		Cannot be aminoacylated under normal conditions. No detectable suppressor activity.	10
U80	C	U	Mischarger; inserts tyrosine and a neutral amino acid *in vivo*. Strong suppressor.	5,2
A81	C	A	Mischarger; inserts tyrosine and a netural amino acid *in vivo*. Strong suppressor.	5,2
U81	C	U	Mischarger; inserts tyrosine and glutamine *in vivo*. Strong suppressor.	11,12,2
G82	A	G	Mischarger; inserts glutamine *in vivo*.	11,12,13,2

a) Mutant denomination in the original literature.
b) Strong suppressors are defined as those having >50% of the wild type suppressor activity. Weak suppressor presents less than 20% of the wild type suppressor activity.

Mutant T4 tRNAs Glutamine (psu^+2, ochre) and Serine (psu^+1, amber)

tRNA	Mutant[a)]	Base change from	to	Interesting properties	Reference number
Gln	U11	C	U	Synthesizes reduced amount of Gln tRNA (49% of psu^+2). No detectable suppressor activity. It accumulates a small amount of Gln-Leu tRNA precursor.	14
	A40	G	A	Synthesizes reduced amount of Gln tRNA (41% of psu^+2). No detectable suppressor activity. It accumulates a small amount of Gln-Leu tRNA precursor.	14
	A62	C	U	Synthesizes reduced amount of Gln tRNA (17% of psu^+2). No detectable suppressor activity. It accumulates a small amount of Gln-Leu tRNA precursor.	14
Ser	U26	C	U	Synthesizes normal amount of serine tRNA. No suppressor activity.	15
	G27	A	G	Synthesizes normal amount of serine tRNA. No suppressor activity.	15
	C48	U	C	Synthesizes reduced amount of Ser tRNA (25% of psu^+1). No suppressor activity.	15

a) Mutant denomination in the original literature.

Mutant T4 tRNAs Serine, Glutamine and Leucine Sequenced at the Precursor Level

tRNA precursor[a)]	Mutant[b)]	Base change from	Base change to	Interesting properties	Reference number
Pro-*Ser*	U1	G	U	Accumulates Pro-Ser tRNA precursor lacking many modified nucleotides. Block RNAase P reaction. Precursor RNA terminates in CCA_{OH}.	15
	O3	A	– (deletion)	" "	15
	C8	U	C	" "	15
	G12	A	G	Accumulates Pro-Ser tRNA precursor lacking many modified nucleotides. Defective in the first step of the biosynthetic pathway, the removal of UAA_{OH} residues at the serine terminus	15
	U17	G	U	" "	15
	A29	C	A	" "	15
	A30	G	A	" "	15
	U58	G	U	" "	15
	A67	G	A	" "	16
	C68	T	C	" "	15
	U70	C	U	" "	16
	U75	C	U	" "	16
	C77	U	C	" "	15

Gln-Leu	U11	C	U	Accumulates Gln-Leu tRNA precursor lacking many modified nucleotides. Partial reduction of RNAase P and cca repair at Gln moiety.	3
	A40	G	A	" "	17
	U62	C	U	" "	17
Gln-*Leu*	U72	C	U	Accumulates Gln-Leu precursor lacking many nucleotide modifications. RNAase P cleavage reduced at both moieties.	17

a) In all cases nucleotide substitutions have been localized in the precursor. The underlying tRNA contains the mutation. The serine tRNA in the Pro-Ser precursor has the anticodon N_2UA (psu^+1, amber). The glutamine tRNA in the Gln-Leu precursor has the anticodon NUA (psu^+2, ochre) and the leucine tRNA has the normal anticodon NAA (su^-).

b) Mutant denomination in the original literature.

References

1. Smith, J.D. and Celis, J.E. (1973). *Nature (London)* **243**, 66.
2. Ghysen, A. and Celis, J.E. (1974). *J. Mol. Biol.* **83**, 333.
3. Celis, J.E., Squire, M., Kaltoft, K. and Riisom, E. (1977). *Nucleic Acids Res.* **4**, 2799.
4. Inokuchi, H., Celis, J.E. and Smith, J.D. (1974). *J. Mol. Biol.* **85**, 187.
5. Smith, J.D., Barnett, L., Brenner, S. and Russell, R.L. (1970). *J. Mol. Biol.* **54**, 1.
6. Abelson, J.N., Barnett, L., Brenner, S., Gefter, M., Landy, A., Russell, R. and Smith, J.D. (1969). *FEBS Letters* **3**, 1.
7. Anderson, K.W. and Smith, J.D. (1972). *J. Mol. Biol.* **69**, 349.
8. Hooper, M.L. (1972). Ph.D. thesis, University of Cambridge.
9. Smith, J.D. (unpublished).
10. Celis, J.E., Miller, D., Piper, P.W., Riddle, D., Sheldon, R., Smith, J.D. and Squire, M. (submitted for publication).
11. Hooper, M.L., Russell, R.L. and Smith, J.D. (1972). *FEBS Letters* **22**, 149.
12. Celis, J.E., Hooper, M.L. and Smith, J.D. (1973). *Nature New Biol.* **224**, 261.
13. Shimura, Y., Aono, H., Ozeki, H., Sarabhai, A., Lamfrom, H. and Abelson, J. (1972). *FEBS Letters* **22**, 144.
14. Seidmann, J.G., Comer, M.M. and McClain, W.H. (1974). *J. Mol. Biol.* **90**, 677.
15. McClain, W.H. (1977). *Accounts Chem. Res.* **10**, 418.
16. McClain, W.H., Barrell, B.G. and Seidmann, J.G. (1975). *J. Mol. Biol.* **99**, 717.
17. McClain, W.H. and Seidmann, J.G. (1975). *Nature (London).* **257**, 106.

SUBJECT INDEX